THE HOCKEY STICK ILLUSION

THE HOCKEY STICK ILLUSION

CLIMATEGATE AND THE
CORRUPTION OF SCIENCE

A.W. Montford

STACEY
INTERNATIONAL

For my family

The Hockey Stick Illusion

STACEY INTERNATIONAL

128 Kensington Church Street

London W8 4BH

Tel: +44 (0)20 7221 7166; Fax: +44 (0)20 7792 9288

Email: info@stacey-international.co.uk

www.stacey-international.co.uk

ISBN: 978 1 906768 35 5

CIP Data: A catalogue record for this book is available from the British Library

Printed in Turkey by Mega Basim

Praise for The Hockey Stick Illusion

'For anyone wanting to understand the mind-boggling complexities of the debate over the "Hockey Stick" I've seen no single better reference than the enaging narrative by Andrew Montford. While not the last word on the Hockey Stick, Montford's recounting leads to some unsettling conclusions about the integrity of a very visible part of the paleoclimate research community.'

Roger Pielke Jnr
Professor of Environmental Studies
University of Colorado

Although the science is not always straightforward, Andrew W. Montford manages to make the story both exciting and accessible to the reader. He uses the Hockey Stick as an example of how manipulation of data and publication routines can change the whole world's view of an important subject. The story is told in such a fascinating way that it is hard to take ones eyes from the page.

Wibjörn Karlén
Professor of Geography (Emeritus)
University of Stockholm

'This is a thriller about codebreaking – not Napoleon's or Hitler's codes, but computer codes that generated a false signal to the world about runaway global warming. Like most codebreaking it was painfully slow but Montford keeps the drama pacy as the years pass, while he explains the intricacies in the plainest possible language. By military codebreaking, the likes of Scovell and Turing helped to change the course of history, and McIntyre and McKitrick should soon do the same, when the statistical fudges that misled the politicians become more widely known.'

Nigel Calder
Former editor, *New Scientist*
Co-author *The Chilling Stars*

CONTENTS

LIST OF FIGURES

LIST OF TABLES

Dramatis personae

The Climate Auditors

STEVE MCINTYRE   Canadian mining consultant who investigated the science behind the Hockey Stick.

ROSS MCKITRICK   McIntyre's co-author. Professor of economics at the University of Guelph.

PETE HOLZMANN   Climate Audit reader who performed the re-sampling of bristlecone trees at Almagre.

DAVID HOLLAND   McIntyre supporter who tried to obtain undisclosed IPCC reviews under British freedom of information legislation.

CRAIG LOEHLE   Ecologist who published a study based on non-tree ring proxies which showed a Medieval Warm Period.

The Hockey Team

MICHAEL MANN   The lead author of the Hockey Stick papers. Initially an adjunct professor of climatology at the University of Massachusetts, later at the University of Virginia. Now at Penn State.

RAY BRADLEY   Co-author of the Hockey Stick papers. Professor of climatology at the University of Massachusetts.

MALCOLM HUGHES   Co-author of the Hockey Stick papers from the University of Arizona.

KEITH BRIFFA   British tree ring researcher. Author of several studies underpinning paleoclimate reconstructions and lead author on the paleoclimate chapter of the IPCC Fourth Assessment Report.

CASPAR AMMANN   Mann supporter whose papers were alleged to have rebutted McIntyre and McKitrick's work.

ROSANNE D'ARRIGO   Paleoclimatologist. Noted for her controversial statement on cherypicking of data.

GABRIELE HEGERL   Paleoclimatologist. Author of an important temperature reconstruction.

TOM CROWLEY — Paleoclimatologist who accused McIntyre of threatening behaviour. Author of an important temperature reconstruction.

PHIL JONES — Climatologist. Maintains the HADCRUT temperature index, author of an important paleoclimate reconstruction.

SCOTT RUTHERFORD — Mann's assistant who delivered the Hockey Stick data to McIntyre. Later author of one of the 'independent confirmations' of Mann's work.

Other scientists

HANS VON STORCH — German climatologist who wrote papers critical of both Mann and McIntyre.

EDUARDO ZORITA — Spanish climatologist who was one of the reviewers of McIntyre and McKitrick's submission to *Nature*.

GERRY NORTH — Professor of climatology from Texas A&M University. Chairman of the NAS panel on paleoclimatology.

EDWARD WEGMAN — Statistician from Rice University. Author of a review of the statistics of the Hockey Stick that confirmed that the study was flawed.

PETER HUYBERS — Oceanographer who wrote a critical comment on McIntyre and McKitrick's GRL paper.

IAN JOLLIFFE — Emeritus professor of statistics from the University of Aberdeen who was a reviewer of one of McIntyre and McKitrick's critiques of the Hockey Stick. Later revealed that he had missed the flaws in Mann's work.

LINAH ABABNEH — PhD student who updated the critical Sheep Mountain chronology.

The bureaucracy

RALPH CICCERONE	Head of the US National Academy of Sciences who drew up the terms of reference for the NAS panel.
SUSAN SOLOMON	Head of the IPCC's Working Group I who threatened to remove McIntyre as a reviewer if he asked for data from study authors.
JOHN MITCHELL	IPCC review editor who is alleged to have performed his review in his spare time.

Journals and journalists

MARCEL CROK	Dutch journalist who discovered that Mann had calculated the verification R^2.
JAMES SAIERS	Editor at *Geophysical Research Letters*.
JAY FAMIGLIETTI	Executive editor of *Geophysical Research Letters* who replaced Saiers as editor in charge of the McIntyre and McKitrick paper and its responses.
SONIA BOEHMER-CHRISTIANSEN	Editor of *Energy and Environment* and global warming sceptic.
STEPHEN SCHNEIDER	Editor of *Climatic Change* and global warming promoter.
DAVID APPELL	Freelance science journalist and Mann's outlet in the media.

Politicians

| JOE BARTON | Texas congressman and chairman of the House Committee on Energy and Commerce who opened an inquiry into the IPCC and the Hockey Stick. |
| SHERRY BOEHLERT | Chairman of the House Science Committee who commissioned the NAS report on paleoclimate. |

Preface

In 2005 I followed a link from a British political blog to Steve McIntyre's Climate Audit site, then the newest addition to the blogosphere. While some of the statistics was over my head, there was plenty to interest a lay reader with an interest in sceptical arguments against the global warming hypothesis. While I was never a daily reader of the site, I found myself returning regularly, learning more and more each time, until I eventually found I could follow most of the postings without difficulty.

From time to time, new visitors to Climate Audit would plead for an introduction to the site and while there were some excellent primers, like Ross McKitrick's *What is the Hockey Stick Debate About?*, there was nothing that explained the story in the level of detail that I felt was required to enable the newbie to get fully up to speed on the intricacies of the science, and from time to time I wondered if my newly-found understanding of the debate would enable me to take on the task myself.[1]

It wasn't until the story of Caspar Ammann's purported replication of the Hockey Stick came to light during 2008 that I finally decided to take the plunge. The antics involved in keeping Ammann's paper alive, despite the catastrophic failure of its verification statistics, was so extraordinary, it seemed almost to be a public duty to make the story more widely known. Over the course of the next two or three days, I summarised a series of Climate Audit postings into a long article on my blog. *Caspar and the Jesus Paper*, as I chose to call the story, briefly turned my sleepy and relatively obscure website – my daily visitor count was probably a couple of hundred a day at the time – into a hive of activity, with thirty thousand hits being received over the following three days alone. To move from ten hits per hour to ten per second was something of a shock. Suddenly, writers I admired and respected were linking to my site and saying nice things about what I had written. There was even an attempt to use my article as a source document on Wikipedia. Perhaps, I thought, there was a demand for this kind of thing.

Many commenters have described *Caspar and the Jesus Paper* as a history of the Hockey Stick, but in truth it covers only a small part of the tale, reproduced here in Chapters 8 and 12. There was so much more to tell. Eventually I was spurred into telling the full story by the sight of the Hockey Stick in the manuscript of a new science textbook that crossed my desk one afternoon. Two years after it had been discredited the Hockey Stick was still apparently being used as the basis of a programme of environmental propaganda for schools. What made it worse was that the author was using the stick in its 'unofficial' guise, the twentieth century instrumental record grafted onto the end, the separate datasets not revealed to the reader. By the autumn of 2008 I was immersed in telling the tale from beginning to end.

With only one or two minor exceptions, there are no new revelations here. Every part of the debate between McIntyre and Mann has been fully documented on their competing blogs and elsewhere, and to some extent my task as a chronicler has been merely to sort their postings into coherent order and to distil the essence of the statistical arguments into something comprehensible by a lay reader. The reader can decide for themselves if I have been successful in doing this.

With my work here being based on the public record, the task of writing this book has been largely solitary. I am, however, grateful to several people who provided help and assistance along the way. Steve McIntyre, Ross McKitrick and Roger Pielke Jnr read the manuscript and provided perceptive reviews. Steve and Ross also provided some source materials that I was unable to locate elsewhere. David Holland sent me some unpublished details of his apparently never-ending search for the IPCC review comments and

Eduardo Zorita allowed me to identify him as the second reviewer of the *Nature* submission. Dr Angela Montford harrassed me over my grammar and spelling and asked many of searching questions, while doubling as an unpaid researcher. Dr Lesley Montford also read the text and made sure I stopped work from time to time.

Notes on Usages

There is considerable discussion in the text of the rather frightening sounding 'Pearson's squared correlation coefficient' (its meaning and importance, which are relatively straightforward, are explained in Chapter 2). In different fields of study this measure is signified by either R^2 or r^2, the former being more common in the social sciences, the latter in the physical sciences. Throughout the text I have preferred the usage R^2, since this is the style adopted by Steve McIntyre.

There is also much discussion in the text of a statistical technique called principal components analysis. The technique is described, hopefully in a non-threatening manner, in Chapter 2, but it is worth explaining here the particular terminology I have chosen to use. The technique of PC analysis, as I will refer to it, comes in one widely used 'vanilla flavour' plus a number of rarely used ones. Much of the story revolves around the use of a novel variant which we will refer to as short-centred PC analysis, although as we will see that its classification as a form of PC analysis is not generally accepted. Elsewhere, short centring has often been referred to as 'decentred' PC analysis, but I use the former style as I think it gives a better idea of the what has happened to the underlying data.

Much of the debate over the Hockey Stick has taken place online, on the blogs of the participants. It is therefore inevitable that much of the argumentation does not involve the checking of spelling and grammar that was normal in the past in print-based disputes. Rather than excuse myself of every error by appending a 'sic', I have preferred to correct each one, except in one or two cases where a mistake impinged directly upon the story.

Quoting as I do, directly from blog postings, I have had to make many simplifications, both for the benefit of a non-technical readership and for reasons of space. All such changes are marked by brackets and/or ellipses.

Throughout the text, I will use 'bristlecones' to refer to both bristlecone pines and foxtails, two closely related species that are critical in the story of the Hockey Stick. The two species are found on adjacent mountain ranges in the USA and, in fact, are so closely related that they interbreed.

1 The Hockey Stick

And thus Bureaucracy, the giant power wielded by pygmies, came into the world. (Honoré de Balzac)

The Hockey Stick was a long time in the making. The idea that manmade emissions of carbon dioxide might cause the Earth to heat up can be traced back to the French scientist, Joseph Fourier, who worked at the start of the nineteenth century.* Fourier is probably better known for his mathematical studies, but in a seminal paper of 1824, he also described how atmospheric gases might be capable of warming the atmosphere. In the 1850s John Tyndall, the Irish head of the Royal Institution, built on Fourier's work, performing a number of experiments that demonstrated the effect in action.

The term 'greenhouse effect' was not itself used until the end of the nineteenth century. The expression was coined by the Swedish physicist and Nobel Prize winner, Svante Arrhenius. Arrhenius was the first to attempt quantitative work on the warming produced by the atmosphere, and was the first to raise the question of whether manmade emissions of carbon dioxide could actually alter the temperature of the Earth. However, Arrhenius, far from being concerned about this possibility, thought that if man's activities caused a rise in temperature, the effects on humankind would be entirely beneficial. Warmer temperatures, he explained, would lead to higher crop yields and so to fewer hungry mouths,

---

*   The early history of the science of global warming was ably documented by Spencer Weart, on whose work much of this section of the story is based.[2]

an issue which was of great public concern at that time as the population of the planet continued to grow. Arrhenius also put forward a theory that carbon dioxide might be behind the cycle of ice ages and warmings that scientists had perceived in the geological record, and even went so far as to suggest that increasing the levels of carbon dioxide in the atmosphere could actually *prevent* the Earth from slipping into another ice age, again demonstrating a rather different set of concerns to those of many people today.

While Arrhenius's theory attracted the attention of his fellow scientists and a certain amount of controversy at the time, it soon disappeared from the mainstream of scientific life. The theory made a brief reappearance in the early twentieth century when a British engineer and amateur meteorologist called Guy Callendar wrote a number of papers expanding on Arrhenius's work, but the subject remained a scientific backwater until after the Second World War.

In the 1950s, the global warming hypothesis received a boost when accurate measurements of atmospheric carbon dioxide levels started to be recorded by the observatory on Mauna Loa in Hawaii. Until then it had been widely assumed that any carbon dioxide emitted into the atmosphere would simply be absorbed by the oceans, but the Mauna Loa results showed a clear and steady upward trend, and scientists started to dust off the work of Callendar and Arrhenius to work out what this might mean for the climate.

Work continued quietly but steadily in the background. Then in 1977 the pace started to quicken. The impulse was provided by the creation of a separate climate bureaucracy under the auspices of the World Meteorological Organisation (WMO). The WMO had organised the first World Climate Conference, which was held in Geneva two years later, and it is to that first meeting that the beginnings of the global warming movement can be traced.

The conference was instructed to review the state of knowledge of climatic change and variability, due both to natural and anthropogenic causes, and also to assess what this meant for humankind. In the way that bureaucracies sometimes do, however, the scientists actually did something slightly but tellingly different to what they had been asked to do. Rather than simply assess the state of scientific knowledge and consider what might happen in the future, they set out the steps they thought policy makers should take in a 'Call to Nations' that was issued at the end of the conference. This statement called for full advantage to be taken of man's knowledge of climate, for steps to be taken to improve that knowledge, and for potential manmade changes to climate to be foreseen *and prevented*. This then was not merely a call for more research, but also a demand for a particular policy outcome – prevention rather than adaptation. One can almost detect the germ of a idea forming in the minds of the scientists and bureaucrats assembled in Geneva: here, potentially, was a source of funding and influence without end. Where might it lead?

A couple of years later there was, to coin a phrase, something of a shift in the climate. James Hansen, a physicist from NASA's Goddard Institute for Space Studies, and a man who has been central to the whole global warming movement, published a breathtaking paper in Science in which he claimed that global warming was going to start happening much sooner than had previously been expected and that temperature records would start to be broken by the 1990s.[3]

Another climate conference, this time held in Villach, Austria in 1985, upped the ante even further. This meeting has been described as the first time that a scientific 'consensus' emerged on the issue of manmade or 'anthropogenic' global warming and the conclusions of the conference were certainly more outspoken than its predecessor. Predictably, the delegates called for more scientific research, but again went rather further than would have been expected from a scientific conference. They also demanded that policymakers fund research into the economic, social and political impacts of climate change and consider what steps could be taken to mitigate any future changes. Climatology was moving quickly from being an obscure backwater of scientific research to being an area of study which could shape policy in almost every conceivable area and affect the lives of millions of people around the world. The man in the street might not know it yet, but there were to be some big changes coming.

The first breakthrough in bringing the global warming hypothesis to public notice came in 1988, when Hansen went to the US Congress to explain how the release of carbon dioxide into the atmosphere was likely to affect the climate in coming years. Fortuitously, or perhaps by design, the hearing was held in mid-summer on a swelteringly hot day. The baking temperatures outside may well have affected the views of the assembled congressmen anyway, but Hansen was certainly not pulling his punches either. He told the Senate Committee on Energy and Natural Resources that the Earth was hotter in 1988 than at any time in the history of instrumental measurements, and that it was possible to point the finger of blame at the greenhouse effect.[4] His models, Hansen explained, predicted violent extremes of weather including, coincidentally, summer heatwaves.

This no-holds-barred warning seemed to have had the desired effect and it was reported around the world. With headlines secured around the world, 1988 turned into a pivotal year for the global warming hypothesis. A few months later, Margaret Thatcher gave a speech to the Royal Society in which she is quoted as having said that 'we may have unwittingly begun a massive experiment with the system of the planet itself'.[5] Thatcher's conversion to the green cause is credited to her ambassador at the United Nations, Sir Crispin Tickell, although Hansen may also have played a part – Thatcher is said to have read his congressional testimony and he is also believed to have made a presentation to her on his findings. The Royal Society speech was not the only time that she spoke out on global warming either. In the heady atmosphere following the finalisation of the Montreal Protocol to ban CFCs and save the ozone layer, the environment was the political buzzword *du jour*, and Thatcher was able to add global warming to the list of green issues she outlined in an address to the UN the following year:

What we are now doing to the world, by degrading the land surfaces, by polluting the waters and by adding greenhouse gases to the air at an unprecedented rate – all this is new in the experience of the earth. It is mankind and his activities that are changing the environment of our planet in damaging and dangerous ways.[6]

The floodgates were open. Politicians were leaping onto the bandwagon and soon the political momentum of the issue would be all but unstoppable as global warming found its way onto front pages and into election speeches around the world. The final step was the formation of a permanent climate bureaucracy and in the same year, 1988, the WMO and the UN together set up the Intergovernmental Panel on Climate Change (IPCC), a scientific body that would report on the state of climate science, advising policy makers on what was known about global warming and what should be done about it. Everything the climatologists had demanded just three years before at Villach had been granted to them.

Climate science

In its First Assessment Report (FAR), the IPCC was rather circumspect in its conclusions about what was happening to the Earth's climate and the reasons for any change that might be perceived.[7] Despite what Hansen had said in his congressional testimony about there being a high degree of confidence in the causal relationship between carbon dioxide and recent temperature rises, climatology was, and to a large extent remains, a science in its infancy. In the executive summary, the report's authors commented:

> We conclude that despite great limitations in the quantity and quality of the available historical temperature data, the evidence points consistently to a real but irregular warming over the last century. A global warming of larger size has

almost certainly occurred at least once since the end of the last glaciation without any appreciable increase in greenhouse gases. Because we do not understand the reasons for these past warming events, it is not yet possible to attribute a specific proportion of the recent, smaller warming to an increase of greenhouse gases.[7]

Their words, and particularly the closing sentence, show the problems that the global warming movement faced. If they were going to persuade policymakers to vote them still more funds and to take drastic action in terms of changing the workings of the economy and the way people lived, it was going to be necessary to persuade the public as well, and the public were unlikely to be convinced by science that was sparse and limited in quality.

There was a bigger problem too. The report included a chart showing how global temperatures had varied in previous ages, according to the scientific understanding of the time. This was something of a dampener for the argument for catastrophic global warming because it suggested that past temperatures had been warmer than today in a long period lasting from the eleventh to the fifteenth centuries. This period had been followed by two or three hundred years of much cooler temperatures, lasting until the eighteenth century. Since then warming had recommenced, but current temperatures were still thought to be well short of those reached during the medieval warming. This then was a huge problem for those promoting the idea of global warming – how would they convince anyone that a rise of a fraction of a degree in temperature portended something dangerous when the climate had been much warmer in the past?

At the time, the FAR graph was pretty much a representation of what might have been considered common knowledge. The so-called 'Medieval Warm Period' was extremely well represented in medieval annals and other documentary sources and it had come to have at least some impact on the public imagination. Every schoolboy knew that the Vikings had taken advantage of warmer

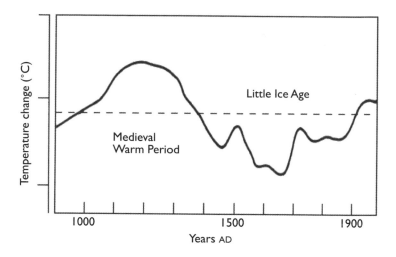

FIGURE 1.1: The Medieval Warm Period as shown in the IPCC First Assessment Report in 1990

temperatures to colonise Iceland and Greenland at the end of the first millennium, and historians had also discovered that grapes had been grown commercially in England at the time. There was a wealth of evidence that the medieval period had been an age of warmth, plenty and a flourishing of culture. The 'Little Ice Age', meanwhile, was equally well known – the Viking evacuation of Greenland at the start of the fifteenth century, the freezing of the Golden Horn in the seventeenth, stories of ice-fairs on the Thames and the winter paintings of Breughel the Elder had all created a strong public perception of years of biting cold winters. So at the start of the 1990s nobody was going to take issue with the story that the IPCC was telling – of Medieval Warm Period giving way to Little Ice Age before another gentle warming was ushered in.

The exact origins of the chart presented by the IPCC were, at the time, obscure; rather strangely, the report did not contain a citation or other indication of its authorship. Although it appeared to be a schematic or cartoon rather than a proper graph, it must have had some basis in scientific research, but quite what this basis was was not discovered until many years later when it was shown to be derived from the work of a British climatologist called Hubert Lamb.[8] Lamb, while an important scientist, was born in 1913 and the chart turns out to have been based on work he did in the 1960s. The relative antiquity of this climate history might explain the reluctance of the IPCC to explain its provenance. What was still more surprising was that Lamb's work turned out to be largely based on the Central England Temperature Record, a long series of instrumental readings, which dated back to the mid-seventeenth century. In other words, the understanding of world climate history propagated to the public by the IPCC was based, not on any understanding of global climate, but on the records for just one part of England: an odd situation to say the least.*

The Medieval Warm Period becomes less warm

In 1994 a pair of tree-ring researchers called Malcolm Hughes and Henry Diaz co-authored a journal review which struck a major blow at Lamb's view of climate history.[9] The two men surveyed the evidence supporting the existence of the Medieval Warm Period, considering all the different types of data that had been used to reconstruct past temperatures since Lamb's time. Their conclusions were that temperatures *had* been higher in some parts of the world – they singled out Scandinavia, China, the Sierra Nevada in California, the Canadian Rockies and Tasmania. However, they emphasised that these warm periods seemed to have happened at slightly different times in different places. This suggested that the warmings had probably had different causes. They also claimed that there was no evidence for any abnormal medieval warming at all in the southeast United States, southern Europe along the Mediterranean, and parts of South America. If they were right, then it would only be possible to conclude that the Medieval Warm Period was, at most, a series of regional warmings.

---

* We should note in passing that the caption to the original FAR graph was unequivocal that it was a representation of *global* temperatures.

On its own, these findings might look interesting but otherwise unremarkable. But put in the context of the temperature history of the last thousand years their impact on the climate debate was potentially explosive. Anecdotally at least, the Medieval Warm Period, represented by the bump upwards in temperatures at the left hand side of the IPCC 1990 graph, was being slowly flattened out. And as it flattened, the current warming started to look more and more significant – if current temperatures were in excess of anything seen in previous times, it would be powerful evidence that manmade global warming had already had a serious and deleterious effect on the world's climate. The flatter the representation of the medieval period in the temperature reconstructions, the scarier were the conclusions.

This was one paper in a single volume of review articles. It would take more than that to overturn an well-embedded paradigm. However, behind the scenes climatologists were busy, and a short time after the Hughes and Diaz paper was released, the public got a brief glimpse of what was happening. It was not at all as it should have been.

The Deming affair

David Deming was a geoscientist from the University of Oklahoma, whose expertise was in boreholes. From these holes, drilled deep into the Earth's surface, it was possible to extract a profile of the temperatures within the rocks all the way down. This profile was a direct record of what the surface temperature had been in the past. The deeper you went, the older the temperature record you could get. Of course it wasn't as simple as that – there were all sorts of confounding factors affecting the reliability of the results but it was one of the approaches being tried as a way of discerning the history of the Earth's climate.

Deming had recently created a temperature reconstruction for the last 150 years, based on boreholes in North America. In his study, he concluded that North America had warmed somewhat in the period since 1850, but had little to say beyond that. This was good, solid science but not the stuff of newspaper headlines. His findings were, however, considered highly important in climate science circles. With the expectation that temperatures were being driven upwards by carbon dioxide emissions, the Deming study seemed like good evidence to support the hypothesis. Because of this interest, Deming was able to get his work published in one of the world's most important journals, *Science*.[10] And with a storyline of rising temperatures published in such a prestigious publication, he also attracted the notice of some of the most influential people in the global warming industry, who thought they saw in Deming a valuable new recruit to the cause. Deming explained what happened in a later article:

> With the publication of the article in *Science*, I gained significant credibility in the community of scientists working on climate change. They thought I was one of them, someone who would pervert science in the service of social and political causes. So one of them let his guard down. A major person working in the area of climate change and global warming sent me an astonishing email that said 'We have to get rid of the Medieval Warm Period.'[11]

This sudden flash of light on a particularly murky shadow of climatological practice is probably unique. Suddenly it was possible to see that the Hughes and Diaz retake on the Medieval Warm Period was not considered enough. The aim was to erase it from the climatological record in its entirety. Although Deming himself did not identify the email's author, Richard Lindzen of MIT has confirmed that the email was written by Jonathan Overpeck of the University of Arizona.[12] It was evident to anyone who was watching that, in some quarters at least, there was a concerted effort to rewrite the Earth's climate history so that the Medieval

Warm Period disappeared. Unfortunately, few people were watching. Those who noticed what Deming was saying, and tried to raise the alarm, were ignored by the media.

Deming had another interesting story to tell too. A couple of years after the publication of his *Science* article he had been the reviewer for another borehole study, this time written by Shaopeng Huang of the University of Michigan. Huang's results had shown a pronounced worldwide Medieval Warm Period, something that was anathema to those in the global warming mainstream. In fact the study suggested that medieval temperatures might have been well in excess of those in modern times. Deming explained what happened next:

> The Huang et al. (1997) study was originally submitted to *Nature*. I was one of the reviewers of the manuscript. I told the *Nature* editors that the article would surely be one of the most important papers they published that year. But it never appeared in print. *Nature* asked the authors to revise the paper twice and then, after a long delay, ended up rejecting it.[11]

This difficulty in getting into print any result that went against the idea of catastrophic global warming was to be a consistent complaint among sceptics, and readers may like to note *Nature*'s treatment of Huang and compare it to later events in this story.

A few months after Deming's revelations about the fate of Huang's paper, the second IPCC report picked up on the changing attitudes towards the Medieval Warm Period. The report's authors noted that:

> Based on the incomplete observations and paleoclimatic evidence available, it seems unlikely that global mean temperatures have increased by 1°C or more in a century at any time during the last 10,000 years.

and went on,

> The limited available evidence from proxy climate indicators suggests that the 20th century global mean temperature is at least as warm as any other century since at least 1400 AD. Data prior to 1400 are too sparse to allow the reliable estimation of global mean temperature.[13]

This represented a significant change in emphasis by the IPCC. The story in the FAR, of a pronounced Medieval Warm Period with temperatures exceeding modern ones, had been replaced by a new narrative, in which it was said that modern warmth was probably unprecedented – or at least as high as anything seen in the last six hundred years. And if anyone were to question how all the historical records of warm temperatures in the medieval period could be wrong, it was explained that these were a regional phenomenon and that overall, the globe appeared to have been no warmer back then than it was at present.

There was one major problem with the case for the Medieval Warm Period having been an insignificant regional phenomenon though. This was the paucity of hard data to support the case – the 'limited available evidence' referred to above. It was simple for critics to point out that any conclusions drawn from this data would have to be highly speculative at best. Climate science wanted big funding and big political action and that was going to require definitive evidence. In order to strengthen the arguments for the current warming being unprecedented, there was going to have to be a major study, presenting unimpeachable evidence that the Medieval Warm Period was a chimera.

Enter the Hockey Stick.

The paper

The Hockey Stick paper made its grand entrance in an article published in *Nature* on 23 April 1998.[14] Its main author was a hitherto relatively obscure scientist based at the University of

Massachusetts (UMass) called Michael E. Mann, and it went by the distinctly unmemorable title, 'Global scale temperature patterns and climate forcing over the past six centuries'. Despite this unpromising opening and a style of writing that has been politely described as 'rather obscure',[15] it was to become one of the most cited scientific papers of that year or indeed of any other year. In fact, when the controversy was at its height, one investigator discovered that it had been cited twice as often as was normal for a scientific paper, and years after its publication it was still being referenced at a startling rate in the scientific literature.

Mann was just starting out on his scientific career, receiving his PhD in 1998 at the age of 33. At the time of the paper's publication he was still only an adjunct member of faculty at UMass. Mann may have been a late developer, but he was ambitious and self-confident and the reception for his paper suggested he was destined for great things.

Apart from Mann, the Hockey Stick paper had two secondary authors: the first was Ray Bradley, a colleague of Mann's from the University of Massachusetts, while the other was Malcolm Hughes of the University of Arizona, whom we have already met as one of the authors of the first serious attempt to 'get rid of the Medieval Warm Period'. In the years since its publication the paper has become known by the initials of its authors' names and we will be referring to it as MBH98 from here on.

MBH98 was novel on a number of levels. Firstly, it had been based on a much greater volume of raw data than earlier studies. Mann, Bradley and Hughes had trawled the archives for anything from which they might extract a temperature signal and had come up with a network of 112 'indicators', as they termed them. These are more normally referred to as 'proxies' (see Chapter 2). Although the majority of the indicators didn't extend back to the critical medieval period, the MBH98 dataset represented a significant advance and struck a blow at those critics who had rejected earlier studies as lacking sufficient data to be reliable. In fact, some of the indicators were actually summaries of larger networks of proxies, so there was even more data backing up their reconstruction than was suggested by the reported number of 112 series. This summarising had been done using a statistical procedure called principal components analysis (PC analysis) and this first application of the technique to temperature reconstructions gave the study an air of great technical sophistication, which would again render it much harder to criticise. With a large dataset and state of the art methodology in place, the authors wanted their readers to be in no doubt as to how good their results were, speaking of its 'highly significant reconstructive skill'. This suggested a study that was going to be hard to refute.

What then of the findings? The abstract of the paper explained that Mann and his team had been able to reconstruct temperatures since the year 1400 and that recent temperatures were warmer than any other year since the start of their records. In the remainder of the paper, they went on to assess possible reasons for the dramatic change in temperatures by testing how the graph of their reconstruction correlated against possible causes ('forcings' in the jargon), such as atmospheric dust, solar irradiance and carbon dioxide. It will be no surprise to anyone that their conclusion was that the only potential culprit was carbon dioxide. The implications were once again clear: mankind was warming the globe. Here then was the beginning of the end of the process of getting rid of the Medieval Warm Period. All that was lacking was a degree of publicity, something that was to be dealt with in fairly short order, as we will see.

The key graphic in the paper was a chart of the reconstruction of Northern Hemisphere temperatures for the full length of the record from 1400 right through to 1980. The picture presented was crystal clear. From the very beginning of the series the temperature line meandered gently, first a little warmer, then a little cooler, never varying more than half a degree or so from peak to trough. This was the 500-year long handle of the Hockey Stick, a sort of steady state that had apparently reigned, unchanging, throughout most of recorded history. Then suddenly, the blade of the stick appeared at the start of the twentieth century, shooting upwards in an almost straight line.* It was a startling change and it was this that made the Hockey Stick such an effective promotional tool, although to watching scientists, the remarkable thing about the Hockey Stick was not what was happening in the twentieth century portion – that temperatures were rising was clear from the instrumental record – but the long flat handle. The Medieval Warm Period had completely vanished. Even the previously acknowledged 'regional effect' now left no trace in the record. The conclusions were stark: current temperatures were unprecedented.

The splice

As presented in *Nature*, the Hockey Stick chart was a dreadful example of scientific graphics, with the authors managing to cram no less than four lines onto the same chart, making it hard for even the attentive reader to see exactly what it was they were looking at (see Figure 1.2). One should charitably point to the space restrictions necessary for publication in *Nature* , and the difficulties of presenting information in black and white diagrams.

Presented on top of each other were four sets of numbers: the reconstructed temperature, a smoothed version of these figures, the error bars, and at the right hand side (and easy to miss for the inattentive), the thermometer record for the twentieth century. The inclusion of the instrumental record was instantly controversial, with global warming sceptics accusing Mann of having spliced two entirely different datasets. The effect of this scientifically dubious presentation was, they said, to make the twentieth century portion look more frightening than the underlying data would warrant. Mann's counter-argument to these accusations was that the data was not spliced, but overlaid, and that its inclusion was justified in order to extend the reconstruction

---

*   Readers outside North America may wonder why a straight upward line on the end of a long flat handle should make the graph look anything like a hockey stick: it is, of course, and with delicious irony, an *ice* hockey stick.

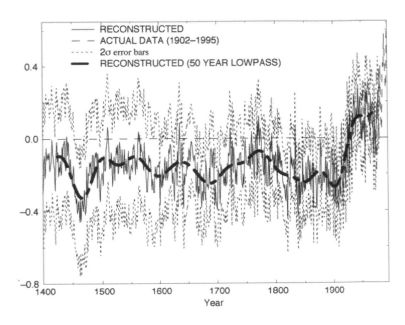

FIGURE 1.2: The Hockey Stick in MBH98

from 1980, which was when most of the underlying data ended, right up to the present day.

While it is difficult to see exactly what had been done from the black and white graph in *Nature*, later versions of the Hockey Stick were much clearer about the splice/overlay, using colour to distinguish the different datasets, although some well-known users of Mann's work did forget to make the distinction, as we will see later in the story. The IPCC's version of the Hockey Stick is shown on the back cover, with the instrumental overlay shown in red at the right hand side. Clearly, a large proportion of the blade of the stick was not from the same dataset as the handle, although there is undoubtedly a rise in the reconstructed temperatures too. Opinions on the issue remained divided; the first Hockey Stick controversy was off and running.

Reactions

The paper was clearly expected to be of huge public interest, and a press release was issued by UMass, timed to coincide with its publication:

> Climatologists at the University of Massachusetts have reconstructed the global temperature over the past 600 years, determining that three recent years, 1997, 1995, and 1990, were the warmest years since at least AD 1400.
>
> The researchers were able to estimate temperatures over more than half the surface of the globe, pinpointing average yearly temperatures in the northern hemisphere to within a fraction of a degree, going back to AD 1400. The study places in a new context long-standing controversy over the relative roles of human and natural changes in the climate of past centuries, according to Mann.
>
> Advanced statistical techniques were used to translate the proxy information into surface temperature patterns, so that past centuries could be compared with the twentieth century.[16]

With the press release so unequivocal, it is hardly surprising that the media found the story irresistible. Just five days after its publication in *Nature*, Mann was given the honour of an article in the *New York Times*, announcing the results of his study to the world.[17] This was a truly remarkable achievement for Mann, who, as we have seen, had only just finished his PhD, although it should be noted in fairness that he had been active in paleoclimate for some years previously. The story was penned by the *New York Times'* science reporter, William K. Stevens and its headline echoed the press release's certainty about the findings:

NEW EVIDENCE FINDS THIS IS WARMEST CENTURY IN 600 YEARS

Interestingly, beneath the headline, much of the article was actually taken up with discussing doubts about the reliability of the study. One scientist quoted in the *New York Times* article wondered if it would ever be possible to get a temperature reconstruction that was reliable enough to tell if the current warming was unprecedented or not. Even Mann himself was quoted as saying that there was quite a bit of work to be done in reducing the uncertainties. However, the headline and another scientist quoted in the study left no doubt that this was expected to be a very significant piece of work.

USA Today was much less equivocal though:

90S WERE WARMEST YEARS IN CENTURIES

> The 20th century has been warmer than the five centuries that preceded it, and 1997, 1995 and 1990 were the warmest years since 1400, says the latest study to relate global climate change to the burning of fossil fuels.[18]

Elsewhere it was the same: NBC website told its readers, 'Millennium ending with record heat',[19] while *Time* magazine went for the jauntier headline, 'It hasn't been this sizzling in centuries'.[20]

MBH99

Buoyed by the success of their first paper, Mann, Bradley and Hughes set about extending the study back to the start of the millennium, publishing their new results in *Geophysical Research Letters*.[21] 'Northern Hemisphere temperatures during the last millennium: inferences, uncertainties and limitations' (which we will refer to as MBH99) was, as the title suggests, a much more cagily worded article. There was so little data available for the first four centuries of the reconstruction that any conclusions could only be extremely tentative. That said, the conclusions were broadly similar – that the twentieth century appeared to be anomalous compared to any other period in the last 1,000 years. The global warming bandwagon was on a roll.

There was no let up on the public relations front either. Once again, UMass made sure that the paper received maximum publicity, with a press release that concentrated on the scary bits. Under the headline '1998 was warmest year of millennium, UMass Amherst climate researchers report', they quoted Bradley as saying, 'Temperatures in the latter half of the 20th century were unprecedented'.[22] Those who read further might have noticed Mann discussing the uncertainties and the sparseness of the data, but this was clearly not the key message, and most media outlets chose not to dwell on the uncertainties when they reported the results to their readers. The newspaper headlines were all written in terms which left no room for any doubt.

IPCC: The Third Assessment Report

The two Hockey Stick papers were good for Mann. Within months of the first paper's publication, he found himself advancing rapidly through the academic ranks with a speed that was simply breathtaking. In 2000, John Daly, a prominent global warming sceptic explained just how dramatic Mann's rise to fame had been, and how influential he had now become in the climatology community.[23]

> At the time he published his 'Hockey Stick' paper, Michael Mann held an adjunct faculty position at the University of Massachusetts, in the Department of Geosciences. He received his PhD in 1998, and a year later was promoted to Assistant Professor at the University of Virginia, in the Department of Environmental Sciences, at the age of 34.
>
> He is now the Lead Author of the 'Observed Climate Variability and Change' chapter of the IPCC Third Assessment Report (TAR2000), and a contributing author on several other chapters of that report. The Technical

Summary of the report, echoing Mann's paper, said: 'The 1990s are likely to have been the warmest decade of the millennium, and 1998 is likely to have been the warmest year.'

Mann is also now on the editorial board of the *Journal of Climate* and was a guest editor for a special issue of *Climatic Change*. He is also a referee for the journals *Nature, Science, Climatic Change, Geophysical Research Letters, Journal of Climate, JGR-Oceans, JGR-Atmospheres, Paleooceanography, Eos, International Journal of Climatology,* and NSF, NOAA, and DOE grant programs. (In the 'peer review' system of science, the role of anonymous referee confers the power to reject papers that are deemed, in the opinion of the referee, not to meet scientific standards).

He was appointed as a 'Scientific Adviser' to the U.S. Government (White House OSTP) on climate change issues.

Mann lists his 'popular media exposure' as including – 'CBS, NBC, ABC, CNN, CNN headline news, BBC, NPR, PBS (NOVA/Frontline), WCBS, *Time, Newsweek, Life, US News & World Report, Economist, Scientific American, Science News, Science, Rolling Stone, Popular Science, USA Today, New York Times, New York Times (Science Times), Washington Post, Boston Globe, London Times, Irish Times,* AP, UPI, *Reuters,* and numerous other television/print media.'[23]

As time went on, prizes and titles flowed his way too, with papers he had written lauded on all sides. In 2002 *Scientific American* selected him as one of the '50 leading visionaries in science'; all the work that went in to preparing the Hockey Stick certainly seemed to have been worthwhile.

As Daly had noted, one of Mann's most most significant accolades after the triumph of the Hockey Stick was his appointment as the lead author of the paleoclimate chapter in the

IPCC's Third Assessment Report of 2001. Again, we can only stand back in admiration that someone who had published his PhD a matter of a year or so earlier could be invited to head the team writing one of the most critical chapters in one of the most important scientific reports written for decades. Mann had certainly made an impact in the climate world.

Mann's position as lead author did present an apparent problem, however, since in that position he had a clear conflict of interest in assessing the published literature – he was going to be considering his own work. It is unfortunate then that the Hockey Stick was given extraordinary prominence in the Third Assessment Report, particularly in Mann's own chapter on paleoclimate. In fact the whole IPCC report started to look like a locker room, it was so full of hockey sticks. As one observer noted:

> [The Hockey Stick] appears as Figure 1b in the Working Group 1 Summary for Policymakers, Figure 5 in the Technical Summary, twice in Chapter 2 (Figures 2-20 and 2-21) of the main report, and Figures 2-3 and 9-1B in the Synthesis Report. Referring to this figure, the IPCC Summary for Policymakers (p. 3) claimed it is likely 'that the 1990s has been the warmest decade and 1998 the warmest year of the millennium' for the Northern Hemisphere.[1]

The IPCC report also 'bigged up' the paper's claims to statistical sophistication, stating that the reconstruction had 'significant skill in independent cross-validation tests'. Whenever the Hockey Stick appeared, it was larger, bolder and more colourful than any other temperature series presented. Mann must have been thrilled with the report. The final icing on the cake was when the IPCC chairman, Sir John Houghton, announcing the publication of the report, sat in front of an enormous blow-up of the Hockey Stick itself. This was Mann's moment of triumph: 1998 was officially the warmest year of the millennium, a stunning recognition of his work.

In the years that followed more and more interest was focused on the Hockey Stick. In particular, it was one of the key arguments used to support the need for the Kyoto treaty. Citations of Mann's work flooded in and its influence and importance grew without restraint, until it came to symbolise the very idea of manmade global warming. As one BBC reporter put it, 'it is hard to overestimate how influential this study has been'.[24] Every home in Canada was sent a leaflet quoting the paper's conclusions and warning of the dangers of climate change. School books told children that the Hockey Stick meant that the world had to change. Politicians told voters that only they could save people from the threat it demonstrated. Insurers, newspapers and magazines, pamphlets and websites were all in thrall to its message; the Hockey Stick swept all before it.

2 Science

Torture numbers, and they'll confess to anything.

Gregg Easterbrook

We saw in the last chapter how Mann and his team created the Hockey Stick and the impact it had on the world. Before we go on to tell the story of how his work was undone, we need to learn some paleoclimatology and a little statistics (really, just a little!), so that you can follow what the arguments were about.

Paleoclimatology

So how do you actually go about measuring the temperature of the past? For recent centuries, it's relatively straightforward. Thermometer records go back at least a hundred and fifty years, and in some places, even further than that. In principle, all you need to do is to take all your thermometer readings and work out an average. Of course, many parts of the world are not covered by a thermometer record, and many of the records may be unreliable, and ways have to be found to deal with these issues. Another problem is that as you go back into the nineteenth century, the thermometer coverage of the globe becomes thinner and thinner. However, compared to the situation in earlier centuries, the twentieth century and the second half of the nineteenth can be said to be fairly well understood.

Before about 1850 though, there are very few instrumental records to speak of, and scientists have to find some other way of assessing the temperature. We saw briefly in the last chapter how attempts have been made to *directly* measure past temperatures from boreholes. This is a procedure which is fraught with difficulty and there are many confounding factors, although the approach is not necessarily worse than any of the other ways we are going to examine. Mostly though, historic temperatures are estimated indirectly using *proxies*. A proxy is simply some quantity that varies with temperature and which leaves some trace after the event that can be sampled and measured. There are lots of different kinds of proxies and we will meet many of these in the course of this story, but the most common ones, and the ones which are of most relevance to the rest of the story, are tree rings.

The basis of the theory of tree rings as a proxy for temperature is that if you pick the correct tree, it can be seen to grow more in a warm year than in a cold one. The annual growth rings will be wider and the wood will be denser. So by taking a core through the tree from the outside towards the middle, it should be possible to extract what is effectively a record of temperatures throughout the lifetime of the tree.

Not all trees respond to temperature in this way though. A tree at the edge of a desert will *not* grow more when it gets hotter because it can't get enough water – scientists say that it is *precipitation limited*. Other trees might be limited by a lack of nutrients or by competition with other species. But, the theory goes, there are some trees which are indeed limited by temperature. These are trees that are located on the upper tree lines on the sides of mountains – where the forest gives way to rock and grass – or perhaps those that are at the northern limit of their geographical range.* To a paleoclimatologist, these special trees are a kind of thermometer. By examining the width (and also the wood density) of the rings of these particular trees, it is thought that you can get an estimate of how warm it was at any point in time in the past – as long as the tree was alive then.

---

* The northern limit applies to trees in the Northern Hemisphere, the majority of trees studied in this way. Temperature-limited trees in the Southern Hemisphere are found at the southern geographical limit.

It goes without saying that you wouldn't want to base your temperature estimate on a single tree, which might be affected by the conditions in its immediate vicinity, or by insect infestation or some other unidentifiable problem. Any of these factors could affect the record in such a way as to completely ruin the temperature reconstruction. Because of this, dendroclimatologists, as the scientists who collect tree ring samples are called, put together the ring samples from a number of trees at a particular site into what they call a *chronology*, which shows the average picture of tree ring behaviour there. The idea is that all the small variations – issues with insects and so on – average out once you sample enough trees, leaving just the temperature signal behind. And from this you can, in theory, measure the temperature of the past.

In fact it is rarely this simple. There are very large numbers of confounding factors, not least the fact that even if you were to keep a tree at a constant temperature it wouldn't actually grow the same amount each year. Researchers have discovered that trees tend to grow quickly at first and then gradually less and less each year, a fact that, if uncorrected, would produce a matching slow decline in the temperatures reconstructed from the rings. To deal with this, the chronologies have to be standardised. This involves working out an expected growth curve for the trees in question, and then expressing the ring width for any given year as a percentage of the expected ring width, essentially leaving only the temperature-related information – or so the researchers hope.

Even then there are other factors that may destroy the effectiveness of the studies. For example, it has been noted by one researcher that trees within a single site can show completely opposite growth patterns – some grow more in times of higher temperatures, but others grow less.[25] If this tendency is widely replicated then the whole approach of calculating historic temperatures from tree rings is thrown into doubt.

While most of the proxies used in paleoclimate reconstructions are tree ring series, there are other types as well. Ice cores are much used, as well as speleothems (more normally known as stalagmites and stalactites and similar rock formations). Ocean sediments and corals are also used in a similar way. The extraction of a temperature record from these other proxy types mostly involves analysis of relative proportions of the isotopes of certain atoms in the proxy. The theory is that, for one reason or another, the isotope ratio will be different in a hot year to a cold one. Find the ratio in a sample of your stalagmite and you can work out the temperature. So long as you can also date the sample correctly then you can create a valid proxy series.

For a long time, temperature reconstructions were created using single proxies, but the trend has been increasingly towards multi-proxy reconstructions, using a mixture of different proxy types, and MBH98 fell into this latter category. As we will see, while using a multiproxy approach means that much more data is available, this advantage is offset by the complex statistics required, but that tale will have to wait until later in the story.

Methodology

Once you have collected together your proxies, how exactly do you go about reconstructing the temperatures of the past? There are different ways of doing this, but they all fall under one overall framework, which is what we need to describe next.

In very simple terms, we can derive temperatures of the past by calculating the mathematical relationship between tree ring width and temperature in the recent past. Once this relationship has been determined, it is quite simple to reverse it, enabling researchers to work out temperatures in the distant past from the widths of the rings of ancient trees. That's the simple explanation. Let's take a look in a bit more detail.

Paleoclimatologists have proxy records stretching back into the past for several hundred years. Let's say that we have a tree ring chronology which goes back to 1400. We can divide this 600-year period into three parts: the calibration period, the verification period and the reconstruction period (see Figure 2.1). In our example, the calibration period stretches from 1900 to 2000 and the verification period from 1850 to 1900. In these two periods we already know the temperatures because we have instrumental records. The objective of the exercise is to estimate temperatures in the reconstruction period, which runs from 1400 to 1850.

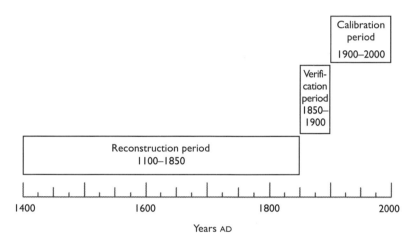

FIGURE 2.1: The periods in a temperature reconstruction

The first step of the process is called calibration. All this means is that the mathematical relationship between the proxy values and the temperature values is calculated, using only the data from the calibration period – i.e. the twentieth century proxy values and the twentieth century instrumental records. By way of a simple hypothetical example, we might work out that ring width in millimetres times ten is equal to average annual temperature in degrees centigrade. In other words a temperature rise of 1°C gives an extra 0.1 mm of ring width. Of course, in reality, it's not that simple, but this is all that is meant by calibration. The relationship is established by a statistical technique called regression analysis, which simply comes up with the linear relationship between ring width and temperature that gives the least errors, which is to say the one that best matches the actual data in the calibration period.

One problem that occurs in calibration is that any relationship we might have been able to calculate could have arisen purely by chance. In other words, just because tree ring width multiplied by ten happened to be equal to temperature in the twentieth century, that doesn't mean that it was *always* that way. For example, twentieth century ring widths might have been lower than normal, say because of insect infestation in that particular set of trees, or maybe even in trees in general. Another possibility is that the tree isn't actually responding to temperature at all, but to something else. If any of these issues were really affecting tree growth, it might be that the *normal* relationship is actually an extra 0.15 mm of growth from a rise in temperature of 1°C.

Assessing whether this kind of problem exists is what the verification period is for. We have to take the tree ring widths in the verification period (the second half of the nineteenth century in our example) and use the mathematical relationship we have just worked out to calculate what temperatures were at that time. In our example, the calibration exercise told us that 10°C gives us 1 mm of ring width. Let's now say that in 1850, the average ring width was 1.25 mm. According to our maths, this means that the average temperature should have been 12.5°C. We can now go and check the value we have calculated against the thermometer records and find out if we were right or not. We can also repeat the exercise for all the years in the verification period, and then by comparing the calculated temperatures to the actual ones we can work out how good our algorithm is. If the insect infestation we talked about above had actually occurred in the twentieth century (and remember, we probably wouldn't know that this period was abnormal) then the results could be wildly different. If the results of the verification exercise were no good, then we would have no choice but to start again.

However, assuming that the verification procedures show that the algorithm is still effective in the verification period, we can repeat the process on the reconstruction period, taking the proxy values for these earlier centuries and working backwards using the algorithm to get to the equivalent temperatures.

How many proxies?

In the simple example above, we have considered a single tree ring series and have shown how this can be used to reconstruct the temperature. But this isn't the only way to go about a reconstruction. Multiproxy reconstructions, in which a variety of different proxies are used, have been increasingly popular in paleoclimate. Once you have more than one proxy, the mathematics involved becomes considerably more complicated, but for the purposes of this story it is not necessary to go into this. The essence of what is done is that the calibration in a multiproxy reconstruction produces a weighting for each proxy series. Put the series and the weightings together and you have a multiproxy reconstruction.

If you have a large number of proxies, there are two main ways in which you can go about the calibration.[26] The first of these has been described as 'the Schweingruber method', or composite-plus-scale (CPS), and involves taking proxies that are expected to be temperature sensitive, calibrating them against local temperatures and essentially taking an average. The other way, the 'Fritts method', or climate field reconstruction, involves taking lots of proxy series, which are sometimes not even responding to their local temperatures, and seeing if some sort of correlation can be found with temperature measurements somewhere in the wider vicinity. What emerges from this latter method is essentially a weighted average of the full proxy set, with the temperature sensitive proxies having a much higher weight than the non-temperature-sensitive ones. It's this Fritts method that is the relevant one for our story. The Fritts method involves a certain leap of faith to trust that trees that are not responding to their own local temperature can nevertheless detect a signal in a wider temperature index. You have to believe in the existence of something called 'teleconnections', whereby temperatures in a possibly distant part of the world affect the climate in the locale of the tree in such a way as to affect its growth, and in a consistent manner. If this sounds implausible to you, then you are not alone. However, the reality of the mechanism is accepted by the paleoclimate community and for the purposes of our story that's what you need to know.

Principal components analysis

History of principal components analysis

We mentioned in the last chapter that Michael Mann used a technique called principal components analysis in MBH98 to summarise some of the proxy records, and it's this that we need to look at next.

Principal components analysis is not a new technique, and it may surprise many readers to read that much of the debate over the Hockey Stick was about a process that, far from being cutting edge statistics, was actually invented over a hundred years ago. The technique was developed by the English statistician Karl Pearson in the first few years of the twentieth century, and it has been in regular and uncontroversial use ever since; in recent years it has found new applications in facial recognition and image compression.

Pearson was a giant of the early development of mathematical statistics, doing groundbreaking work on regression and correlation analysis. In fact, he is often credited with turning statistics into a true science and he was the founder of the world's first university statistics department at the University of London. It is perhaps only his aggressive support for eugenics that has lead to his relative obscurity in modern times, at least outside his own specialism; he was for a time the holder of the Galton chair of Eugenics at the University of London.

What principal components analysis does

Principal components analysis (which we will refer to as PC analysis for the rest of the book) sounds complicated, and if you're a layman and you are presented with a page of matrix algebra showing how it works, it certainly looks scary as well. The good news, for this story at least, is that the essence of what it does is actually rather simple.

PC analysis simply extracts key patterns from the data in a database. Imagine you have lots and lots of data series, let's say a database of tree ring chronologies. Let's also say there are a hundred chronologies, each one being 400 years long. The first thing to notice is that there is a lot of data. How do you get your head round what is going on in there? One possible approach would be simply to average all the data. This is fine, and useful and will give you an overall picture, but it's quite possible that there are interesting things going on in the data that will be obscured by averaging. Let's say that half of the trees have a sharp uptick in ring widths in the twentieth century while most of the others show a gentle decline. In the average, you might well see pretty much nothing at all, maybe a gentle rise in the twentieth century, but nothing that will grab your attention. You would almost certainly miss the important information about the different behaviour in different trees. As we saw above, this is a real issue in paleoclimate.

So, averages can hide interesting patterns. This is where PC analysis comes in: PC analysis will extract from the database the most important pattern in the data, which is called the first principal component or PC1. A principal component is somewhat analogous to a stock exchange index. The FTSE or Dow Jones indices are simply weighted averages of the underlying share values, and stand in for those underlying datasets. In just the same way, a principal component is simply a weighted average of the underlying proxy data series, the weights calculated in such a way as to explain as much of the underlying variability as possible. When the PC1 is calculated, as well as getting the pattern, you also get a set of weightings that explain the relationship between each data series and the pattern. In the example we have just looked at, the PC1 might show the twentieth century uptick, and the weighting for each underlying proxy series showing such an uptick would be positive. The weighting for series showing a twentieth century *downtick* would be negative, indicating that they were, broadly speaking, mirror images of the PC1.

Once you have got the PC1 out, you can go on to extract the second most important pattern, the PC2, and this would be accompanied by another set of weighting coefficients explaining how each proxy series related to *that* pattern. In fact, you can go on extracting more and more patterns (and more and more weightings) from the data, right up to 100 PCs; the PC3 might show a relatively small rise in ring widths in the seventeenth century for example, the PC4 would be something different again. However it is important to note that each extra PC that you extract from the database explains less and less of the total variance in the underlying data. So while the PC1 might explain 60% of the total variance, by the time you get to PC4, you might be talking about only 6 or 7%. In other words, the PC4 is not telling you anything of much significance at all, in this example at least. How many PCs you want to extract and retain depends on how much variance each one explains and the exact nature of the data you are analysing.

The PCs are often described as being like the shadow cast by a three-dimensional object. Imagine you are holding an object, say a comb, up to the sunlight, and it is casting a shadow on the table in front of you. There are lots of ways you could hold the comb, each of which would cast a different shadow onto the table, but the one which tells you the most about the object is when you expose the face of the comb to the light. When you do this, the sun passes between the teeth and you can see all the individual points. You can tell from the shadow that what is being held up is a comb. This shadow is analogous to the first PC. Now rotate the comb through a right angle,* so that you are pointing the long edge of the comb

---

*　PCs are all at right angles to each other; they are 'orthogonal', in the jargon. In a database of 100 chronologies, you can have 100 PCs, all at right angles to each other, in a 100-dimensional space. I've explained this in a footnote for fear of some of my readers' heads exploding at the idea of a space with 100 dimensions.

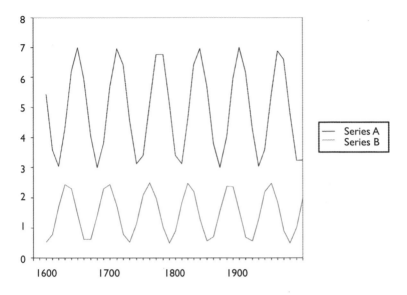

Figure 2.2: Raw series

to the sun. If you do this, the shadow cast is just a long thin line. You can see from the shadow that you are holding a long thin object, but it could be just about anything. This would be the second PC. It tells us something about the object, but not as much as the first PC. You can rotate through a right angle again and let the sunlight fall on the short edge of the comb. Here the shadow is almost meaningless. You can tell that something is being held up, but it's impossible to draw any meaningful conclusions from it. This then, is the third PC.

Centring the data

I said earlier that the way PC analysis works is difficult for the layman to follow, and this is true. Nevertheless, if you are going to follow the story you will need to know just a couple of things about the actual mechanics of performing PC analysis. Again, these are not terribly complicated, so bear with me. But make sure you understand these next few paragraphs because they are critical to the later story. Here goes.

We have our database of tree ring series. Each column in the database represents a single chronology, the average of all the trees sampled in that particular site. Let's say we have a hundred chronologies and therefore a hundred columns. Each row represents the years; let's say we have 400 years and therefore 400 rows. Each cell contains the average ring width for that chronology in that particular year.

Because of the way the underlying mathematics of the PC calculation works, before you can start crunching the numbers, every chronology in the database of tree ring chronologies has to be adjusted to a mean of zero – a process called 'centring'. This is mathematically quite simple – high school level maths in fact. Take your first chronology. First, you calculate the average tree ring width for that chronology. Then one by one, you go down each of the 400 ring measurements for the chronology, and you take away the series average you have just calculated. And that's it. See? That was easy wasn't it? Now repeat the process for the other 99 series and you are ready to start the PC calculation.

Let's look at it what happens to the tree ring data on some graphs so we can see the effect of the centring. We will look at two dummy tree ring series (they're nothing like real tree rings at all, but the effect of the centring is clearer if you do it this way). Figure 2.2 shows the uncentred data. There are two series, with Series A having a slightly higher mean than Series B – it's higher up the chart.

The effect of taking away the relevant series average from each value is to slide both lines down the page until they hover above the x-axis. In other words, the two resulting 'centred' series have an average of zero. Let's see the effect on our two tree ring series (Figure 2.3).

What you should now see is that the effect of centring the data is simply to move all the series down the chart until they are varying around an average of zero.

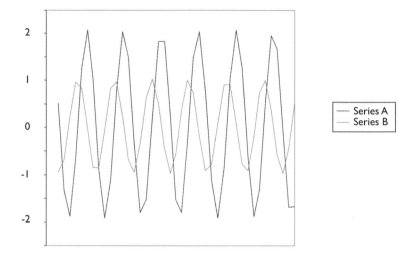

FIGURE 2.3: Centred series

And that's it: that's all you need to know. Throughout the months and years of bitter argument over Mann's Hockey Stick, this simple step was the only part of the PC analysis that was in dispute. For the purposes of the story it is not really necessary to understand anything else about how the subsequent calculations work. You can think of PC analysis as a big black box which takes the centred data and churns out as many patterns as are felt necessary. It is, however, useful to understand just a little of the detail of what happens to the centred data, as it will help explain just why centring is so important.

As it processes each proxy series, the PC algorithm calculates the square of each value in the series and sums the resulting figures down the column. The resulting number, the 'sum of squares' has a very specific meaning: it is a multiple of the series variance. This is a very important result because, of course, we are interested in series with large variances: they are the ones which will form the basis of the dominant patterns in the dataset. In other words, by centring the data, the PC algorithm points us automatically to the most important series. It is important to realise, however, that this result is only achieved if the data is first centred. Because of this, centring is considered an integral part of PC analysis.

3   Face-off

Man is in error throughout his strife.

Johann Wolfgang von Goethe

Climate Skeptics

At the start of 2003 a short comment was posted to an Internet forum for global warming sceptics by a Dutchman called Hans Erren. Erren had developed an interest in Mann's 1999 update to the Hockey Stick and had decided to dig into the study in a little more detail. It quickly became apparent to him that it was not a simple task. Mann's paper was opaque and difficult to understand. Also, some of what Erren read seemed a little odd; he could make little sense of the way the proxies were calibrated against temperature, for example. So, as people in the sceptic community would often do in these circumstances, he threw the question open to the forum to see if anyone could help or shed any light on the problem.

The readers at Climate Skeptics were a diverse bunch. People from all over the world and from very different backgrounds used the forum to exchange news and views on the development of the global warming scare, from both scientific and political viewpoints. Teachers and engineers were common on the site, and there were other interested amateurs from a huge variety of other specialisms. There were some eccentrics, of course, but there were also climatologists and meteorologists, and even the editors of a couple of scientific journals. There was a steady stream of postings covering subjects as diverse as urban heat islands, radiative physics and historic sea levels in Tasmania, all discussed in the mixture of erudition and outrage that you can find on any Internet site.

There had been a great deal of excitement on the forum in recent months. A new study by two Harvard astrophysicists, Willie Soon and Sallie Baliunas, had just been published and its appearance had caused a huge furore in the world of paleoclimate.[27] Soon and Baliunas had reviewed a large dataset of paleoclimate proxies to see how many showed the Medieval Warm Period, the Little Ice Age and the modern warming. They had concluded that the Medieval Warm Period was in fact a real, significant feature of climate history. The paper had been extremely controversial, contradicting the mainstream consensus that the Medieval Warm Period was probably only a regional phenomenon. Climatologists from around the world had fallen over themselves to attack the Soon and Baliunas paper, mainly on the grounds that many of the proxies used in the study were precipitation proxies rather than temperature proxies. So great was the uproar, in fact, that several scientists resigned from the editorial board of *Climate Research*, the journal which had published the paper in the first place. In the face of all this opposition, the paper had gained little traction in terms of changing mainstream scientific opinion on the existence of the Medieval Warm Period. It had been a huge disappointment for the sceptic community.

Another topic which had been visited and revisited by the forum over previous months, was, of course, Mann's Hockey Stick and what the climate sceptics saw as its use as a 'sales tool' in the IPCC report – pushing the idea of dangerous global warming in the face of what otherwise would have amounted to a distinct lack of hard evidence. There was no doubt in the minds of anyone posting there that it was the Hockey Stick that was driving the relentless momentum of the global warming movement. It was this centrality in the debate that had attracted the interest of Erren.

McIntyre

It's perhaps a little surprising then that Erren's posting received just a single response. Perhaps the subject had been debated too hard already; maybe it was just a quiet day, but the only person who had anything to add was Steve McIntyre, a semi-retired mining consultant from Toronto. His response though was rather encouraging:

> Hans – Just to note that I think the dissection of MBH1999 (also 1998) is very important. I've spent a fair bit of time on this and intend to reply to your posts, but plan to spend some attention on radiative physics for a little while.

McIntyre was one of the mainstays of the Climate Skeptics site, posting comments on a wide array of subjects. In recent weeks he'd spent a great deal of time discussing radiative physics, trying to understand how the IPCC came up with an expected temperature rise of 2.5°C every time atmospheric carbon dioxide doubled. He'd not really got anywhere with it so far (and in fact it remains a mystery to this day), but he was far from giving up hope. If there was an explanation to be had, he fully expected to find it.

McIntyre had always been a talented mathematician. At high school in his native Ontario, he'd been a prize winner, going on to come top in the maths exams at provincial and national level. It was pretty much a given that he'd continue his studies at university – a goal which he achieved when he graduated in pure mathematics from the University of Toronto in 1969. Postponing an opportunity to study mathematical economics at MIT, he decided to broaden his horizons and head for England, where he spent a year studying PPE (Philosophy, Politics and Economics) at Oxford.

With family commitments preventing him from taking up the MIT position when his Oxford studies were complete, he had embarked upon a career in the mining industry, where he was to spend the next thirty-odd years. He'd mainly been involved with small mineral exploration companies, his work involving preparation of listing prospectuses, analysing prospects for acquisitions and the like. This work involved a great deal of heavy duty data analysis, and gave McIntyre a facility with statistical tools which was to stand him in good stead during the later controversies over climatology.

By 2002, McIntyre was in comfortable semi-retirement in Toronto, the easy routine of children and grandchildren and visits to the squash court only interrupted by occasional scraps of consultancy work for old contacts in the mining industry. Climatology was not even on his radar. He knew nothing of the IPCC or the Hockey Stick or Michael Mann. But 2002 was to be the year in which all that began to change. We have seen that in the drive to promote the Kyoto treaty, every home in Canada was sent a leaflet about the risks of global warming. When McIntyre read in his copy that it was the warmest year in the last millennium, instead of shrugging his shoulders as most of us would do, he found himself wondering just how it was that scientists knew this. Canadians were brought up on stories of the Viking explorations during the Medieval Warm Period and the opening up of Canada by fur traders during the Little Ice Age. The claims in the leaflet seemed to be overturning a very well-embedded paradigm.

Within a matter of days, McIntyre was absorbed in the world of climatology, his reading quickly taking him to the IPCC's Third Assessment Report with its prominent display of the Hockey Stick. To someone with long experience of mining promotions it immediately struck McIntyre that someone had spent a great deal of time and money making an effective sales tool out of the Mann's graph. This observation seemed even more pertinent when McIntyre saw for the first time the confusing way the Hockey Stick had been presented in the original *Nature* paper. Improving the presentation of the graph for inclusion in the IPCC report didn't make it wrong, of course, but it was clearly intended to be persuasive.

While McIntyre's readings in climatology broadened, he also began discussing the IPCC's claims of unprecedented warmth with friends and acquaintances. His contacts in the mining industry were particularly interesting on the subject. Familiar as they were with the long-term history of the Earth, many of the geologists McIntyre spoke to had strong opinions on claims that recent temperatures were unprecedented and most were highly sceptical of the idea. When it came to the Hockey Stick itself, mining people – geologists, lawyers and accountants – were openly contemptuous. Hockey sticks were a well known phenomenon in the business world, and McIntyre's contacts had seen far too many dubious mining promotions and dotcom revenue projections to take such a thing seriously. The contrasting reactions to the Hockey Stick of politicians and business people – on the one hand doom-laded predictions of catastrophe and on the other open ridicule – acted as a spur to McIntyre, who flung himself headlong into the world of climatology.

Proxies

By the time Erren posted up his initial questions in April 2003, McIntyre was therefore already thinking far ahead of the Dutchman. Like Erren, he had discovered that getting to the bottom of the Hockey Stick was no easy task. McIntyre was used to the corporate world, where tight regulation and nervous investors meant that clear explanations of methods and results were an absolute necessity. In the mining industry, a garbled explanation of what a drill core contained would send potential investors running for the door, so McIntyre had been completely bemused by the obscure language and vague allusions that littered Mann's papers. It was very hard for an outsider to get any purchase on the slippery slope of Mann's narrative. Working out what he had done was clearly going to take a great deal of effort.

Despite the difficulties, McIntyre had made a reasonable start on understanding the two Mann papers. He could see that Mann had used a network of 112 proxy series, and in fact behind the scenes there was even more data than this. For some parts of the world, particularly North America, the archive of tree ring data was very large and if it had been used directly in the calculations, these parts of the world would have been grossly over-represented. So in order to avoid this problem, Mann had used PC analysis to distil these series down into a few key patterns. Therefore, while some of the 112 series that went into the final calculation were single site chronologies, others were PC series, representing a summary of many sites. This was unobjectionable and must have seemed an eminently reasonable step.

MBH98 was a multiproxy study, using a variety of different proxy records to reconstruct temperature. As we saw in Chapter 2, there are two main approaches to temperature reconstruction: either find the relationship between the individual proxies and their local temperature and calculate an average (the Schweingruber method) or find the relationship between the full set of proxies and some regional temperature index (the Fritts method). Mann had taken the Fritts approach although he referred to it in the paper as a 'climate field reconstruction'.

The proxy series used were a real mixture. The majority were tree-rings, but among the 112 there were also ice cores, corals and ice melt records, together with a few oddities. Several of his series were rainfall records, included presumably because of a possible correlation between rainfall and temperature. This was surprising in view of the furore over the Soon and Baliunas paper. With so many voices of outrage having been raised at Soon's use of precipitation records, it is amazing that nobody in the paleoclimate mainstream seemed to have spotted that Mann had done the same thing a few years earlier. In fact, at the time of the Soon controversy, Mann and the group of colleagues who would later be known as the 'Hockey Team'* had themselves written a critique of Soon and Baliunas in which they said,

---

*   It has been suggested that the use of the expression 'Hockey Team' to describe Mann and the group of climatologists associated with him is derogatory and amounts to accusation of a conspiracy. Its earliest use in this context appears, however, to have been due to Mann himself.[28]

In drawing inferences regarding past regional temperature
changes from proxy records, it is essential to assess proxy data
for actual sensitivity to past temperature variability . . .

. . . and went on to note that it was 'patently invalid' to fail to do so.[29]

As well as using proxies which he had previously deemed
invalid, Mann had also used some 'proxies' that weren't actually
proxies at all, such as the Central England Temperature Record.
This, as we saw in Chapter 1, is a long series of thermometer
readings extending back to the seventeenth century. Sceptics
unkindly noted that the eccentric mix of proxies, some of which
had an unproven relationship with temperature, was hardly a
reliable methodology – merely throwing everything bar the kitchen
sink into the database was not necessarily a recipe for success. In
fact, quite the opposite.

Temperature series

As we have seen, the first step in a temperature reconstruction is to
calibrate the proxies against instrumental temperatures. The most
reliable surface temperature record is agreed by many researchers to
be the series prepared by Professor Phil Jones of the Climatic
Research Unit (CRU) at the University of East Anglia; the record
is generally known by the acronym HADCRUT.* This database
consists of figures compiled from a host of weather stations and
other sources of temperature data such as ocean buoys and ships, all
put together in tabular form. Imagine a huge spreadsheet, rather
like the one for the proxies. Each column represents a single
weather station and each row represents a period of time. In each
cell is the temperature for that period for each station. It's pretty
simple stuff. There are plenty of problems with the temperature
data, but all the raw figures were cleaned and adjusted for known
issues. The data that Mann used was the CRU's best stab at what

---

*   The acronym is derived from the names of the land temperature series
      (CRUTEM) and the sea temperature series (HADSST) used in its
      preparation.

the actual temperatures had been for the previous 150-odd years, and as we've noted, CRU's data was reckoned to be the best. Whether these adjustments were correct, or valid, is another question entirely, and it was one which had already attracted the attention of climate sceptics, but it is not one with which we need concern ourselves as it does not bear directly on this story.

Taking all the temperature data, Mann had used PC analysis to summarise the records into its key patterns, reducing 1000 or more individual temperature series into just 16 PCs. In other words, PC analysis was used twice in the study – once on the proxy records and once on the instrumental temperature records. The temperature PCs were the climate patterns, or 'climate fields' as Mann called them, that he was going to try to recreate in the reconstruction by calculating the mathematical relationship between them and the tree ring data. Having recreated the temperature PCs he would then be able to reverse the PC analysis using a procedure called *expansion*, taking the PCs and their respective weightings to recreate an approximation of how the underlying temperature data would have looked in the past. From there it was a relatively simple step to recreate the northern hemisphere temperature average for the last 600 years.

Regression

With his data ready, Mann's next task was to calibrate the proxies against the temperatures to establish the mathematical relationship between them. As we saw in Chapter 2, this is done using regression analysis.

In the simple case of performing a regression on a single set of data points, this is a relatively straightforward exercise in fitting a line through a cloud of points. However, MBH98 involved a multivariate calibration – in other words there were multiple sets of data needing to be calibrated: the 112 proxies and the 16 temperature PCs. But quite how Mann had gone about this much more complex process was a mystery that was not revealed in the text of the paper. That would have to wait for another day.

McIntyre's conclusions on the first analysis

So, after spending a good few hours perusing Mann's papers McIntyre had a sense of how the studies had been performed, at least at a perfunctory level. But already questions were emerging that suggested that there might be more to MBH98 than met the eye. Something was not quite right with the Hockey Stick papers. As he said to the sceptics:

> I am not able to comment at present on his methodology, but my sense is that there are weaknesses to it, which deserve careful auditing.

On this score at least, McIntyre was not mistaken.

First look at divergence and Briffa

Meanwhile, there was another issue that was nagging at the back of McIntyre's mind. This concerned some comments made by another prominent paleoclimatologist called Keith Briffa about tree ring records in the twentieth century. McIntyre couldn't recall the reference but distinctly remembered that Briffa had said that twentieth century ring widths hadn't gone up alongside the warming that had been seen in the instrumental record. But if this was the case, then how had Mann got a sharp rise in twentieth century temperatures from a study which was dominated by tree rings? And if tree rings and temperatures didn't move in tandem, wasn't the very basis of paleoclimatology thrown into doubt?

The more he looked at the paper, the more McIntyre found questions that were not answered in the text and issues that needed to be checked or clarified. But even at this point, the Hockey Stick remained a secondary consideration for the Canadian. Over the next couple of months he continued to post comments at Climate Skeptics on a range of different topics – climate models, radiative physics and so on – and very little about the Hockey Stick. He even commented that he felt that the best contribution that the sceptic community could make would be to concentrate on publicising the lack of any adequate disclosure from the IPCC. But in the background he was still working away at Mann's paper. He had spent quite a lot of time looking at some of the available proxy records, particularly those not based on tree rings. So far as he could see, *none* of these showed the uptick that would have been expected from rising temperatures in the twentieth century. On the tree ring front, moreover, he had located all the Briffa papers that discussed the divergence of tree ring growth and temperatures, and their conclusions were just as he had recalled them:

> Averaged around the Northern Hemisphere, early tree growth . . . can be seen to follow . . . trends in recorded summer temperatures, tracking the rise to the relatively high levels of the 1930s and 1940s and the subsequent fall in the 1950s. However, although temperatures rose again after the mid-1960s and reached unprecedentedly high recorded levels by the late 1980s, hemispheric tree growth fell consistently after 1940, and in the late 1970s and 1980s reached levels as low as those attained in the cool 1880s . . . The reason for this increasingly apparent and widespread phenomenon is not known but any one, or a combination, of several factors might be involved.[30]

This was a rather remarkable finding, given the prominence accorded to tree ring reconstructions in the IPCC's report. It seemed simply inconceivable that a 'widespread phenomenon' – and one that could potentially undermine much of the science of paleoclimatology – should barely warrant a mention* in the main study in which the findings of climate science were reported to the public. As McIntyre observed, Mann should presumably have observed this divergence effect too, but if he had, it was not

---

* The section on tree rings mentioned Briffa's work, together with a possible explanation for it from another researcher, but given the criticality of the question, two sentences seem rather inadequate.

mentioned anywhere in his papers. His failure to observe the divergence, or worse, if he had failed to report it at all, must seriously undermine the credibility of the Hockey Stick papers.

McIntyre starts to concentrate on paleoclimate

By the start of April, McIntyre was starting to rein back on his other climatological interests in order to concentrate his efforts on the paleoclimate papers, and in particular the Hockey Stick. He announced this intention to the Climate Skeptics forum. There was, he said, 'an opportunity for some quite provocative analyses'. At the time though, the reaction from his fellow sceptics was only one of mild interest. There was no great sense of expectation or excitement. There was no inkling of the controversy that was to be unleashed in the coming months and years.

First detailed analysis of Mann's work

Within a matter of days of his announcement, McIntyre was posting findings to the Climate Skeptics forum. He had now worked through Mann's explanation of his methodology and he had soldiered his way through the matrix algebra. It was still very strange. The use of PC analysis was new in the realm of paleoclimate and Mann had made no attempt to prove the validity of the technique in the field, instead relying on a bold assertion that it was better than the alternatives. In view of this and given the surprising results – with no Medieval Warm Period or Little Ice Age visible in the reconstruction – one might have expected that experts in the field would have questioned whether Mann's novel procedures might have been a factor in his anomalous results. But despite a thorough search of the literature, there was no sign that anyone else had seen fit to probe the issue further. Nor had any other researchers adopted Mann's methodology in the five years since his paper had been published. Given how often the Hockey Stick had been cited in the scientific literature, these were very surprising observations, which seemed to suggest that paleoclimatologists liked Mann's results rather more than they liked his methodology.

Lies, damned lies and calibration statistics

Another issue was also attracting McIntyre's attention. During his calibration exercise, Mann had assessed how well the temperature data matched up against the proxies by calculating various statistical measures – in other words, numbers that acted as a score of how good the match was. The main way he did this was using a measure that he called the beta (β), which he described as being 'a quite rigorous measure of the similarity between two variables'.[14]

This was a somewhat surprising choice since the beta statistic was virtually unheard of outside climatology circles. (It also goes by the names of the 'resolved variance statistic' or the 'reduction of error (RE) statistic' – the latter being the term we will use to refer to it henceforward.) With his experience in statistics, McIntyre was aware that there was great danger in using novel measures like these, whose mathematical behaviour hadn't been thoroughly researched and documented by statisticians. The statistical literature was littered with examples where particular statistical measures gave results which misled in certain circumstances. Mann had left no clue as to why he had preferred the RE rather than the more normal measures of correlation, such as the correlation (r), the correlation squared (R^2) or the CE statistic. The behaviour of all of these measures under a wide range of scenarios was well documented, so McIntyre was surprised not to see an explanation.

Mann indicated in the paper that the r and R^2 *had* also been calculated, which might have provided some reassurance to McIntyre but for the fact that the results of these calculations were not presented for the calibration step anywhere in the paper or in the online supplementary information. However, by now McIntyre had got hold of the data for the second Hockey Stick paper, MBH99 – the extension back to the year 1000 – so he was able to start to make some significant progress in answering some of these questions. Because the number of proxies used in MBH99 was so small (there being very few proxies that extended so far into the past) it was a relatively straightforward task for McIntyre to recreate

Mann's calibration and to calculate some of the correlation statistics for himself. The results were eye-opening, to say the least. As he reported to the climate sceptics:

> The R^2 . . . ranges from −0.006 to 0.454; on this basis, only 2 of 13 proxies have R^2 adjusted over 0.25, and 7 of 13 have values under 0.1 . . .

To put this in perspective, R^2 will normally vary between 0 and 1. A score of 0 indicates that there is no correlation at all, and 1 indicates perfect correlation. So what McIntyre was seeing was that the proxies and the temperature PCs didn't really match up very well, according to a standard measure of correlation. The best among them were not even halfway good, and some simply showed no correlation at all. Could this explain why Mann was so enthusiastic about the RE statistic, the climatologists' own measure of correlation?

Struck by this result, McIntyre repeated the calculation in a slightly different way. Previously he had measured the correlation of each proxy against the full set of temperature PCs – sixteen in all. This time he restricted himself to only the temperature PC1, which was the only temperature PC that was used to recreate temperatures in the early part of the MBH99 reconstruction – it was the only one that really mattered. And when McIntyre saw the results, they turned out to be even worse, with only one proxy achieving an R^2 score of more than 0.2. As McIntyre noted:

> The low correlations against [the temperature] PC1 need to be carefully noted. Some/most of the datasets are essentially uncorrelated to [it]. There is no mention in MBH98 or MBH99 of these low correlations.

Many people might have sat back and stopped at this point but McIntyre decided to take his analysis a step further. He started to swap the proxies for some completely unrelated datasets in order to see what sort of correlation scores he could get. For example, 79 consecutive values from a table of Eurodollar six month *interest rates* achieved an R^2 of 0.595, far in excess of the proxies. Concentrations of potassium from an ice core dug out of a glacier on Mount Everest scored 0.444. He also tried regressing nineteenth century proxy data against *twentieth* century temperatures and found no great difference in the R^2 score to those achieved when the correct proxy data was used. The conclusion was clear: if you could get such high correlations from obviously unrelated data, what meaning could there be in the proxies, whose scores were so much lower? Taken literally, the implication was that it would be better to recreate historical temperatures with Eurodollar interest rates than with the proxies, a conclusion which was obviously nonsense.

As we saw in Chapter 2, after calibrating the proxies and the temperature records, it is necessary to demonstrate that any correlation between the two is real rather than spurious by checking the temperature reconstruction against historic temperatures in the verification period. McIntyre explained the details to the sceptics:

> Mann's verification was to show that he could get similar correlations in a withheld period 1854–1901. However, since all of the correlations are at rather low levels and similar correlation levels are obtained with completely spurious series or against unrelated time periods, it is not clear to me that this verification exercise shows that statistical significance has been achieved, or alternatively and perhaps more importantly, that there has been any material narrowing of error bands.

In essence then, a whole new set of correlation statistics had been generated, just like the RE statistic mentioned above, this time measuring the correlation between the actual temperatures and the reconstructed temperatures generated by Mann's mathematical model in the period 1854 to 1901. What McIntyre was pointing out though, was that if the correlations were insignificant during calibration (according to the standard R^2 measure), what was the point of even following the analysis through to verification?

MBH99 data

On an even simpler level, there was a great deal about the data used in the MBH99 reconstruction that was peculiar. Of the 13 proxy series which had generated the reconstruction for the period between the years 1000 and 1399, four were ice cores from a single small ice cap area of Peru, called Quelccaya, while a further three were the first three PCs from a PC analysis of tree rings in the south-west of the USA.

You will remember that in the first Hockey Stick paper, Mann had noted that his roster of proxy series was too heavily weighted to the USA, and so he had distilled down the data using PC analysis, creating a summary of the main patterns in that area. Here, in the second paper, exactly the same problem seemed to exist, with 4 of 13 series being from a single location. Why then, had he not summarised the data in the same way? If having too many series in a single area was a problem in the first Hockey Stick paper, why was it not a problem for ice cores in the second?

The southwest USA tree rings also didn't seem quite right. Why had Mann retained the first *three* PCs from this analysis? Surely if you were trying to recreate only the first temperature PC (the Northern Hemisphere mean temperature) you would only need the first proxy PC1? Remember, the tree growth was supposed to respond to temperature, so the temperature signal should have been right there in the first PC.

McIntyre decided to examine what would have happened if Mann had prepared his proxy data along the more logical lines this analysis was suggesting. He prepared a PC analysis of the Quelccaya ice cores and eliminated the second and third PCs of the south-west USA tree rings. This left him with eight series, which he regressed against the temperature PC, creating a completely new calibration.

By now he wasn't expecting there to be any significant correlation between the temperatures and the proxies, and he wasn't disappointed. The R^2 score reached a measly 0.385, virtually indistinguishable from what you would get from random numbers. In fact, the proxies didn't even exhibit any sort of correlation with *each other* – in other words they were all wiggling up and down apparently independently, making a nonsense of the idea of extracting some sort of a common temperature signal from them.

Reaction from Climate Skeptics

The surge of alarming results from McIntyre's analysis was starting to attract the attention of the other members of Climate Skeptics. There was a flurry of comments from the regular readers, some of whom started pressing McIntyre to submit his findings for publication. One less enthusiastic commenter pointed out that there were at least five independent studies that had arrived at the same conclusions as Mann and that on the face of it, it seemed extremely unlikely that they had *all* used flawed approaches. However McIntyre had already been thinking about this and, as he pointed out, there seemed to be a certain amount of commonality of the data used, his point being that perhaps the independent analyses weren't *quite* as independent as they seemed. But in the meantime, he was beavering away, collecting the data from the other papers, ready to see just what a careful study of their findings might reveal.

With his long experience of the mining industry, McIntyre was well equipped to get to the underlying truth of a compelling graphic like the Hockey Stick. In a posting on Climate Skeptics, he pondered some similarities between the work of a mining analyst and a climate auditor.

> [A]n individual time-series has much the same function as a drill-hole. Where there is an ore-body (i.e. a significant 'signal'), the information in the individual drill-holes is not subtle. Any analyst recommending a mining stock has to

look at the drill holes – not just the compilations. The application of valid statistical methods to invalid data can result in fiascos like Bre-X [a famous mining scandal in which drill-hole results had been 'improved']. 'Adjustments' are always something to be suspected.

Mining promoters would often come up with carefully manipulated sets of data which they would summarise in a compelling graph in order to convince potential investors to part with their money. In his mining days McIntyre had dealt with this kind of promotional graphic by turning to the raw data in graphical form to 'get a feel for the numbers'. Only by looking at *raw* data could he be sure that he was seeing what the rocks were saying rather than the results of some statistical shenanigans overlaid on the raw data specifically to fox the unwary.

As he got his hands on more and more proxy data, McIntyre became frustrated by the fact that most of the proxies stopped at around 1980. This meant that the dramatic warming of the 1980s and 1990s, which should have vastly inflated the ring widths, couldn't be seen. As he tartly observed:

> If the IPCC were a feasibility study for a mere $1 billion investment in a factory or a mine, you can be sure that the engineers would bring all this type of data up to date. The casualness of the IPCC process in respect to not bringing the data up to date (but relying on it for sales presentations) is really quite awe-inspiring.

However, where there *were* up-to-date numbers, it became increasingly clear that the raw proxies actually showed twentieth century trends that were, broadly speaking, absolutely normal. There was no sign of the proxies breaking new ground. How was it then, that the Hockey Stick showed twentieth century temperatures that were unprecedented in a thousand years, when they were based on the same proxy data? Was it something to do with the particular proxies that Mann had used? Or was it perhaps an artefact of the PC methodology? Only time would tell.

Making contact

At the same time as doing this work on MBH99 McIntyre had also made a start on the Hockey Stick proper – the original MBH98 paper. His first step was to contact Mann directly in order to get hold of the proxy data. On 8 April 2003 he wrote the first of what was to be many emails to Mann.

> Dear Dr Mann,
> I have been studying MBH98 and 99. I located datasets for the 13 series used in [MBH99 . . .] and was interested in locating similar information on the 112 proxies referred to in MBH98 . . . (the listing at [the official website] is for 390 datasets, and I gather/presume that many of these listed datasets have been condensed into PCs, as mentioned in the paper itself). Thank you for your attention.
> Yours truly,
> Stephen McIntyre, Toronto, Canada[31]

Mann's response was almost immediate, but while it was quite courteous, it also contained something of a surprise.

> Dear Mr McIntyre, These data are available on an anonymous FTP site we have set up. I've forgotten the exact location, but I've asked my colleague Dr Scott Rutherford if he can provide you with that information.
> Best regards,
> Mike Mann [31]

So apparently, the author of one of the most important scientific papers in recent decades didn't know where the data he had used in that study was located. This seemed a little odd, but the quick reply and its businesslike tone boded well.

A couple of days later, on 11 April, Mann's assistant, Scott Rutherford, had finished his search for the data and emailed McIntyre to tell him what he had come up with. Again the response was not what would have been expected:

> Steve,
> The proxies aren't actually all in one FTP site (at least not to my knowledge). I can get them together if you give me a few days. Do you want the raw 300+ proxies or the 112 that were used in the MBH98 reconstruction?
> Scott[31]

So, according to Rutherford, and somewhat contrary to what Mann had said, the data wasn't even in one place. Stranger and stranger. The fact that the data had never been compiled into a single record also strongly suggested that *nobody* had ever asked to see the figures before. Nobody had ever tried to replicate Mann's study. However, McIntyre didn't raise the question with Rutherford or Mann, instead indicating that the 112 distilled proxies would suffice, and offering to organise the data to make things easier for anyone who might want to use it in future. (The 300+ series that Rutherford was referring to were the raw proxy data, some of which would be summarised down using PC analysis, leaving just 112 to be put into the calibration.)

Arrival of the data

After a couple of weeks, and following a few gentle reminders, an email from Scott Rutherford popped into McIntyre's inbox, indicating that the proxy data was now available on Rutherford's FTP site at the University of Virginia.

> Steve,
> OK, I think I have it all straight now. You can get the data via anonymous FTP at holocene.evsc.virginia.edu/pub/sdr
> Regards,
> Scott [31]

The data took the form of a simple but fairly large text file called pcproxy.txt which contained the values for the 112 proxies for each of the 600 years of the MBH98 paper: each column of the file represented the values for one proxy, and the rows were the years. At the topmost rows of the file were the data from the oldest proxies, the first starting in the year 1400. They ran down through the centuries, with the most recent figures being from 1980. Rutherford had helpfully attached a list of descriptions of each series and provided a link to a website that would enable McIntyre to work out what each one was. At last having the actual data in his hands, McIntyre was ready to start the audit.

A strange shortage of hockey sticks

With all the amazing findings from his work on MBH99 he must have been intrigued as to what he would find in the MBH98 data. Once again, the reality was to be every bit as surprising as he had expected, although the sheer number of issues must have taken even him aback.

After loading up the data into a spreadsheet McIntyre plotted all 112 series in separate graphs. In this way he hoped to be able to see clearly which series were driving the twentieth century warming in the reconstruction. He quickly noticed that the more prominent anomalies were coming, not from the individual proxy series, but from the PC series – where multiple proxy records had been summarised to stop their geographic area being overrepresented. This didn't seem quite right: if the output from the PC analyses showed significant twentieth century warming (i.e. wider tree rings) then the tree ring series that went in as raw proxies must have shown the same warming too. But why then did the other raw proxy series – the ones that hadn't been put through a PC analysis – show nothing of the sort?

Qualitatively then, there was a problem, but McIntyre needed to get a firmer grasp on the scope of the issue. He needed to define which series had a hockey stick shape. In order to do this, he had to define 'hockey stick shaped' in mathematical terms and the definition he came up with was this: any series where the value in the year 1975 was greater than one standard deviation from the series average. This was a crude measure, but it would help to make sense of the full roster of proxies. Armed with his new definition he analysed the full dataset and discovered that hockey stick shaped series constituted just 13 of the original 112. In other words, most of what went into the reconstruction was essentially just noise and had no effect on the final result. The shape of Mann's temperature reconstruction emanated from a short list of hockey sticks, which is shown in Table 3.1.

TABLE 3.1: Hockey stick shaped series in MBH98

SERIES NO	NAME	1975 EXCURSION
53	Gaspé	3.05
96	Australia PC1	2.47
65	Mongolia, Tarvagatny Pass	1.50
17	West Greenland Ice Melt	1.44
84	North America PC1	1.39
93	South America PC1	1.35
91	North America PC8	1.28
58	Coppermine River, Canada	1.28
54	Arrigetech	1.22
85	North America PC2	1.18
60	Churchhill Canada	1.17
94	South America PC2	1.14
61	Castle Penin, Canada	1.14

The 1975 excursion is the 1975 distance from the series mean measured in standard deviations.

All but one of the series were derived from tree rings, and no less than six were PC series. This raised the question of why was it, if all of the proxies could carry a temperature signal, it only seemed to be

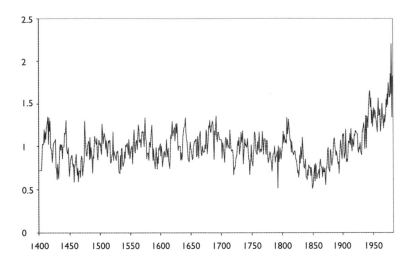

FIGURE 3.1: The Gaspé series, showing the dramatic twentieth century uptick in ring widths.

the tree rings which were picking up the rapid warming seen in the twentieth century? Were the non-tree-ring series actually temperature proxies at all?

At the top of the list, the enormous deviation from the mean of Series 53 was striking, but here, once again, McIntyre's mining background proved to be surprisingly useful. As he pointed out to the climate sceptics:

> Series 53 is from Gaspé, Québec. At the time that this series was making its big excursion, I happened to be working for the Canadian company which owned the Gaspé. It's certainly hard to think of a reason why trees in Gaspé were making a 3 [standard deviation] excursion in the 1970s.

McIntyre smelt a rat, and it wasn't to be the last time either. Might not the database be infested with them?

Principal components

The trouble started while McIntyre was trying to replicate one of the PC analyses. As we have seen, because certain parts of the world were overrepresented in the tree ring archives, Mann had used PC analysis to reduce this mass of records to a few key patterns. He had identified five regions as needing to be distilled down in this way: Texas–Oklahoma, Texas–Mexico, International Tree Ring Database (ITRDB US), South America and Australia–New Zealand. Even without considering the underlying data or calculations, these groupings looked a trifle strange. Why should Texas–Oklahoma and Texas–Mexico be worthy of a PC compilation in their own right when there was a US-wide compilation available – the ITRDB US series? Surely it would have made more sense just to lump them all together? There may of course have been a rational explanation for this, but Mann's paper was silent on the subject.

Meanwhile McIntyre's roving eye alighted on one of the *individual* proxies, series 106, which Mann had given the designation MEXI001. As the name suggested, this was a series from Mexico. The question this raised in McIntyre's mind was why Mann had not included this series in the Texas–Mexico PC compilation? It made no sense. He investigated further and it turned out that this wasn't an isolated issue either. Series 49–64 were all from North America, 46 and 47 were South American, while 43 and 45 were from the Australia–New Zealand region. It would have made far more sense to put all of these into the relevant PC compilations. Again, Mann gave no rationale in the paper for what appeared to be an entirely illogical approach to the question of data compilation.

Mann had indicated that the individual sites for each of these regions could be found in his online Supplementary Information. However, it turned out that this information wasn't actually complete although, with only one or two exceptions, McIntyre was able to identify the missing sites by other means.

While he was suspicious of some of the individual proxies used, McIntyre was expecting the PC replication to be straightforward, but it turned out that it was far from simple. After downloading clean data from the World Data Center for Paleoclimatology, and having centred it ready for analysis, McIntyre immediately ran into problems in replicating what Mann had done. When he looked at the Australia–New Zealand series, he had data going back to 1625, but Mann's PC series only started in 1750. Why was that? And worse, in Texas–Mexico, the data only went back to 1760 and yet Mann had PC calculations going back to 1400. From data which went back 220 years, Mann had extracted a main pattern that went back nearly 600. Now this really was a mystery; PC analysis will not work if the series used have missing data – its default reaction to a missing value would be for the algorithm to fail. So if these were the series that Mann had actually used, how had he been able to get his PC analysis to work? He had obviously done *something* – used alternative versions of the data perhaps – but what? Again, the paper was no help. Scratching his head somewhat, McIntyre put the problem aside and decided to take a look at the data for the 112 series used in the calibration, that is, the figures as they were *after* the PC analysis. Perhaps seeing some actual Mann data – some cold hard numbers – would shed some light on the issue.

Dodgy data

To a lay reader, the columns of proxy series are pretty much indecipherable – rows of numbers, columns of numbers, like so many grains of sand. But to an experienced eye, used to picking out patterns from dense screeds of data, certain things can jump out and demand to be examined more closely. So when McIntyre started to study Mann's proxy data series, it wasn't long before he noticed something odd.

The Texas–Mexico chronologies had been reduced to nine PCs, which appeared as proxy numbers 72 to 80 out of the 112. What McIntyre noticed was that for the year 1980, the values for each of these series were the same. This wasn't a case of rounding making them appear the same; the value was *identical*, to seven decimal places: 0.0230304. This simply could not be correct. It looked almost as if someone had copied the data from one series and pasted it over the others.

When auditors of companies' financial statements find errors, they have no alternative but to extend their testing and see if what they have found is an isolated error, or whether they are scratching at the surface of something more serious. McIntyre's climate audit was no different, and he commenced a careful examination of all of the 112 proxy series. It wasn't long before more and more oddities of the same kind were tumbling out of the woodwork. The copying of 1980 values that had infected the Texas–Mexico PCs was also seen in the three PC proxy series known as the Vaganov PCs, as well as four of the nine PC series derived from the International Tree Ring Database (ITRDB). Again, each one was identical to seven decimal places. Another strange feature was also observed in these PC series. All but two started in a year ending either in 99 or 49. For example, Series 73 started in the year 1499, 74 and 75 in 1599 and 76 to 80 in 1699. The PC series were meant to have started either on the century or half-century, so it looked as if what had happened was a simple clerical error – some of the data appeared to have been copied into the file at the wrong row and then the missing data at the bottom of the column had been infilled by copying from an adjacent series.

These three groups of affected records – Vaganov, Texas–Mexico and ITRDB – amounted to a total of 16 series. Assuming one in each group was actually correct (that is to say, the one from which the infilled value had been copied) then that left thirteen which were incorrect, or more than ten percent of the series used in the reconstruction.

Infilling

More digging into the proxy records turned up a different kind of error: proxy number 45 had the same value in every year from 1978 to 1982.

Series 46, on the other hand, was identical from 1974 to 1980. And as McIntyre looked across the columns he saw similar problems in still more of the series. Series 51 was the same. So was 52. And 54, 56 and 58 as well. It looked as though some of the numbers had been missing from the series, and rather than discard the proxy or locate the missing data, someone had infilled the missing numbers with the final available figure. Further across the file, the same thing could be seen in another sequence of series, numbers 93, 94, 95, 96, 97, 98 and 99. And number 6 too. All of these series had their final values infilled in the same way: by simple means of copying the last available value into the empty cells. Elsewhere though, Series 53 was infilled for four years at the beginning, and Series 3 showed every sign of having been infilled for a period during the 1950s.

The *pièce de resistance* though, was Series 50. Here, the values for the entire period from 1962 to 1982 were copied from Series 49. With a little digging, McIntyre was able to work out that, although Mann had attributed both 49 and 50 to a study by Fritts and Shao, Series 49 was in fact derived from an entirely different study, by Keith Briffa.

Most of the infilling was happening, as you can see, during the modern era, which is when you would expect it to be easiest to obtain complete data, but more importantly it was during the calibration. Inaccurate results here would have a direct knock-on effect on the reconstruction of historical temperatures. To be fair to Mann, he had said in the online supplementary information to the original paper that there had been some infilling in the data:

> Small gaps have been interpolated. If records terminate slightly before the end of the 1902–1980 training interval, they are extended by persistence to 1980.[14]

but it must also be said that this didn't cover the copying of data from adjacent series and nor did it really give the reader a sense of the sheer amount of infilling that had seemingly gone on. In all, more than a third of the series had been affected in this way, and for Mann to have been fair to the reader this should have been disclosed, together with some assessment of the potential impact on the reconstruction.

Where did the data come from?

It got worse. Series 10 and 11 were two instrumental records – the Central England Temperature Record (CETR) and the Central Europe Temperature Record, (you will remember that not all of the inputs into the regression were proxies – some were actual temperature readings). When McIntyre checked these back to the original publicly archived data he found that the figures didn't match. Where was Mann getting his numbers from? With a little digging the answer turned out to be that the figures were *actually* based on the average of June, July and August for each year, rather than the full year average. The problem with this was that Mann was trying to recreate an annual average temperature, not a summer average, so why choose the summer figures? All the other instrumental records were full year averages, so why should CETR be different?

CETR is the one of the oldest uninterrupted temperature records in the world. It measures the average temperature for an area roughly corresponding to the English Midlands, but also includes areas which an Englishman would normally consider 'the North'. It was started in the year 1659, giving it the best part of 350 years of uninterrupted measurements. It is hard then to understand why Mann should have truncated the record at 1730, reducing the length of the series to 250 years. Cynical observers might, however, have noticed that the late seventeenth century numbers for CETR were distinctly cold, so the effect of this truncation may well have been to flatten out the Little Ice Age.

When McIntyre transferred his attentions to the Central Europe series, he came across a similar problem – the data had been truncated at 1550, when the full series actually went back to 1525. Here the warmest part of the record was removed from the series, and the effect was presumably to flatten the Medieval Warm Period somewhat. In neither case were these truncations disclosed or justified.

Mislocations

The MBH98 reconstruction included 11 precipitation (rainfall) series, which Mann had referenced to a paper by Jones and Bradley 1992 (the same Phil Jones who prepared the temperature data).* However, when McIntyre tried to check the precipitation numbers back to original data in the public archive he immediately ran into problems. He was able to check the matches *en masse*, by calculating the correlations between the archive version and Mann's version of the same series. His best scores were above 0.9, indicating a close but not exact match, but many of the proxy series barely matched at all, with correlation scores of less than 0.5. Where had Mann got this data?

There were more problems with these series too. Series 37 was identified as being the rainfall records for Paris, France, and the numbers Mann had used had a high correlation with the archive figures. The start dates of the two sets of numbers were the same as well. On the face of it, this looked to be correct. However, in Mann's reconstruction the series had been located at 42.5N, 72.5W, which is just outside Boston, Massachusetts, an error which prompted McIntyre to quip cruelly that 'The rain in Maine falls mainly in the Seine', much to the amusement of the sceptic community.

Two other precipitation series were located in India according to Mann's paper, but the authors of the study which Mann had quoted as his source, Jones and Bradley, didn't actually have any Indian series in their paper, so it simply couldn't be correctly located. Certainly the figures didn't match actual Indian rainfall figures, and the best match McIntyre could find in the archived precipitation records turned out to be Philadelphia, although not

---

* See page 61.

with a high enough correlation to make the identification definitive, or even likely.

The rest of the precipitation series were either unarchived or were from unreported sources or had been manipulated in some way prior to use in the reconstruction, otherwise they would have been identifiable by correlation analysis.

And last, but not least, Series 20, an ice core from Greenland, was materially mislocated, and the locations of Series 46 and 47 had been swapped. It was all fairly amazing, but if McIntyre thought that was the end of the story, it was an idea of which he was shortly to be disabused.

Old data

McIntyre's comparison of the data Mann had used to the figures in the public archives, which had identified the origins of the precipitation figures, also revealed another puzzling aspect of MBH98. When McIntyre checked Series 51 to 61 to the archive, it turned out that all of these series had more up-to-date figures available – Mann had been using *old versions of the data*. Of course, it was quite possible that some of these might have been the current versions at the time MBH98 was originally written, so McIntyre made some enquiries at the World Data Center for Paleoclimatology who maintained many of the records. Their response was that the updated figures had been available since 1991 and 1992, more than six years before the publication of MBH98.

In all, there were 24 series where more up-to-date figures were available in the public archive, and some of the differences between the two versions were far from trivial. One in particular was astonishing: Series 56, was a tree ring-width chronology called Twisted Tree, Heartrot Hill (see Figure 3.2). Mann had used an old version of the data, which ended in 1975. Needing data to run up to 1980, he had therefore infilled up to 1980 by simply repeating the 1975 value for the final five years of the series – this can be seen as the tiny plateau at the right hand end of the record. The

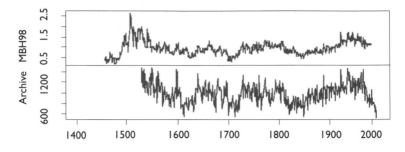

FIGURE 3.2: Twisted Tree, Heartrot Hill
Top: Mann version; *Bottom*: Up-to-date version

overall trend in Mann's data was upwards. However, the updated version in the archive now included figures right through to 1992, and these showed that during the 1980s the trend in the ring widths had dramatically reversed, with all the gains from earlier years being lost.

This decline confirmed exactly what Briffa had said about the divergence between tree ring widths and temperature in the modern era; it is therefore not surprising that McIntyre found that this divergence was not unique among those series where updated figures were available, with Series 51, 54 and 59 all showing declining ring widths while the versions used by Mann showed increases.

More on PCs

Having finally exhausted, for the time being, the possibilities for error in the MBH98 database, McIntyre returned to the subject of the principal components calculation. Where Mann had got a PC that extended back in time further than the underlying data, nothing could be done until McIntyre was able to unravel the mystery of how the PC calculation was made to work with sections of data missing. But, as we've seen, this situation didn't apply to the Australia–New Zealand PC analysis, where the raw data went back to 1625 but Mann's calculations had run only from 1750.

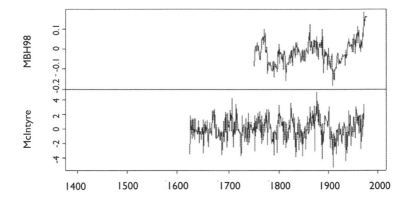

FIGURE 3.3: The Australia PC1

Top: Mann version; *Bottom*: McIntyre's version

McIntyre had all the data, so there was nothing to stop him replicating at least this small step – there were only 16 proxies in the compilation, after all.

With a statistics add-in for his spreadsheet, running a PC analysis was very straightforward, particularly with such a small dataset. The results, which are shown in Figure 3.3, were once again spectacular; they told an entirely different story to the one that Mann had apparently read from the same data.

The top chart shows Mann's figures for the first PC from the Australia–New Zealand compilation, with a sharp twentieth century uptick at the right hand side, suggesting widening tree rings, apparently the result of global warming. The bottom chart is the equivalent from McIntyre's analysis of the same dataset, calculated using a standard PC algorithm and showing, well, not much of anything.

This was starting to look like dynamite: Mann had got his PC calculation wrong, and in the Antipodean proxy compilation at least, the effect of putting it right was to make the hockey stick shape disappear completely.

Reconstruction

How would this pan out in the Northern Hemisphere reconstruction? Remember, the tree ring PC calculation distils groups of raw tree ring proxies down into their main patterns. You then have to calibrate these tree ring PCs (and the individual proxies) against temperature PCs ('This tree ring pattern behaved like that temperature pattern during 1902– 1980'), validate them ('My tree ring pattern still behaved like the same temperature pattern in 1850–1901') and then use the tree ring PCs to recreate temperature PCs of the past ('Since I know my how my tree rings were behaving in 1400 to 1850, I can work out what the temperature was in the same period'). It was quite possible that the Australia–New Zealand series could be completely trendless and yet the Northern Hemisphere reconstruction could still show a hockey stick, driven by another series. The reconstruction, remember, would be driven by those proxies which correlated best against temperature in the calibration period.

Now that the proxy database had been cleaned up, all the mistakes corrected, and up-to-date data collected, putting together a recalculation of the reconstructed temperature should have been easy, but with Mann's description of his methods being so vague, it was still a hard task to work out exactly what he'd done.

Making little headway, McIntyre emailed Mann again to try to get clarification of some of the ambiguous calculation steps, but received little useful information in return. Without Mann's input it was almost impossible to get an exact replication of the Hockey Stick, and without an exact replication it was hard to be certain how correcting the errors in the database would affect the final result. However, if his findings were ever going to see the light of day, he was going to have to reconcile his own work to Mann's. There was nothing for it except to use trial and error to try to discover the exact combination of methodological steps that Mann had used.

Over the next few weeks, McIntyre laboured at his task. Returning again and again to the text of the paper and using what further clues there were in the online Supplementary Information, he threw every conceivable methodological permutation at the data, noting anything that brought his own results closer to Mann's. The problem was that small changes in the methodology could have a dramatic effect on the outcome, a finding that suggested Mann's results were far from robust. However, slowly but surely McIntyre's take on the Hockey Stick edged closer and closer to Mann's, until by September he had something that he felt was good enough. It wasn't exact – something was still not quite right – but if he hadn't arrived at Mann's algorithm, he certainly had something that looked very like it.

With a good approximation of Mann's methodology at hand, McIntyre now reached the moment of truth. How would correcting all the errors in the database affect the results? McIntyre pointed the program at the corrected data and set the calculations in train. In a minute he had the answer: when he saw the results, it was clear that the hunches he'd had when looking through the graphs of the proxies were entirely borne out. With the database corrected, the handle of the Hockey Stick was warped – that is to say, there was a pronounced Medieval Warm Period. In fact the temperatures of the reconstructed fifteenth century were even higher than those reached in the twentieth century.

Now McIntyre really had something. The Hockey Stick looked as though it was bent. But he can have had little inkling as to how long it would be before it would be broken once and for all.

4 Energy and Environment

It has been said that though God cannot alter the past, historians can; it is perhaps because they can be useful to Him in this respect that He tolerates their existence.

Samuel Butler

The audit team

McIntyre's Hockey Stick postings on the Climate Skeptics forum had garnered him a great deal of support and encouragement, including many kind words from some of the professional scientists in the sceptical community, such as physicist David Douglass and geologist Bob Carter. Two other academics who were supportive of McIntyre's work went on to play significant parts in the story that followed.

Ross McKitrick is a professor of economics at the University of Guelph, not far from McIntyre's own home in Toronto. A prominent global warming sceptic, he had already written books and articles critical of the scientific basis for the theory.[32–34] He also hung out at Climate Skeptics from time to time and in the middle of July 2003, McIntyre sent him a short email suggesting that they get together to discuss his work on MBH98. McKitrick explained in his reply that he was en route to a summer holiday in British Columbia but suggested that they should meet up in early September, when he expected to have some free time. Ahead of the meeting, however, McIntyre sent through his notes, together with an explanation of how his thinking had developed and the kinds of calculations he was doing. As McKitrick pored over the intracacies of the mathematics he realised that McIntyre had made discoveries that were of undeniable importance. However, it was one thing to make a scientific step forward, but quite another to turn it into a paper fit for publication. McIntyre had a lot to do if he was to show the world what he had found.

At around the same time, Sonia Boehmer-Christiansen, a geographer based at the University of Hull, also got in touch. Boehmer-Christiansen was the editor of a controversial journal called *Energy and Environment*. *Energy and Environment* was, and remains, a rather obscure journal with (rightly or wrongly, depending on your point of view) a reputation for publishing climate science papers that more mainstream journals would rather not touch (for valid or invalid reasons, again depending on your point of view). Its circulation was tiny, and it could probably be best described as a social science journal, specialising as it did in policy matters rather than physical sciences. However, it did publish occasional scientific papers, and since it adhered to the scientific norm of peer review it was almost impossible for mainstream climatologists to ignore something published there.

In the middle of 2003, Boehmer-Christiansen was planning a special issue of the journal dealing with climate issues and, aware from the Climate Skeptics forum that McIntyre had made some interesting discoveries about MBH98, invited him to submit an article. This was something of a surprise for McIntyre who had never published anything in an academic journal before, and he admitted to being rather flattered that anyone should approach him in this way. He accepted the invitation and set to work on a draft, although time was very short: the special issue was due to be published in the autumn, just a few months away.

Writing the paper

By early September, McIntyre had completed a draft of the paper and arranged to meet McKitrick on 19 September at a small restaurant near Toronto's Pearson airport, midway between their homes. This was to be their first meeting, no suitable occasion having arisen since they had first mooted the idea back in the summer. Over lunch, McKitrick explained that he was impressed by the content of McIntyre's draft paper but not by how it was presented. He later said of the paper that 'the discussion was unfocused and the conclusions unclear',[35] and the two men spent a long time considering how it might be tightened up and clarified. With McKitrick's experience of drafting academic papers and his detailed knowledge of statistics, McIntyre realised that the economist was a very useful man to know. Before the day was over, the two had agreed to continue their work together.

The paper was clearly going to be very critical of Mann; McIntyre had found too many errors in MBH98 for a reader to come away with anything other than an unfavourable impression. The big worry was that something essential had been missed. It would be humiliating if they published a damning criticism that turned out to be flawed in some way. Having checked and rechecked their work, they were certain that they had made no mistakes. This almost certainly meant that the mistakes they had identified were real and the critique they had prepared was valid. However, there was one other possibility. Could it be that somehow they had been working with the wrong data? Of course, they had got the numbers from Mann's colleague, Scott Rutherford, but it was just possible that an inadvertent mistake had been made and that the data he had supplied was not the same as the data Mann had used. It seemed best to check out this possibility before they went to print and it was agreed that McIntyre would draft a letter to Mann to confirm they had indeed received the right dataset. At the same time, he could probe the nagging issue of how Mann had got his PC calculations to work with the gaps in the data.

> Dear Prof Mann
> Here is the pcproxy.txt file sent to me last April by Scott Rutherford at your direction. It contains some missing data after 1971. Your 1998 paper does not describe how missing data in this period is treated and I wanted to verify that it is the correct file. How did you handle missing data in this period? In earlier periods, it looks like you changed the roster of proxies in each of the periods described in the

Supplementary Information using only proxies available throughout the entire period. I have obtained quite close replication of the [reconstructed PCs] in the 20th century by calculating coefficients for the proxies and then calculating the [reconstructed PCs] using the . . . procedures described in MBH98 and the . . . Supplementary Information. The reconstruction is less close in earlier periods . . . The description in MBH98 was necessarily very terse and is still very terse in the Supplementary Information; is there any more detailed description of the reconstruction methodology to help me resolve this? Thank you for your attention.

Yours truly,

Steve McIntyre,

Toronto, Canada[31]

The reply from Mann was brief, but evasive on the questions:

Dear Mr. McIntyre,

A few of the series terminate prior to the nominal 1980 termination date of the calibration period (the earliest such instance, as you note, is 1971). In such cases, the data were continued to the 1980 boundary by persistence of the final available value. These details in fact, were provided in the supplementary information that accompanied the *Nature* article . . .

The results, incidentally, are insensitive to this step; essentially the same reconstruction is achieved if a calibration period terminating in 1970 (prior to the termination of any of the proxy series) was used instead.[31]

Mann also seemed very keen to end any further enquiries into his work (brackets in original):

Owing to numerous demands on my time, I will not be able to respond to further inquiries. Other researchers have successfully implemented our methodology based on the information provided in our articles [see e.g. Zorita, et al 1998]

I trust, therefore, that you will find (as in this case) that all necessary details are provided in the papers we have published or the supplementary information links provided by those papers.

Best of luck with your work.

Sincerely, Michael E. Mann [31]

Mann's reference to the Zorita paper was not particularly helpful, since the authors of this paper had actually applied a variation on the MBH98 methodology to a completely different dataset. With Mann giving nothing more away, McIntyre and McKitrick had little choice but to go ahead with the data and methodological information they had already.

Publication

McIntyre's findings were clearly going to be highly controversial and it was therefore important to make the paper scientifically watertight. To make sure nothing had been missed, he and McKitrick recruited a number of external reviewers to examine the findings and the draft paper. Then, to check that their work was unassailable from a statistical perspective, they also commissioned a review from a professional statistician with paleoclimate expertise. All of this input greatly improved the paper, but it also took time and there was precious little of that left. The deadline for *Energy and Environment* was 30 September 2003, now just a matter of days away. Work continued on the final drafts at a furious pace, McIntyre and McKitrick exchanging emails and drafts on an almost hourly basis. Even then, it looked as if their efforts were not going to be enough to have the paper ready on time, but fortunately Boehmer-Christiansen was prevailed upon to extend the deadline by a few valuable days.

Finally, at the start of October, McIntyre and McKitrick were ready to go and sent the paper, with its provocative title, 'Errors and defects in Mann et al. (1998) proxy data and temperature history', on its way to *Energy and Environment*. With the publication deadline already passed, Boehmer-Christiansen had been struggling to get the paper included, but by dint of persuading the peer reviewers to do their work in a fraction of the normal time* and with McIntyre and McKitrick responding equally quickly, she was able to scrape the paper home just in time to go to press at the end of October. MM03, as the paper became known (after its authors' initials and the year of publication), was about to make a splash.

The paper was published in Volume 14, Issue 6 of *Energy and Environment* on 27 October 2003 with a revised title of 'Corrections to the Mann et al. (1998) proxy data base and northern hemisphere average temperature series'.[37] As it was being printed, it was also posted online on the *Energy and Environment* website and, unusually, because of its political importance it was made freely available. Publication was accompanied by the launch of a dedicated website called Climate2003, which contained background information on McIntyre and McKitrick and more details of the findings. The website also contained all of the data and code used in McIntyre's research. This was done quite deliberately, mindful of possible accusations of hypocrisy, as the two authors had been very critical of Mann and his team for failing to make all of their data and methods available. To fail to have their own supporting material available at the time of publication would have left them looking very foolish.

News of McIntyre and McKitrick's findings hit the media within hours of publication. First out of the blocks was *USA Today* which declared on 28 October, just 24 hours after publication:

---

*　Mann's supporters later questioned whether the peer review can have been adequate because the process appeared to be very brief. The review appears to have taken between three and four weeks.[36]

An important new paper in the journal *Energy and Environment* upsets a key scientific claim about climate change. If it withstands scrutiny, the collective scientific understanding of recent global warming might need an overhaul.[38]

Mann's mouthpiece

Even more startling was the fact that Mann managed to shoot back almost as quickly. The very next day, through the website of a sympathetic journalist called David Appell, he fed an extraordinary story about why McIntyre and McKitrick's results were so different from his own.

Mann's first claim was, almost predictably, that McIntyre had used the wrong data. Appell reported that what had happened was this:

> [McIntyre and McKitrick] asked an associate of Mann to supply them with the Mann et al. proxy data in an Excel spreadsheet, even though the raw data is available [on Mann's University of Virginia FTP site]. An error was made in preparing this Excel file, in which the early series were successively overprinted by later and later series, and this is the data [McIntyre and McKitrick] used.[39]

McIntyre and McKitrick were taken aback. Mann's explanation of what had happened bore no resemblance to what had actually happened. McIntyre had certainly made no request for the data to be delivered in spreadsheet format and when the data was eventually delivered, it was as a text file.* As for the rest of the claims, it was a mystery how Appell and Mann expected McIntyre to have checked the data to the FTP site.

This was the first time McIntyre had even heard of the site's existence, let alone that it contained a data repository. There was

---

* Readers may like to refer to McIntyre's original email request on page 72.

certainly no link to the site on Mann's homepage. Neither Mann nor Rutherford had made any mention of it in their correspondence with McIntyre. A few days after McIntyre's original request, Mann had said he didn't know where the data was. Rutherford had said that he would have to compile the figures from different locations, suggesting that he too was also unaware of the FTP site's existence. If Mann didn't know where the data was and Rutherford didn't know that a single compilation of the data existed at all, why should McIntyre be expected to know about it?

In a second posting later that day Appell reproduced McIntyre's original email request for the data, together with Mann's response that he would get Scott Rutherford to look up the FTP location. Appell didn't seem to notice that this first email made no mention of a spreadsheet and therefore contradicted the story he had posted a few hours earlier. However, he also posted some new details of Mann's side of the story.

> Mann says that the crux of [McIntyre and McKitrick]'s error is their use of a Excel dataset with only 112 columns (where each column represents one set of proxy data–tree rings, ice cores, historical temperature data, etc.), when in fact *the full paleoclimatic data series requires 159 to be used properly.* [40] [Emphasis added]

If the story about a spreadsheet was a surprise, this new claim was truly bizarre. Appell explained to his readers that McIntyre was aware that there were 159 series used in MBH98 rather than 112, and pointed his readers to McIntyre's original email to Mann.*[40] The problem was that this email referred specifically to 'the 112 series' (the same figure that was mentioned in the original paper) and to the 390 raw series, which had been summarised down using PC analysis. The figure of 159 series was completely out of the blue, appearing nowhere in either of Mann's papers or the online

---

*   See page 72.

supplementary information that went with them, nor could mention of it be found in Rutherford or Mann's correspondence with McIntyre, or in any other scientific papers which referred to Mann's work.

Appell continued:

> I have asked McIntyre and McKitrick if they had checked the data they received from Mann and associate against [Mann's] raw data, as you'd think you would if you were truly trying to double-and triple-check an important established scientific conclusion (especially if you were going to seriously slam it), but haven't received a reply.[40]

Appell's idea that McIntyre should have checked the data back to another dataset run and controlled by the same research group was odd. The validity of the data could only be checked by matching it to the original sources in the scientific archives and this was something which McIntyre had done; he had set out the results in all their gory detail in his *Energy and Environment* paper. He had also contacted Mann to check if he had supplied the correct data, but Mann had failed to direct him to the FTP site and in fact rebuffed him in no uncertain terms. How could he now complain that McIntyre had failed to check their dataset sufficiently?

McIntyre and McKitrick couldn't allow Appell's story to stand unchallenged and decided to make a considered reply. They explained that they were reluctant to engage the argument in this way since a full Mann response was apparently on its way, but they set about the task with a certain relish anyway.

To his credit, Appell posted another article the next day, pointing to McIntyre's response, which had been posted on the Climate2003 website.[41] It is probably fair to say that Appell was taken aback by the ease with which McIntyre was able to rebut Mann's story since his tone was considerably milder than in his earlier postings, and he seemed to avoid exploring their responses in detail.

It was relatively easy for McIntyre to refute the idea that he had requested a spreadsheet, simply by pointing to his correspondence with Mann. Likewise, it was simple to cite MBH98 itself, where the text referred to 'the full multiproxy network of 112 indicators'.[14] Mann and Appell had also claimed that the data McIntyre used contained meaningless splices from the earlier and later centuries. Clearly, since Mann had supplied this data, the splices couldn't be McIntyre's fault, but were presumably attributable to Scott Rutherford, who had compiled the numbers on Mann's behalf. But, as McIntyre pointed out, he and McKitrick had checked each series to the archives and while there were plenty of errors, there was absolutely no sign of the kind of splicing errors Mann described. Perhaps Mann would consider telling them which particular series were affected in this way?

As McIntyre's response went on, it became worse and worse for Mann, whose accusations were simply opening more avenues of enquiry:

> Why did the data file have to be assembled from scratch? Did he not have a copy for his own work? Has no one ever asked for it before? Is he accusing his associate, Scott Rutherford, of inserting all the fills? And if what we received was 'a complete distortion', and bears 'no relation' to the dataset he used, how were we able to replicate his original results so closely?
>
> While the claim implicit in Professor Mann's defence is that he actually did work from correctly collated data file in his 1998 paper, this still fails to address the substantial problems of obsolete series, mislabelled locations, truncation of sources, extrapolations of missing data, use of [summer] data where annual are available etc.[41]

The point about having been able to replicate Mann's work using the allegedly erroneous data was key. This meant that Mann's claims that the MM03 result of a pronounced Medieval Warm

Period had been due to the use of incorrect data couldn't be true. It was when McIntyre had used clean, up-to-date data direct from the scientific archives that he had got a Medieval Warm Period. When he used the erroneous data that Rutherford had supplied, he was able to replicate Mann's Hockey Stick closely. Mann *had* used the erroneous data, or at least something that looked very much like it.

Investigating the FTP site

As well as preparing their response to Mann, McIntyre and McKitrick were busy checking out some of his mysterious claims. Mann's attempted rebuttal was the first time either McIntyre or McKitrick had heard of Mann's FTP site at the University of Virginia, which should not be confused with Rutherford's FTP site, from which McIntyre had first downloaded the proxy data back in April. The obvious step was to examine the site to see if the information it contained in any way supported Mann's defence of his work.

After checking that the data they had originally downloaded was still up on Rutherford's site, McIntyre visited Mann's website and downloaded the copy of pcproxy.txt, the equivalent of the original data file they had received from Rutherford. After verifying that the data on Rutherford's site was unchanged and noting that the file carried a date of 8 August 2002, he checked off each series against the Mann version and discovered that the files were identical. The file creation date being in 2002 meant that the data they had been sent must have been prepared well before his request for it in April 2003.

As well as examining Mann's FTP site, however, McIntyre also spent some time looking at Rutherford's website and here he chanced upon further evidence that the pcproxy.txt file had been around for several years. On a graph comparing different temperature reconstructions, Rutherford had made reference to the pcproxy.txt file, and when McIntyre traced the heritage of the web page using the Wayback Machine,* he discovered that Rutherford

---

* The Wayback Machine is a website that archives the whole of the Internet. See www.archive.org.

had originally posted this page in 2001, well before he had even started looking at Mann's paper, let alone requested the data. This and the file creation date together amounted to definitive evidence that the file hadn't been prepared especially for him, as Mann was now claiming.

McIntyre also discovered that even if he had been aware of the existence of the directory on Mann's FTP site, he wouldn't have known which of the proxy series there were to be used. You will remember that in Mann's initial response through Appell's website, the number of proxy series used in the calibration had suddenly changed from 112, as reported in MBH98, into 159. In the directory on Mann's FTP site there were 430 raw proxy series, some of which would have been summarised down using PC analysis, while others were non-PC series which would have gone into the calibration as they were. These latter series would presumably have been quite simple to identify, simply by comparing the values in pcproxy.txt to the files on the FTP site. However this would then leave the task of working out how the remaining series were put through the PC calculations. Which PC rosters did they go into? In which periods? And at the end of the day, how many PCs were to be extracted from each PC calculation? McIntyre was working towards a total of 112 (81 non-PC series plus 31 PCs), but according to Mann, he should have been trying to get to 159 and should therefore have produced 78 PCs on top of the 81 non-PC series. The task was essentially impossible.

The visits of McIntyre and McKitrick to the websites of Mann and Rutherford didn't go unnoticed. A few days later, things started to get very strange. Without warning, the copy of pcproxy.txt on Mann's FTP site was deleted. Fortunately, a vigilant member of the sceptic community noticed the deletion, and on 8 November emailed McIntyre to tell him what had happened. Since the data had been there on the FTP site shortly after the publication of MM03, the deletion must have occurred in the previous two weeks.

A few days later, it got stranger still. As we saw in Chapter 1,* shortly after the publication of MBH98 in 1998, Mann had left UMass and had taken up a postion at the University of Virginia. However, some of the data that McIntyre had used in MM03 was still located on Mann's old website at UMass. On 12 November, just four days after the disappearance of pcproxy.txt, the entire MBH98 directory at UMass was suddenly deleted, again without notice. What was worse was that the deletion of the data had happened before McIntyre and McKitrick had a chance to copy the full contents. Fortunately, the disappearance of the data was again quickly picked up in the sceptic community and an email was dispatched to the webmaster responsible, requesting the restoration of the data from backups. By a stroke of good fortune, he was obliging and the data was restored and made secure again. The deletion, the webmaster alleged, had been made in order to save server space and the timing was, apparently, coincidental.

This was not the end of the deletions though. A few days later, all reference to pcproxy.txt was systematically removed from Rutherford's website. It now appeared that these simultaneous deletions were no coincidence: could it really be that the evidence that the file sent to McIntyre had contained the original MBH98 data was being carefully erased? In some ways this would have been futile, because, as noted above, McIntyre had been able to replicate the Hockey Stick using the data he'd been sent and the methods Mann had described, which was strong evidence that they had both used the same data. Nevertheless, the disappearance of the data made it pretty clear that Mann was intending to argue on regardless.

Mann's reply

Mann's formal reply was published online a few days later at the website of Tim Osborn, a colleague of Phil Jones at Britain's Climatic Research Unit.[42] As expected, it was an aggressive

---

*   See page 37.

defence of his position. After an initial shot across the bows of McIntyre and McKitrick for failing to allow him to review MM03 before publication, Mann got to the meat of his arguments. His first line of defence was based around an alleged failure by McIntyre and McKitrick to include all the data in their calculations.

> It seems clear that [McIntyre and McKitrick] have made critical errors in their analysis that have the effect of grossly distorting the reconstruction of MBH98. Key indicators of the original MBH98 network appear to have been omitted for the early period 1400–1600, with major consequences for the character of the [MM03] reconstruction of Northern Hemisphere temperatures over that interval.[42]

Mann's claim that McIntyre and McKitrick had missed out key data from the early part of the reconstruction was two-pronged. Firstly he was disputing the validity of the corrections. He pointed first to Twisted Tree, Heartrot Hill, which you may remember from Chapter 3 had been used in MBH98 with an obsolete version.* McIntyre had shown in MM03 that a more up-to-date version had declining temperatures in the late twentieth century. Mann's objection was that the more up-to-date version only started in 1530, while the obsolete one went right back to 1459. He was arguing that McIntyre had effectively thrown away over 70 years of data. Of course, it hadn't been McIntyre who had thrown away the data at all, but the scientists who had entered the revised data onto the archive. Assuming they hadn't made an error, these researchers had presumably removed the early years because the data failed quality control measures in some way. Whatever the reason, McIntyre was quite happy to stand on a position of using the most up-to-date numbers available, and leave it to Mann to explain why he thought the older version was more valid.

---

*   See page 83.

Mann's second line of attack was to accuse McIntyre and McKitrick of missing out data by not following the same procedures that had been used in MBH98. Mann explained that, as you went back in time, fewer and fewer proxy series were available in the pcproxy database. As we saw in Chapter 3,* standard PC analysis will fail if there are missing values, and of course, as series dropped out of the MBH98 record in the earlier centuries, there were more and more gaps. In order to get round this, Mann explained, he had adopted a 'stepwise' procedure. He first reconstructed the temperature for 1850 to 1980 using the full roster of proxies. Then he repeated the process for 1800 to 1980 using only those series that were available for the full 180 year period. Continuing in this vein, he could calculate 1750 to 1980, 1700 to 1980 and so on, right back to 1400. Obviously, he now had several reconstructions for each period, each one based on a smaller set of proxies than the last. It was therefore necessary to take the most reliable reconstruction for each period – the one with the most proxies in its roster – and splice it to the most reliable reconstruction for the previous period. So the final Hockey Stick was actually a patchwork of 'steps' or sections of several different reconstructions that had been spliced together.

This stepwise process was how he was able to avoid any failures of the PC algorithm, Mann said, and because McIntyre had failed to use the same procedure, great swathes of data had been dropped from the calculations – in particular the Stahle PC1 and the North American PC1. (This was the series referred to in Chapter 3 as IRTDB US, but it is usually known as 'NOAMER' and this is the way we will refer to it from now on.) This failure explained much of the discrepancy between his results and McIntyre's.

The problem with Mann's argument was that there was no word of this kind of stepwise procedure having been used anywhere in MBH98 or in the online Supplementary Information, although we can see that McIntyre had been guessing that this might have

---

*   See page 78.

been the case from his email to Mann shortly before publication. In fact, Mann had stated in MBH98 that he had used 'conventional principal components analysis' and it was a moot point as to whether what he had done was actually 'conventional' at all. And because Mann had refused to answer his questions, it was impossible for McIntyre to have ascertained what had in fact been done.

Mann may well have felt that he had done enough to fend off McIntyre's criticisms but McIntyre's perspective was quite different. Without realising that he'd done it, Mann had inadvertently shone a little light on another murky corner of his famous paper. To McIntyre, what made Mann's response most interesting was not the fact that Mann had used an undisclosed methodology, but the fact that if you left out just two of the proxy series – the Stahle and NOAMER PC1s – you got a completely different result – the Medieval Warm Period magically reappeared and suddenly the modern warming didn't look quite so frightening. What this meant was that Mann's result – that the Medieval Warm Period didn't exist – seemed to rest on just a tiny fraction of his data. The rest of the series were just 'noise'. Mann may well have been justified in using a stepwise procedure, but if his conclusions depended on just two PC series, then they could hardly be considered robust.

McIntyre and McKitrick shoot back

A few days later, McIntyre and McKitrick responded with their own broadside, a formal response to Mann.[43] As well as detailing the issues around when the FTP site became available, and repeating the responses they had made to Appell, a couple of other points were addressed. Firstly, they noted in an aside that they had discovered, buried deep within the site's directory structure, a subdirectory with a rather surprising name: BACKTO_1400-CENSORED. To what purpose or in what way this directory had been 'censored' was not clear but, the two Canadians noted, with apparently straight faces,

> In light of the identified sensitivity of early 15th century
> values to very slight variations in proxy indicators and the
> evidence elsewhere of truncation (censoring?) of important
> temperature series, we believe that disclosure of the
> censoring process would be helpful.[43]

Although they didn't realise it at the time, the CENSORED directory
was to play an important part in the subsequent story.

The response to Mann also included a long appendix, looking
at the failure by the Hockey Stick authors to disclose materials and
methods in an adequate fashion. The gaps Mann had left had raised
a whole raft of new questions about the decisions he had taken in
designing the study. For example, although he had now revealed
that he had used a stepwise approach to the proxy PC calculations,
it was still impossible to work out exactly which proxies had been
used in which steps and how many PCs were retained from each
calculation. Another question was why so many of the 159 (or
perhaps 112) tree ring series were derived from North America –
more than half of the total. Surely the proxies sampled should have
had a more even spread across the globe? It was also unclear why
the Australian series were only used in the PC calculation from
1750, when they were actually available from 1625. Why was the
South American PC1 not used for its full extent either? The Central
England Temperature Record too? It was a shambles and McIntyre
was not inclined to give Mann the benefit of the doubt. With a
paper like MBH98, which was of such huge public importance,
nothing other than full disclosure was acceptable.

> Regardless of the merits of their methodology, until MBH
> provide the long overdue public disclosure of their PC
> rosters, one is still involved in an ongoing guessing game,
> which is completely unedifying for a paper on which there
> is considerable public reliance. The disclosure within
> MBH98 and the Supplementary Information to MBH98 is
> inadequate and further disclosure was not given upon

private request. Material differences may result from a reconstruction using stepwise PC calculation rather than conventional PC calculation. The non-disclosure in MBH98 of the use of stepwise PC methods is accordingly a material non-disclosure. More adequate disclosure by [Mann, Bradley and Hughes] in MBH98 may well have resulted in a more searching examination of their methodology by statistical specialists long before now.[43]

A new approach to Mann

It was now clear to McIntyre and McKitrick that the only way they were ever going to get to the bottom of MBH98 was to get hold of the actual computer code Mann had used and to obtain complete details of the data, including the crucial information about which series were used in which PC calculations and in which periods. A decision was taken to approach Mann once more and on 11 November McIntyre wrote an email as follows:

> You have claimed that we used the wrong data and the wrong computational methodology. We would like to reconcile our results to actual data and methodology used in MBH98. We would therefore appreciate copies of the computer programs you actually used to read in data (the 159 data series referred to in your recent comments) and construct the temperature index shown in *Nature* (1998) ('MBH98'), either through email or, preferably through public FTP or web posting.[44]

Mann's reply, however, was unresponsive. He made no mention of the code and pointed again to the FTP site as the location of the data:

> To reiterate one last time, the original data that you requested before and now request again are all on the indicated FTP site, in the indicated directories, and have been there since at least 2002. I therefore trust you should

have no problem acquiring the data you now seek.[45]

Still without the details of the proxy rosters and the computer code, McIntyre decided to press the point once more, but unfortunately for everyone, Mann chose to bring the correspondence to an end:

> I am far too busy to be answering the same question over and over for you again, so this will be our final email exchange.[46]

This particular line of enquiry, at least, looked as though it had gone cold.

5   Line Brawl

What a book a devil's chaplain might write on the clumsy,
wasteful, blundering, low, and horribly cruel works of nature!

Charles Darwin

More investigations

The possibility of getting more information from Michael Mann might have come up against something of a dead-end, but there was still plenty to explore on the FTP site, and also the intriguing question of the significance of the CENSORED directory.

McIntyre was still chipping away at the apparently intractable problem of trying to replicate Mann's PC calculations. It seemed that no matter what he did, he just couldn't produce exactly the same results as Mann – he was close, but there was still something missing. Mann had said that McIntyre's attempt at replication had failed because he hadn't used stepwise methods, leading to data from three key 'indicators' dropping out of the reconstruction: the NOAMER and Stahle PC1s and the Twisted Tree, Heartrot Hill series. We also saw that this claim set the alarm bells ringing for McIntyre, suggesting as it did that these series were key to the whole study. Knowing they were key was one thing, but being able to do anything about it was another. That was until the release of the FTP site. When this became available, many of the mysteries which had bamboozled McIntyre finally started to unravel.

McIntyre had been unable to work out from the data repository on the FTP site which series had been used in which steps of which PC calculations. Without this information, he was certain that he would never be able to produce an exact replication of Mann's study. However, as he explored the rest of the site, he suddenly made a breakthrough when he discovered a number of files which contained the unspliced PCs – essentially the results of the individual steps which would then have been spliced to give the final reconstruction. Being able to see these intermediate steps in the calculation was enough to finally enable him to work out how the whole thing had been put together.

Discovery of the PC code

Having made this massive step forward, McIntyre thought that he would finally be able to replicate the Hockey Stick but to his surprise he found he still couldn't get the same answer as Mann. In desperation, he started to go systematically through Mann's FTP site to see if by chance there was something, *anything*, which might provide even a small clue to what the missing step was. File after file was examined and checked for clues until finally, after days of searching, he found what he was looking for. Buried deep in the directory structure of the site, he chanced upon a small fragment of a Fortran computer program, which turned out to be key to the whole Hockey Stick reconstruction.

For those familiar with computer programming languages, Fortran is generally considered rather antediluvian – something that no serious programmer would use these days, there being much more efficient and powerful alternatives available. However, it still appears from time to time in legacy applications and programmers can therefore still chance upon examples of the code 'in the wild'. McIntyre was therefore pretty surprised to find Fortran in use in Mann's group, but he had studied the language in his university days and, with a certain amount of brushing up of his skills, he was able to decipher the code and work out what it did. As he worked his way through the dense text, he realised that he had found just what he was looking for.

The code fragment turned out to be a copy of the actual program used in the tree ring PC calculations. Having grasped this fact, McIntyre set about working out exactly what it did, transcribing the whole thing, one line at a time, into a more modern language. By that time, McIntyre had long since abandoned his use of a spreadsheet, which had proved too clumsy for manipulating the large datasets he had now collated. Instead, he was using a specialist statistical programming language called simply 'R'. R had a number of huge advantages for someone like McIntyre – it could drill through the data in any direction, it allowed him to easily chop and change between different versions of the data series in his endless search for the correct identity of the 159 series, and it had a dizzying array of statistic functionality built in as standard, including PC analysis. This was a hugely powerful tool, available as freeware, and in widespread use by statisticians all over the world. Quite why Mann would be using something as antiquated as Fortran when he could have used R or an equivalent package was something of a mystery, particularly because, with Fortran, you had to program in PC functionality from scratch. It just seemed hugely amateurish.

As he processed the Fortran program into R, slowly building his understanding of its workings, McIntyre finally came across a handful of lines of code that looked as if they might be . . . not quite right. You will remember from Chapter 2 that before performing the PC calculation, you have to centre the data by subtracting the series average from each data point.* That is what happens in conventional PC analysis. It turned out, however, that Mann had done something different. Only slightly different, mind you, but the effect looked significant: instead of subtracting the mean of the *whole* series from each data point, he had subtracted the mean *of the calibration period*. Then, he had 'standardised' by dividing the answer by the standard deviation of the calibration period and then standardised them again in a slightly different way.

With a possible answer finally in their hands, McIntyre and McKitrick worked furiously at deciphering the meaning of what they had found. The standardisation steps were odd, to say the least. Back in Chapter 2 we saw that before tree ring data is

---

*   See page 52.

archived it has already been standardised.* This involves expressing each ring measurement as the ratio of its width to the expected width for a tree of that age. Standardisation in this way makes every tree ring chronology directly comparable to every other one.

The standardisation process that Mann had adopted – dividing by the standard deviation – is normally used when data series are not directly comparable, for example if the data is recorded in different units. It therefore appeared to be an entirely superfluous step, given that the data series were already directly comparable. Why he should choose to standardise *twice* more was a real mystery, particularly because each time he did so he potentially removed the very variance that he was trying to observe in the dataset.

But while the restandardisations looked odd to the two Canadians, it was Mann's decision to centre his data using only the twentieth century mean that looked the most intriguing. The step was wrong, of that there was not a shadow of a doubt, but what was the effect of this error on the final Hockey Stick result? Was it this that was causing the Mann methodology to produce a hockey stick? The more they looked at it, the more certain McIntyre and McKitrick became that they had the answer. Mann's incorrect method of centring (which we will refer to as *short centring*) created a bias towards hockey stick shaped series – any series with either a twentieth century uptick (or a downtick) would be heavily weighted in the PC1, forcing it into the same shape. Although this may seem a little counterintuitive, it is actually extremely simple to show how this happens. The critical step is the subtraction of the twentieth century average from each data point, rather than the series mean, and this effect is shown in Figures 5.1–5.3.

Figure 5.1 shows two dummy tree ring series. Series A, the black line, has an uptick in the nineteenth century, while Series B, the grey line, has an identical uptick in the twentieth century. Now see what happens when you centre these series the correct way – by

---

*   See page 43.

subtracting the full series average from each data point, just like we saw in Chapter 2. This is shown in Figure 5.2.

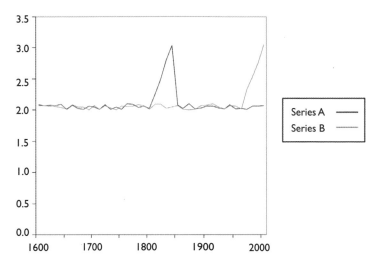

FIGURE 5.1: Two raw tree ring series

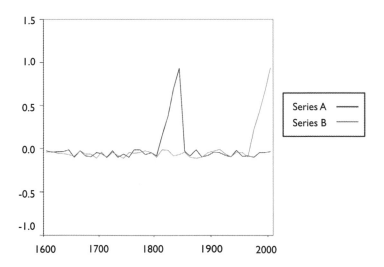

FIGURE 5.2: The same series correctly centred

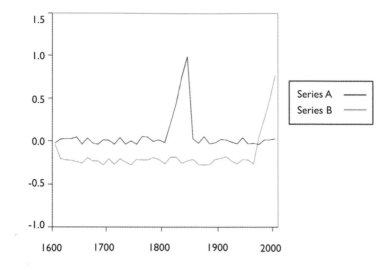

FIGURE 5.3: The series under Mann's short-centring

As you can see, the effect on the two series is identical, shifting them both down so that they are 'centred' around a zero mean – part of each series is above the zero line and part below. Remember that PC analysis will determine how much weight to give each series by calculating the sum of the squares of each variance from the mean. Since both the series here are centred on the mean, the sum of squares is not a large number for either series, both of which should receive approximately equal weightings in the final reckoning.

Figure 5.3, on the other hand, shows the effect on the same series of Mannian short centring. You can see that Series B, with its twentieth century uptick, has been shifted down below zero – in the jargon, it has been 'decentred'. Now, almost every data point has a big variance from the mean, and the sum of squares rapidly inflates to a very large number. Because of this, the weight it receives in the first PC is extremely high compared to Series A. The short centring regime was effectively 'mining' the database for series where the twentieth century diverged from the long-term mean –

hockey sticks in other words – and was then loading all the weight onto them in the final result.

Implications of short-centred PCs

With the answer in their hands, McIntyre and McKitrick needed to set out the problem and its implications for everyone to see in a clear and unassailable manner and after giving it some consideration, they came up with a three-pointed plan of attack. Firstly McIntyre contrasted the effect of the short centring on some actual MBH98 data. He picked two series from the NOAMER proxy roster known as Sheep Mountain and Mayberry Slough (see Figure 5.4). Sheep Mountain is a hockey stick shaped series with a sharp twentieth century uptick. Mayberry Slough, on the other hand, had its growth peak in the early nineteenth century.

Remember that PC analysis assigns weights to each component of the dataset, so that those series accounting for the most variance get the most weight. Both of these series show a spike in growth, although at different times, rather like the artificial example we have just looked at. Standard PC analysis would be expected to give them similar weights in the first PC. However, the effect of short-centred standardisation was to grossly

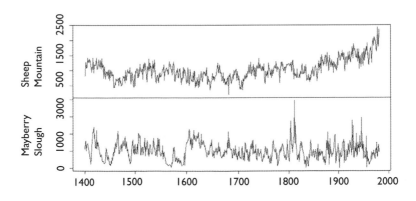

FIGURE 5.4: Sheep Mountain and Mayberry Slough

overweight Sheep Mountain, with its recent growth spurt, over Mayberry Slough. In fact, in the final calculation, Sheep Mountain had *390 times* the weighting of Mayberry Slough, making it look vastly more significant.

In order to reinforce the point, the second line of attack was to prepare some simulations of what Mann-style standardisation would do to random data. When we talk about random numbers we are usually referring to a particular kind of randomness called 'white noise'. In white noise, each additional data point is independent. The throw of a dice is a familiar example: if you get a six with the first throw, it makes no difference to your chances of getting a six with your second. The probability is just the same – one in six. There are, however different kinds of randomness and the one relevant to the story of the Hockey Stick is called 'red noise'. Red noise is distinguished by the fact that each data point is *not* independent of the last one. An example of a red noise process, in slightly more mathematical terms, would be one where the value is given by 'the last point plus or minus a random amount'. Red noise is best described as a 'random walk', which can be envisaged on a graph as a line which wiggles up and down without ever going anywhere in particular. It might wander off in one direction for a while, but eventually it will turn round and head back towards the mean. White noise on the other hand would look just like a mess of dots.

Red noise processes appear to be very common in nature, and in particular are observed in weather and climate systems and in biological processes. So in order to test Mann's algorithm, it was necessary to see what it would do to red noise series rather than to white noise. To make absolutely certain he had headed off any potential objections, McIntyre was careful to ensure that the red noise series had exactly the same statistical characteristics as the noise in the tree ring series actually used in MBH98. And when he fed the results into the Mann PC routines – bingo! Hockey sticks appeared. You could feed pretty much any group of red noise series into Mann's algorithm and, provided there was a rising or falling trend in the twentieth century it would give you a PC1 shaped like a hockey stick. The short centring was simply overweighting any series with twentieth century upticks. Meanwhile, any with twentieth century downticks were given large *negative* weightings, effectively flipping them over and lining them up with upticks. Figure 5.5 shows the result from processing the same red noise series using conventional centring (top) and short centring (bottom), the latter with a distinct hockey stick shape.

McIntyre repeated this process ten times and every time he got a hockey stick. If even random numbers would give you a hockey stick shaped PC1, there could be no doubt that Mann's methods were fundamentally flawed.

There is a subtlety to this result which needs to be understood because there was considerable confusion at a later date. McIntyre and McKitrick were not suggesting that the short centring algorithm produced hockey sticks from nothing – it couldn't conjure hockey sticks out of white noise, for example. But if there was even one hockey stick shaped series in a database containing dozens of series without any significant trend, it would overweight it, yielding a PC1 that suggested that a hockey stick shape was the dominant pattern in the data.

Another possibility was that a database really did have a hockey stick shape as its dominant pattern. But even then the short centring algorithm had to be treated with immense care. Since short centring could produce hockey sticks from red noise, it would have to produce a hockey stick with a much more pronounced blade from tree rings before the result could be seen as statistically meaningful. This was an issue that was to cause considerable controversy later in our story.

The third plank of the argument was to demonstrate the effect of the short centring on the NOAMER PC1. This, if you like, was the *pièce de resistance*, a reconciliation of his work with Mann's. In essence, the significant differences between MBH98 and MM03 boiled down to just a few series: NOAMER, with its problematic centring, the Stahle series, and one which Mann had not flagged up: Gaspé.

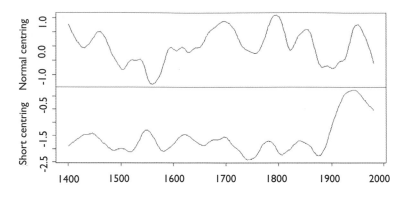

FIGURE 5.5: The effect of short-centring on red noise

We met the Gaspé series in Chapter 3 where we saw its extraordinary hockey stick shape. Since making that observation, McIntyre had discovered that the series had actually been used *twice* in Mann's paper. It appeared once in the NOAMER PC series and once as a single proxy. According to the record in the archive, the series extended back to the year 1404. Strangely, however, when used as a single proxy, Mann had managed to get the data to go back to 1400. It turned out that the value for the year 1404 had been repeated in the three earlier years, a step that was not disclosed in the paper and was unique in the MBH98 dataset. While this might seem a relatively innocuous procedure, its impact was significant. With the data now extending back to the start of the fifteenth century, the Gaspé series, with its dramatic hockey stick shape, could be incorporated into the critical AD 1400 step where it would push down the Medieval Warm Period.

Nature paper
With these amazing findings in the bag, McIntyre and McKitrick could at last make a full response to the critics whose denunciations were still echoing loudly around the Internet. The obvious step was to submit their new work to *Nature*, which had published the original MBH98 paper. This had one major disadvantage in that until

the findings were in print they could not make any public response to their Internet opponents – the findings would have to remain under embargo. However, this appeared to be a price worth paying and so, resigning themselves to having to endure the barbs of the critics, they set to writing once more.

The immediate problem the two men faced was that the information they wanted to convey didn't fit neatly into the categories of article that *Nature* usually accepted. 'Letters' and 'Articles' were categories reserved for the publication of new work, while the final possibility, 'Communications Arising', was for short comments on other articles. The problem was that there was a lot of information to convey and the word limit for a Communication Arising was likely to be too short. They decided however to go with this category since it was reserved for 'exceptionally interesting or important comments and clarifications on original research papers', which seemed to be a pretty good description of their new paper.

The paper was submitted on 14 January 2004, together with a covering letter explaining its importance and also outlining their difficulties with categorising the content and asking for editorial advice on how best to proceed.[47] The paper, they said, was longer than a normal Communication Arising, but shorter than a normal Letter. They also asked that when peer reviewers were appointed, at least one of them should be an expert in PC analysis, with no connection to the climate field.

When a critical comment like McIntyre and McKitrick's is received, a scientific journal will normally approach the targets of the critique for a response. Then both the comment and the response can be published alongside each other. Towards the end of January, *Nature* contacted McIntyre to indicate that this process was underway. There were, however, some worrying signs in the email:

> You will also see from our guidelines that Communications
> Arising are only published as such if they represent a

scientific advance over the original paper; critical comments such as yours may instead be addressed in the form of a published correction/clarification from the criticised authors, if this is recommended by our referees. In this event, your contribution is likely to be acknowledged.[48]

As McIntyre noted in his reply, the words 'scientific advance' were not actually mentioned in Nature's published policies on Communications Arising, this requirement only being relevant for Articles. Besides, he added, eliminating a material error in a hugely important scientific paper must surely count as a scientific advance. He went on:

We have put a lot of time and effort into this investigation, and along the way have taken a great deal of public criticism on earlier findings published elsewhere, including some from high-profile scientists and commentators. The present submission is a substantial advance from our earlier findings and serves as a complete response to these criticisms. Although we would have obviously liked to have already responded to these attacks, doing so would have required divulging the contents of our submission to Nature. Thus, at some personal sacrifice, we have let some attacks go unchallenged, so as to allow the Nature review process to operate confidentially. If our findings are correct, the originating authors not only applied incorrect methodology, but their disclosure of methods and data to the editors, reviewers and readers of Nature was inaccurate, and substantially influenced how the paper was received. It would be inequitable if our findings are upheld, but the originating authors are allowed to present a 'clarification', without simultaneous presentation of our findings.[49]

Favourable revise and resubmit

At the start of March, *Nature* emailed to say that they had now received the reply from Mann that would be published alongside their article if it was accepted, together with comments from the two external peer reviewers. The comments from the reviewers were overall quite complimentary, and the paper had received what is known as a 'favourable revise and resubmit' – in other words, some amendments were required but both the paper and Mann's reply to it were recommended for publication. McIntyre and McKitrick therefore set to work to incorporate the amendments asked for within the two week deadline set by the journal.

Mann's reply – another mystery

The request for a resubmission meant that everyone involved had to repeat the whole review process again. This gave McIntyre and McKitrick as well as the Hockey Team a chance to hone their arguments in the light of their contributions the first time round, as well as addressing themselves to the points made by the peer reviewers.

Mann's reply was just as forthright as his previous pronouncements on McIntyre and McKitrick's work. He took issue with almost all of the paper, saying that the two Canadians were 'incorrect with respect to each major point' they raised and that their work did 'not demand serious consideration'. However, he quietly conceded a great deal.[50]

Much of his response reiterated the position he had taken in response to MM03, namely that McIntyre's data and methods gave rise to a reconstruction that eliminated much of the data in the early periods – the three key indicators. He went on to point out that when normal standardisation was used, the reconstruction scored very poorly on the RE statistic, which he said showed that McIntyre's reconstruction was spurious. This was rather misleading, as McIntyre and McKitrick never claimed to have produced a reconstruction of their own – only to have shown that Mann's was not robust. He also dismissed McIntyre's red noise simulations out of hand, saying that the statistical properties of the actual series were not emulated by red noise, and that it was not therefore an appropriate model to use.

On some of the data issues, like Stahle, Mann argued that even if McIntyre was right, it was immaterial to his results. As far as the Stahle PCs were concerned, McIntyre was happy to accept the argument that their inclusion was irrelevant to the fifteenth century results, and he indicated that he would amend the paper accordingly, but noted that it was actually Mann who had described it as a 'key indicator' in his early responses to MM03. It was a bit rich of him now to claim it as immaterial.

Meanwhile, on Twisted Tree, Heartrot Hill,* Mann put up no substantive defence of his use of an obsolete data version, while still apparently objecting to McIntyre's preference for the up-to-date data. His position appeared to be that a longer record was better than a shorter one, without addressing McIntyre's argument that the early part of the record had been eliminated from the archives for a good reason. In fact, by the time the revised paper was ready to be resubmitted, McIntyre had discovered that the early part of the series was based on only a single tree, which appeared to be the reason this part of the record had been removed.

Mann also tried to dismiss the extrapolation of the Gaspé record back to the start of the fifteenth century, saying that is was a mere 'technicality'. In his reply, McIntyre noted that 'This is hardly correct', a rather polite choice of phrase in the circumstances.[51] As the only series in the MBH98 corpus that was extrapolated in this way, and with the fact that the extrapolation allowed its inclusion in the AD 1400 roster, it was surely deserving of *some* discussion. And as McIntyre noted, its inclusion had a dramatic impact on the Medieval Warm Period. This being the case it was incumbent on Mann to explain to the readers the reasons for his adjustment.

---

*   See page 83.

Bristlecones

Much of the argument though was over the first two PCs of the North American tree ring database, NOAMER. Of the nine NOAMER PCs, only these two were used in every reconstruction step back to the start of the reconstruction, including the critical AD 1400–1450 step, where the Medieval Warm Period would have been expected to appear. McIntyre and McKitrick had set out the effect of Mann's short centring on the results, noting the overweighting given to Sheep Mountain, which has been described above. Mann, however, objected to McIntyre's focus on this one series; there were, he said, *lots* of other series contributing to the NOAMER PC1. If McIntyre had previously thought he had got to the bottom of the Hockey Stick, he was now set right, because this apparently simple claim was about to open yet another can of worms that would keep him busy for years to come.

Intrigued to discover exactly which series were involved, McIntyre prepared a table outlining the sixteen top-weighted series in the NOAMER PC1. This table was eventually included in the letter which accompanied the resubmission McIntyre and McKitrick sent to *Nature* and it is reproduced here as Table 5.1.[52]

As McIntyre explained,

> It immediately stands out that 15 of the 16 sites are high-altitude sites due to Donald Graybill. We identified a presentation of 12 of these sites in Graybill and Idso (1993) (which was also discussed in Mann et al. (1999)).[52]

The implications of this discovery were enormous. Not only did Mann's Northern Hemisphere reconstruction depend largely on just one of its PC series, namely the first PC from the NOAMER network, but the shape of NOAMER itself depended on just a tiny selection of trees from one corner of the western USA. They were all from one of two closely related species – bristlecone pines and foxtails – and all but one had been collected by a single researcher, Donald Graybill. These records all had a distinct hockey stick

TABLE 5.1: Sixteen heavily weighted series in the NOAMER PC1

ID	Name	Species	Elevation (m)	Author	Graybill-Idso (1993)	Exclusion MBH98 Censored
ca528	Flower Lake	PIBA	3291	D.A. Graybill	13	TRUE
ca529	Timber Gap Upper	PIBA	3261	D.A. Graybill	14	TRUE
ca530	Cirque Peak	PIBA	3505	D.A. Graybill	12	TRUE
ca533	Campito Mountain	PILO	3400	Graybill & Lamarche	5	TRUE
ca534	Sheep Mountain	PILO	3475	D.A. Graybill	11	TRUE
ca555	Yolla Bolly	PIBA	2460	B. Buckley		FALSE
co523	Windy Ridge	PIAR	3570	D.A. Graybill	4	TRUE
co524	Almagre Mountain	PIAR	3536	D.A. Graybill	1	TRUE
co525	Hermit Lake	PIAR	3660	D.A. Graybill	3	TRUE
co535	Frosty Park	PILF	3218	D.A. Graybill		TRUE
co545	Niwot Ridge	PILF	3169	D.A. Graybill		TRUE
nv510	Charleston Peak	PILO	3425	D.A. Graybill	6	TRUE
nv511	Mount Jefferson	PILF	3300	D.A. Graybill	7	TRUE
nv512	Pearl Peak	PILO	3170	D.A. Graybill	9	TRUE
nv513	Mount Washington	PILO	3415	D.A. Graybill	8	TRUE
nv514	Spruce Mountain	PILO	3110	D.A. Graybill		TRUE

PILO: *Pinus longaeva* (Intermountain bristlecone pine); PIFL: *Pinus flexilis* (Limber pine); PIAR: *Pinus aristata* (Bristlecone pine); PIBA: *Pinus balfouriana* (Foxtail pine) shape with a dramatic growth spurt in the twentieth century but unremarkable growth levels in the fifteenth century.

As we have seen, the short-centring routine picks up any deviation from the norm in the twentieth century and vastly overweights it in the PC1. The hockey stick shape of the bristlecones was therefore imprinted on the PC1 as apparently being the dominant pattern in the database. Then, during calibration, when the PC1 was matched up against the temperature records, the blade of the stick – the twentieth century uptick – correlated well with the instrumental uptick and so the PC1 was highly weighted again in the reconstruction. This effectively passed the hockey stick shape onwards from the PC1 to the final temperature reconstruction and lo and behold, no more Medieval Warm Period.

What made it all much more problematic was that Graybill had stated that the twentieth century growth spurt in these trees had nothing to do with temperature changes. Graybill's co-author, Sherwood Idso, was a prominent global warming sceptic who had been seeking evidence that any twentieth century increase in ring widths was due to carbon dioxide fertilisation – increased growth due to higher levels of carbon dioxide. He and Graybill had sampled these particular trees because they thought that bristlecones would be responding to carbon dioxide and not temperature. In other words, it was going to be impossible to use these trees as temperature proxies because the growth pattern was being contaminated by the effects of carbon dioxide. As McIntyre explained:

> Graybill and Idso specifically stated that the 20th century growth in these sites were not accounted for by local or regional temperature and hypothesized that these trees ... contained signals of direct 20th century CO_2 fertilization.[52]

Indeed, even the Hockey Team had agreed that the bristlecones were not indicators for temperature. In an article in 2003, Malcolm

Hughes had described their twentieth century growth spurt as 'a mystery',[53] and the Team had apparently gone on to adjust for the effect in the second Hockey Stick paper, MBH99.

There was more though. You will remember* that McIntyre had found a directory on Mann's FTP site called BACKTO_1400-CENSORED. In this, he had discovered that Mann had created a new set of NOAMER calculations: a sensitivity analysis that excluded certain proxy series. Now he was finally able to see the significance of the data in this directory. In their resubmission letter, he and McKitrick explained to the editors and reviewers, just how important the sensitivity analysis was:

> We also analyzed the 'censored' version of the NOAMER PC calculations at the MBH98 FTP site to determine which sites were excluded (finding out in the process that the PC1 was virtually identical to our own calculations). We found that 19 of the 20 sites so 'censored' were Graybill sites and that all of the above 16 sites were so censored. We believe that this analysis sheds a great deal of light on the critical NOAMER PC1 and also on how much significance can be placed on RE and other verification statistics, and have added a discussion of this effect.[52]

In other words, Mann had created a revised NOAMER PC calculation which excluded all sixteen of the Graybill sites listed in Table 5.1 (together with a handful of others). By doing this, Mann would have removed the few hockey stick shaped records from the database. The rest of his data series, however, amounted to little more than noise, which meant that they would not be picked up by the short-centring algorithm. As a result the hockey stick shape disappeared from the NOAMER PC1 and in turn from the final temperature reconstruction. So Mann's revised reconstruction would have had a new PC1 looking just like McIntyre's, with an

---

*   See page 103.

elevated Medieval Warm Period. *But he hadn't reported these findings in the paper.* McIntyre's discovery demonstrated conclusively that the Hockey Team had been aware that their result – of flat fifteenth century temperatures – depended on just 20 mostly related series out of the 400-odd that were fed into the calculations. And these were series that were known to be contaminated by carbon dioxide fertilisation as well. This was, to say the least, at odds with the statement they made in MBH98:

> On the other hand, the long-term trend in [the Northern Hemisphere] is relatively robust to the inclusion of dendroclimatic indicators in the network, suggesting that potential tree growth trend biases are not influential.[14]

In other words, Mann had claimed that you could take out any trees you liked and it wouldn't make much difference to the reconstruction. It was now clear that this was not the case.

Mann's reply: PCs
Much of the rest of Mann's argument was an attempt to defend his use of short-centred standardisation. In it, he provided a recalculation of the Hockey Stick in which all of the series in the NOAMER PC1 were artificially given an equal weight. In essence he simply bypassed the PC calculation entirely and fed all the NOAMER series one by one into the calibration. This, he showed, still gave hockey stick shaped results, with no Medieval Warm Period. However, even in the new scenario, the carbon dioxide-induced spurt in bristlecone ring widths was still going to match up well against the temperature records and so would dominate the final reconstruction. So the hockey stick shape would still be imprinted on the final reconstruction even though this shape was not due to temperature changes. This wasn't the only problem with Mann's proposal. As McIntyre pointed out in his reply to *Nature*, what Mann had done merely changed a situation in which the Graybill sites were two out of 22 series used to reconstruct the early fifteenth

century, to one in which they were 20 out of 95. The PC methodology had been introduced in order to prevent particular geographical areas being overrepresented. If he was to discard the procedure in this way, Mann would also be throwing out one of the putative strengths of his original study.

Peer review

The comments of the two external peer reviewers were of course much less adversarial.[54] Both were anonymous, as is normal for peer review, although they both inadvertently left clues to who they were, and over subsequent years their identities have been established with some certainty. In his comments, the first reviewer had identified himself as an expert on PC analysis, suggesting that the *Nature* editors had taken up McIntyre's suggestion that they use someone with expertise in this area. It was widely assumed afterwards that he was Professor Ian Jolliffe, emeritus professor of statistics at the University of Aberdeen, and a man who could fairly claim to be one of the world's leading experts on PC analysis. In fact Jolliffe has since publicly acknowledged that this supposition is correct.

From his comments, Jolliffe appeared to be favourably disposed to McIntyre and McKitrick's arguments although he was clearly being even-handed. His initial comments had certainly been complimentary to the two Canadians:

> I find merit in the arguments of both protagonists, though
> [Mann] is much more difficult to read than McIntyre and
> McKitrick . . . [Mann and his colleagues'] explanations are
> (at least superficially) less clear and they cram too many
> things onto the same diagram.[54]

On the subject of the PC procedures though, he had been somewhat bemused.

> [PC analysis] is an area where I have expertise . . . [I am]
> uneasy about applying a standardisation based on a small

segment of the series to the whole series, if that is what is being done.[54]

It was, of course, *exactly* what had been done, and McIntyre took the opportunity to confirm this to Jolliffe in the covering letter which accompanied his response. Jolliffe had also criticised Mann and his team for attempting to reject McIntyre's work on the basis of the RE statistic, and for dismissing the red noise simulations. Surely, he wondered, Mann must have been slightly concerned that the algorithm produced hockey sticks from red noise when none would have been expected, regardless of whether it was precisely the correct statistical model?

Jolliffe had gone on to address some of the issues over the quality of the data in the early periods:

> I am not qualified to say much on [data quality] but it seems to be the crucial point. Both sets of authors agree that the omission of some [medieval] data changes the early reconstruction considerably. [Mann et al] say that the omitted data are reliable; [McIntyre and McKitrick] say they are not. Does anyone know who is correct? If there is disagreement among experts, then the true behaviour of the series must be very uncertain.[54]

This was certainly something that McIntyre and McKitrick could agree with.

The second reviewer was Eduardo Zorita, the head of the department of paleoclimate at the Institute for Coastal Research in Geesthacht, Germany, and the scientist whom Mann had indicated had independently implemented the Hockey Stick methodology. But while he was familiar with Mann's work, Zorita had never worked with Mann and was not associated with the Hockey Team in any way. In many ways he was an ideal candidate as the second peer reviewer.

---

*   See page 92.

Zorita had also had difficulties in assessing the paper and explained that to do so really required an in-depth look at the data and code, perhaps unaware that Mann was refusing access to the code and that there was no definitive description of the dataset. But his overall impression was favourable:

> In general terms I found the criticisms raised by McIntyre and McKitrick worthy of being taken seriously. They have made an in-depth analysis of the MBH98 reconstructions and they have found several technical errors that are only partially addressed in the reply by Mann et al.[54]

In the rest of his comments, Zorita made relatively minor criticisms of both McIntyre and McKitrick and of Mann and his team, and suggested a number of amendments that he thought should be made to the paper before publication. In essence he was saying that the answer to the question of who was right would only be reached by the scientific community examining the arguments of both sides in more detail, something that a peer reviewer couldn't be expected to do. Providing some amendments were made, publication of McIntyre's comment should go ahead alongside a reply from Mann.

Nature problems

With the amendments demanded by the peer reviewers all incorporated and the manuscript safely resubmitted,[55] McIntyre might have expected a trouble-free path to publication, but as was starting to be a habit, the Hockey Stick had a way of screwing things up. On 26 March 2004, *Nature* editor, Rosalind Cotter, emailed McIntyre out of the blue, asking for further changes to be made to the manuscript.

> Before we can proceed further, I am afraid that it will be necessary for you to shorten your manuscript substantially, in accordance with our author guidelines . . . You will have seen that submissions to this section of the journal have a

strict length limit (up to 800 words, with one multipanelled figure and no more than 15 references), although you are welcome to include supplementary material for reviewing purposes only.[56]

This was strange indeed. The manuscript had been submitted once already and had passed muster on its word count at that point. What could have possessed *Nature* to suddenly declare the manuscript was too long? McIntyre's paper was of pressing scientific and political importance and it is hard to credit the idea that Cotter can have been unaware of this fact. It would be hard not to start to become a little paranoid in the face of this sudden *volte-face*. However, there was little alternative but to comply and without further ado, McIntyre and McKitrick set about the almost impossible task of distilling their paper down to just 800 words.

The revised paper, which was submitted just two weeks later, was terse to the point of bluntness, with all erudite discussion cast aside in favour of direct presentation of the facts, but somehow they managed to scrape in under the word limit with their central arguments – short centring, bristlecones and Gaspé – all intact.[57] It may not have been the most elegantly worded paper in history but it remained one of the most important. Now, surely, they had crossed the final hurdle.

A month passed without any word from Cotter beyond an automated acknowledgement of the revised submission. May likewise passed by without a word from *Nature*. Mann was busy, meanwhile, publishing a new paper which said that McIntyre's findings in MM03 should be 'dismissed as spurious'.[58] He also cited another paper he was to publish shortly in *Climatic Change*, which he said demonstrated that there were 'critical flaws' in McIntyre's work. Mann was clearly determined to keep the pressure up.

By mid-June McIntyre was becoming increasingly frustrated with the lack of progress and emailed *Nature* again to enquire after the paper's whereabouts. In the wake of Mann's new paper, he had been receiving questions from a number of media outlets, including

New Scientist, asking after the *Nature* submission. The problem was that while the paper was still going through the publication process, he was unable to discuss its contents with anyone.

When Cotter finally provided a few words of explanation, she said that not only had the journal had to get another comment from Mann but they had felt obliged to add a *third* reviewer, because McIntyre had raised the issue of the suitability of tree ring data in the revised submission. This was indeed the case – it had been suggested by the peer reviewers in their comments. She hoped, however, that matters would be resolved in short order.[59]

Three more weeks passed and another McIntyre email was fired across to Cotter, wondering how the review of a single 145-word paragraph (which is all the discussion of data quality that had been added) could have taken more than 110 days. Cotter remained apologetic but otherwise unresponsive. It would all have to wait until the reviewers reported back. They had only received, she said, one set of comments so far.

Rejection

The *coup de grace* was delivered ten days later, when McIntyre received an email from *Nature*.

> Thank you for your revised comment on the contribution by Mann et al., which I am afraid we must decline to publish. As is our policy on these occasions, we showed your revised comment to the earlier authors, and their response is enclosed. We also sent the exchange to 3 referees, whose comments are attached.
>
> In the light of this detailed advice, we have regretfully decided that publication of this debate in our Brief Communications Arising section is not justified. This is principally because the discussion cannot be condensed

into our 500-word/1 figure format* (as you probably realise, supplementary information is only for review purposes because Brief Communications Arising are published online) and relies on technicalities that do not bring a clear resolution of the underlying issues.[60]

Perhaps this rejection should have been seen as inevitable, with all the delays and obfuscation from the *Nature* editors. It may also be enlightening to compare McIntyre's treatment at the hands of the journal to the way they had dealt with Huang's borehole study a few years earlier.**

Mann was still holding nothing back. McIntyre's comment, he said in his covering letter, did not meet the standards for publication in a 'Communication Arising'. Their work was 'specious' and 'spurious' and their claims were 'false'.[61] As McIntyre and McKitrick looked through the review comments it was easy to identify Jolliffe and Zorita again.[60] Jolliffe was still striving to be even-handed but he was struggling with Mann's rhetorical style.

> I started my original review by saying that I found merit in the arguments of both [Mann et al and those of McIntyre and McKitrick]. To rewrite this, I believe that some of the criticisms raised by each group of the other's work are valid, but not all. I am particularly unimpressed by [Mann's] style of 'shouting louder and longer so [he] must be right'.[60]

Jolliffe agreed that Mann-style standardisation was not PC analysis, but didn't want to venture an opinion on whether it was a useful technique or not without having worked through the calculations

---

*   The change from 800 words in Cotter's email of 26 March to 500 as shown here is odd, as there appears to have been no communication of a further reduction in word count to McIntyre. It may have been a typing mistake. McIntyre has pointed, however, to a subsequent comment published by *Nature* where the author was allowed 1500 words and two figures.

**   See page 29.

himself. Neither would he be drawn on a claim by Mann that he had been unable to reproduce McIntyre's 'hockey stick from red noise' experiment. Without seeing the code and testing it for himself, Jolliffe couldn't tell who was right, and this was something that a peer reviewer just didn't have time to do. He concluded:

> Regarding publication, I think it is all or nothing. Either you publish neither, or both. In the latter case, the main thing that would be achieved is to highlight that a serious disagreement exists. Only a reader with several days to spare (longer if they are unfamiliar with the area), to chase references and probably the authors, could hope to come close to a full understanding of the arguments.[60]

Zorita's position had, meanwhile, shifted away from McIntyre and McKitrick, as he felt the manuscript had weakened 'considerably'. The focus of his concerns was the fact that both Mann's reconstruction and McIntyre's corrected version of it scored poorly on their verification statistics:

> [A] reader of these manuscripts will be led to think that both reconstructions are not trustworthy . . . This . . . conclusion seems to me rather weak for a manuscript.[60]

In some ways this rather missed the point, as this was exactly the argument that McIntyre was trying to make. He had never claimed to be making an alternative reconstruction – he was merely demonstrating that Mann's wasn't robust. That the verification statistics were poor when the data was properly centred didn't in any way detract from the case that they were poor when short centring was used. However, as a climatologist, Zorita was familiar with the RE statistic, which was much used in the field, and he declared that McIntyre required strong justification for preferring R^2. Of course, with the absurd word limit now imposed by *Nature*, there was no room for this kind of discussion and with the paper already having been rejected, the point was moot anyway. As we will see in later chapters, however, the reviewers' comments on the verification statistics opened up another extraordinary line of enquiry.

Concluding, Zorita said

> In summary, judging from the present version of the manuscript and the response by [Mann], I now think that basis for [the critique of McIntyre and McKitrick] has wavered and that further work, or further convincing evidence, would be needed to present a more solid case.[60]

The new reviewer had some strong opinions, and indeed said that he felt McIntyre and McKitrick had some preconceived notions that affected their audit, although he didn't explain what these were. He did, however, note that this didn't mean they were wrong. His conclusions were generally unfavourable:

> Generally, I believe that the technical issues addressed in the comment and the reply are quite difficult to understand and not necessarily of interest to the wide readership of the Brief Communications section of *Nature*. I do not see a way to make this communication much clearer, particularly with the space requirements, as this comment is largely related to technical details.[60]

As McIntyre noted, it was odd that the readers of a leading scientific journal might not be interested in technical details but observations like this would get him nowhere. The *Nature* paper was finished.

Materials complaint

At the same time as the Communication Arising was crawling through the submission process at *Nature*, McIntyre was also having to deal with *Nature* on another front. With Mann's blunt refusal to release his computer code still ringing in their ears, he and

McKitrick had little choice but to try other means: they were going to have to approach *Nature*, who had published the original article, and ask them to obtain the data on their behalf.

Many scientific journals have a policy on data archiving, either requiring authors to make data and methods information available to interested third parties on request, or to place it in a public archive prior to publication. In some of the more politicised social sciences like econometrics, even stronger measures are in place, with authors routinely required to submit all of their data and code with the manuscript they are submitting for publication.

Nature's policy at the time was for authors to make data and materials available to interested parties on request. In theory at least, if an author failed to make the requested information available, the journal could withdraw the article, although this was unlikely to happen in practice. However, *Nature* still had strong powers of moral force as well as a contracted right to the data and code, so it was reasonable to think that they might be able to extract the information McIntyre needed. With this in mind, McIntyre and McKitrick penned a joint letter to *Nature*, telling the whole sorry story of their work on MBH98: the truncations, the duplicate use of series, the obsolete data, the discrepancies between what Mann had actually done and what had been reported in the paper, and Mann's repeated refusals to make data and code available. They also refused to shy away from the issue of the file deletions from Mann's websites, calling it 'very disquieting'. Concluding the complaint, they stated:

> Under the circumstances, we believe that the full data set and accompanying programs for MBH98 should now be included in the *Nature* Supplementary Information, along with an accounting of any discrepancies between what has been listed at nature.com to date and what was actually used in MBH98.[62]

In early December, the editorial staff at *Nature* replied in positive manner:

> . . . we have already been in touch with Professor Mann's group, who have indicated their willingness to supply us with the various materials pertaining to your complaint. Once we have these in hand, we intend to seek external independent advice on the issues that you raise; and on the basis of such advice, we will decide on any actions that need to be taken.[63]

The second materials complaint

This was immensely encouraging, and so, in a follow-up request, McIntyre and McKitrick decided to spell out in more detail exactly what it was they were after.[64] They also took the opportunity to ask for the disclosure of futher information relating to the calculation of the confidence intervals R^2 and RE statistics.

Again, a prompt reply was returned by *Nature* and it was just as encouraging:

> We are putting the points that you raise here to Professor Mann (as we did with those from your original communication) and will await his response. I hope that you will understand that, given both the seriousness of your concerns and the time of the year (our office being closed for several days over [Christmas]), it may take us longer than normal to bring this matter to a conclusion. But we are nevertheless anxious to do so, and I hope that you will bear with us.[65]

The Hockey Team's purposes

As February drew to a close, the long-awaited reply from Mann and his co-authors arrived at the London offices of *Nature*, and was forwarded to McIntyre by Heike Langenburg, the editor responsible for handling the complaint. As might have been expected, it conceded very little and was aggressive in its defence of the Hockey Stick.[66]

McIntyre and McKitrick had raised several new issues that they had discovered since the publication of the MM03, some of which were rather surprising. During their examination of Mann's FTP site, the two men had spent a considerable time trawling through the directories trying to discover which tree ring series had been used as inputs into which PC calculations and in which periods. The problem was that there were over 400 series archived on the site and these didn't match the series that Mann claimed he had used originally. For example, when McIntyre looked at the series for the South America PC calculation, he discovered that there were 18 sites listed in the Supplementary Information for MBH98 but only eleven appeared at the FTP site – and these were what Mann now said had actually been used. It was incredibly frustrating. However, while they were struggling to match claim with reality, they had, during their searches, chanced upon the text of an unusual email, inadvertently saved among a mass of data files. It was written by Mann's co-author Malcolm Hughes and was addressed to Mann himself. (The email is reproduced below, with emphasis added.)

> Mike – the only one of the new S.American chronologies I just sent you that already appears in the ITRDB sets you already have is [ARGE030]. You should remove this from the two ITRDB data sets, as the new version should be different (*and better for our purposes*).
> Cheers,
> Malcolm

It was possible that there was an innocent explanation for the use of the expression 'better for our purposes', but McIntyre can hardly be blamed for wondering exactly what 'purposes' the Hockey Stick authors were pursuing. A cynic might be concerned that the phrase actually had something to do with 'getting rid of the Medieval

Warm Period'. And if Hughes meant 'more reliable', why hadn't he just said so? By any stretch of the imagination, it was a strange choice of words.

The existence of the email was too important to withhold from the journal, even if McIntyre *had* felt that there was no nefarious intent. He had therefore concluded his remarks to *Nature* on the data issues by saying that there was 'evidence of intentional exclusion of a disclosed South American site'.

In his reply to *Nature*, Mann was apparently outraged by the suggestion that the two records had been swapped for anything other than valid scientific reasons. It was, he said, 'distasteful' and 'deeply offensive'.[66] What had happened, he explained, was that Hughes had been using a screening process to weed out proxy series that weren't of adequate quality. This process involved looking at the mean segment length (the average number of rings in the series – remember that a series will be an average of many trees) and replication of the chronologies (whether the trees on the site were all telling the same story) as well as some other criteria. The series involved, he claimed, had simply been excluded on this basis, but unfortunately, they had then been inadvertently included in the Supplementary Information, thus leading to the confusion over what was actually used in the final calculations. Mann also pointed to a subsequent paper he had written, which discussed the screening process and its use in MBH98.

Mann's explanation did rather concede the point – namely that the Supplementary Information and the FTP site didn't match up – but it didn't really explain series ARGE030. In the email, Hughes had said that Mann should remove the chronology from the database because the revised one should be 'different' and better for their 'purposes'. But when a new version of ARGE030 was received, it might have been expected that they would have included it as a matter of course, the newer data presumably being more up-to-date? In fact, though, it appeared likely that they *didn't* do this as a matter of course since, as we have seen, McIntyre's audit of the MBH98 database had uncovered many series that were obsolete. But if they had been screening new series using the method Mann described, why wouldn't Hughes have told Mann to remove the old data because the new data was *better*, full stop? Why tell him to do it because the new data was 'different' and better for their 'purposes'? In the event, the series didn't actually seem to have been used in the final calculations, despite Mann having listed it in the MBH98 Supplementary Information as having been included. It was all very strange.

Regardless of these apparent weaknesses in Mann's story, McIntyre and McKitrick set about examining Mann's quality control procedures and testing to see how they would apply to the MBH98 data series in practice. Assuming the process was valid (and this appeared reasonably likely, given Mann's reference to a discussion of it in another paper) McIntyre expected to be able to reproduce the discrepancies between the FTP site and the Supplementary Information. The differences should be only those series which failed the screening tests.

By now, readers will probably not be surprised that the screening test didn't appear to explain the discrepancies at all. For example, one of Mann's tests was that the series should have commenced by the year 1626. But of the series in the FTP site, there were no less than 39 which didn't pass this test. By 1680, a series had to have at least 8 trees in it to be considered valid; 22 series failed this test. 171 sites that had gone into the final database failed a test of the minimum correlation between the individual trees and the site average. In fact, one of these sites failed the test so spectacularly that McIntyre emailed the author of the original study, Professor Rosanne D'Arrigo, who discovered that the wrong site chronology had been archived. If Mann had actually applied the tests he claimed to have used, McIntyre asked, how was this failure not picked up earlier?

A *wild guess*

There had been a second Hughes email too, this time relating to the Vaganov PCs,* a set of tree ring chronologies from Siberia. In the email, Hughes had explained to Mann a little bit about the format of the data and some of its oddities, before going on to discuss which of the records he felt they should use. He listed four that he felt should be excluded straight off, as they were better covered by another series already in their possession. His next statement was, however, rather odd.

> Now, we do not know what the internal replication is, but, at a wild guess, I would hope that series starting by 1570 should be reasonably replicated by 1625. I would include these, and their file numbers are: 26, 6, 31, 32, 41, 10, 11, 12, 15, 42, 44, 46, 47, 49, 50, 51, 52, 53, 16, 17, 22, 23, 25, 55, 56, 57, 59, 61.
> For the present (1625 on) exercise I would exclude all the others. Cheers,
> Malcolm

The problem with this was that a 'wild guess' is not a scientific way of deciding if a series should be included or not. What was even more strange, the series that he said should be excluded didn't seem to have been deleted anyway. Again, there may have been less to this than met the eye, but together with the earlier email, it did start to present a somewhat alarming picture of the way the study had been put together. Mann had little of substance to say in his defence, beyond claiming that the procedure was as objective and rigorous as possible, and accusing McIntyre of quoting Hughes out of context (although McIntyre had reproduced the email in full).

---

*   The collection is named after Eugene Vaganov, the researcher who collected the data.

Truncations

Mann's tactic in defending the truncations of the proxy series was to fire off an aggressive denial, but then effectively to concede the point, while explaining it away. He said that each of McIntyre's claims was 'either false or disingenuous' and the accusations were 'distasteful'. For example, McIntyre had complained that the Central England Temperature Record (which, you may remember, was used as one of the 112, or perhaps 159, proxy series), was deleted for its first 70 years without notice to the reader.* In his response, Mann adopted the politician's trick of ignoring the actual accusation and mounting a defence against another charge altogether: he subtly changed the wording of McIntyre's points, suggesting that the accusation was one of *unjustifiably* eliminating this data and then merely set about justifying it. Once this is seen, it is clear that he was effectively conceding the point – he didn't deny that the data was deleted, or that he had not provided notice of this fact, but tried to rationalise it away by saying that other scientists did the same thing.

Duplicate versions

McIntyre and McKitrick had pointed out to *Nature* the fact that many of the series were used more than once in MBH98, among them Gaspé, which as we have seen had also been extrapolated to allow its inclusion in the AD 1400 proxy roster. In their complaint, the two men had provided an appendix in which they listed all the series used more than once, pointing out that Mann and his team should have explained why they did this (if indeed there was a rational explanation). The response from the Hockey Stick authors was blunt: 'this claim is incorrect with one exception'. That exception was Gaspé. Mann went on to say that if you left Gaspé out of the North American PC series, it made no material difference to the final results. Alert readers may wonder though what would happen if the *other* copy of Gaspé, the one used as a single series,

---

*   See page 81.

were removed instead. This is a story that will be told in a later chapter.

Obsolete data

McIntyre had claimed that many data versions used were obsolete when the paper was published, an accusation that Mann in his response called absurd.

> We listed the specific data used by us (albeit with some typos, and incorrect references, as noted) in the supplementary information, and provided all of the data on our data site. We did not indicate there, or elsewhere, that all of the tree ring data used were available in the NOAA databank.

Again, he was denying something slightly different to the actual accusation. The specific allegation McIntyre had made was that the versions were obsolete. Mann was claiming that the versions he had received were donated by fellow researchers and therefore not to be found in an archive. This was a classic rhetorical sleight of hand by Mann. McIntyre had *not* said that the proxy versions Mann had used were not in the archive, he was saying that they were in the archive but that they were obsolete: there were more up-to-date versions available.

Principal components

With the new paper still wending its way through *Nature*'s peer review process at the time, McIntyre and McKitrick had to be a little careful what they said about the problems with Mann's use of PC analysis. It was enough to point out that, contrary to what Mann had said in MBH98, the PC analysis was not 'conventional'. Mann had not only used an *unconventional* tree ring data standardisation – the short centring discussed earlier – but had also done something odd when he dealt with the temperature PCs: the raw data series had many gaps in the record and so conventional PC calculations should have 'fallen over'. Mann would had to have done something to overcome this problem, but had not disclosed either the existence of the problem or the steps he had taken to overcome it.

Once again, Mann responded to McIntyre's claims with bravado and counter-accusation while tacitly conceding the point. With respect to the missing data in the temperature records, he said that McIntyre and McKitrick had made a fundamental mistake in using a different version of the temperature data to that used in MBH98. With this fighting talk out of the way, he went on to explain how the missing data had been infilled: he had interpolated from known to missing data points, thus confirming that there *had* been a genuine problem and a hitherto undisclosed procedure.

When it came to the tree rings, Mann dismissed the whole of McIntyre's claims as incorrect, and now, almost predictably, went on to answer a different point to that originally made. In response to McIntyre's claim that there had been an unconventional standardisation of the data (short centring, in other words) he said that McIntyre and McKitrick had failed to implement the stepwise procedure. This, as we've seen is a whole different ballgame and one which represents a story in its own right, but it was no defence to an accusation of standardising the data in an invalid way.

Draft Corrigendum

While Mann had made fighting defences of his work, the editors at *Nature* seemed to have their doubts about what he was saying: at the end of February Langenburg emailed to say that the journal was going to ask Mann, Bradley and Hughes to issue a corrigendum.[67] A corrigendum is a published correction to a scientific paper, which is required when serious errors are uncovered. *Nature*'s publishing policies explained that it was '*notification of an important error made by the authors* that affects the publication record or the scientific integrity of the paper, or the reputation of the authors or the journal' (*Nature*'s emphasis). Clearly then, this would necessitate a major admission by Mann and his colleagues that there were indeed serious flaws in their papers. Or would it? Only time would tell.

In the meantime, McIntyre and McKitrick realised that being given space in the pages of *Nature*, even for a corrigendum, would give Mann and his team the opportunity to fire some more shots in the ongoing war of words. McIntyre therefore sent off an email to Langenburg seeking assurances that this would not be permitted, and she quickly confirmed that this would indeed be the case. Here at least was one concern put to rest.

Towards the middle of March, the editors at *Nature* sent McIntyre and McKitrick the draft text of the Corrigendum. When McIntyre cast his eye over the wording though, he immediately realised that this was not the capitulation they might have hoped it would be. It failed to address many of the errors uncovered by McIntyre and although it was exceedingly short, still managed to include a whole host of *new* mistakes.[68]

As well as publishing the Corrigendum, Mann was to prepare a new Supplementary Information website to accompany it. When he discovered this, McIntyre immediately requested that he be able to view its contents, but was swiftly rebuffed by *Nature*, who declared that they did not edit supplementary information. This was something of a surprise as, on the surface, it seemed to contradict their declared policy of including supplementary information in the peer review process.

McIntyre's biggest concern with the Corrigendum itself was what was missing from it. Anyone reading it would have come away with the impression that there were a few issues with data citations and not much else. For example, there was no explanation of why the original paper had claimed that 112 series were used when the real number had apparently been 159. There was no mention of the use of decentred PCs and there was not a word about the stepwise application of PC analysis either. As far as *Nature's* publication record was concerned (and their policies suggested they were keen to protect this), Mann had used 'conventional principal components analysis', the explanation given in the original paper, despite both Mann and McIntyre having agreed that a non-standard stepwise process had been used.

Many of the data collation errors that McIntyre had uncovered were simply ignored in Mann's Corrigendum. As far as Mann and *Nature* were concerned the rain in Maine was still falling in the Seine – the geographical errors in the precipitation series remained uncorrected. Even simple data citation errors went only half-fixed, with a vague reference to an alternative location given the nod by *Nature* as an adequate response. And Mann's approach to explaining the infilling and truncation of tree ring series were terse statements that left the reader unaware of exactly what had been done. For example, on the Gaspé series, where the years from 1400–03 had been infilled by copying the value from 1404 back into the earlier years, Mann's simply said that:

> For one of the 12 'Northern Treeline' records of Jacoby et al. used in ref. 1 (the [Gaspé] series), the values used for AD 1400–03 were equal to the value for the first available year (AD 1404).[69]

There was absolutely no indication of the importance of this seemingly trivial adjustment. As we've seen before, this allowed Gaspé, a series with a dramatic hockey stick shape, to be used in the early years of the reconstruction. For such an important change Mann should, by rights, have discussed the impact and the reasoning for it.

For the deletion of the first 25 years of the Central Europe temperature series, Mann's Corrigendum explanation was merely to state:

> The start year for the 'Central Europe' series of ref. 1 is AD 1525.[69]

So once again the reader was left in the dark as to the fact that values prior to the year 1550 had been removed, let alone the reasons for doing so or the effect on the final reconstruction. Meanwhile the similar truncation in the Central England Temperature Record wasn't mentioned at all.

All these failings and many others were outlined in a long email which McIntyre sent off to *Nature*. As he pointed out, unless all the issues were resolved satisfactorily, nobody would be able to replicate what Mann had done. He presented a list of all the information that Mann needed to file on the new Supplementary Information website in order that this fundamental step in the scientific process could be met. This included the series identities, the results of the screening tests, the number of PCs retained from each PC calculation, details of the stepwise PC calculations, the actual temperature series used for the temperature PCs and above all the actual computer code used. With this last piece of information, a huge amount of misunderstanding and bickering would be avoided, because anyone trying to replicate Mann's work would be able to see exactly what he had done.

Path to issuing of the Corrigendum

Once again, *Nature*'s reaction was not unfavourable and they agreed to take the Corrigendum out of production while McIntyre's criticisms were digested. However as emails were exchanged over the next few weeks, it became clear that *Nature* were backtracking somewhat. On 26 March Langenburg emailed again with an amazing set of statements. McIntyre's criticisms of the misleading claims Mann had made on PC analysis – namely that they were 'conventional' – were ruled out of order. The consistency of the methods used, she said, was 'not the subject of a corrigendum'. She went on:

> You also make a number of additional comments to the Corrigendum, but for reasons of space constraints, we insist that such publications are as concise as possible. We feel that the current version, together with the Supplementary Information explicitly listing the data sets and methods used, clearly establishes which data were used in the paper.[70]

Nature was allowing Mann to have his way. The Corrigendum would be about data issues alone.

By the middle of June and with the Corrigendum still not in print, there were some new developments. Mann had been just as busy as McIntyre and McKitrick, and two new papers were working their way through the publication process. The first of these was the submission to the journal *Climatic Change*, which was mentioned previously. McIntyre had been asked to act as a reviewer of this paper by *Climatic Change*'s editors, enabling him to see the content pre-publication, and he had picked up on a reference to a new MBH98 page on Mann's website in the manuscript. It looked as though this was the new Supplementary Information that was to sit alongside the Corrigendum. As might have been expected, the contents still failed to provide sufficient information to enable someone to reproduce the study – for example the 159 series remained unidentified, the data citations were still inadequate and the computer code was still a closely guarded secret. However, there were some surprises. There was a new description of the PC methodology which confirmed the short centring of the tree ring series, thus vindicating the claims made in McIntyre's *Nature* submission. Given that this information didn't appear in the text of the draft Corrigendum, it was clear that *Nature* were going to allow Mann to retain as much face as possible. The correction was to be kept out of sight in the Supplementary Information, the journal itself remaining uncorrected.

There was another kick in the teeth for McIntyre in the Supplementary Information: after explaining the short centring methodology, Mann had added the words 'The results are not sensitive to this step', citing the forthcoming *Climatic Change* paper as evidence. Discussing the PC methodology in this way also spoke directly against McIntyre's own *Nature* submission, which Mann had seen as a reviewer. This appeared to be in direct contravention of *Nature*'s policies for reviewers, which required Mann to keep McIntyre's paper entirely confidential and not to use the contents for his own purposes. It also appeared to breach Langenburg's undertaking that Mann would not be permitted to use the Corrigendum as an opportunity to attack McIntyre. McIntyre immediately issued a complaint to Karl Ziemelis, the physical sciences editor at *Nature*, but it was all too late. The decision was already made.

The Corrigendum

The Corrigendum was published on 1 July 2004, and even after over six months of email exchanges, claim and counterclaim, there were still some surprises.[69] *Nature* had quietly decided to allow Mann to make a late change to the Corrigendum, namely the addition of a sentence at the end stating 'None of these errors affect our previously published results'. So Mann's claim had been carried forward from the online Supplementary Information to the main body of the Corrigendum in the printed journal. McIntyre had shown the journal that the claim was false, a position confirmed by the peer reviewers' who had accepted that the errors mattered in their written reports. And yet *Nature*, the world's premier scientific journal published Mann's claim regardless.

This was pretty outrageous, and McIntyre sent off yet another email to *Nature* the same day, protesting formally about the breach of *Nature*'s own policies and the undertakings made by Langenburg.

> In March 2004, we were shown page proofs of the Corrigendum, which did not contain this sentence. It appears that it was inserted after the review process had closed. Like the above sentence on the [Supplementary Information] web site, we have shown in our [*Nature* Submission] that it is untrue. By publishing it while the Communications Arising is under embargo, Professor Mann has attempted to pre-empt our submission and has breached the embargo under which we continue to withhold our own material. This sentence is not merely incidental; it is already being cited and circulated by Professor Bradley and perhaps others.[71]

In response to these points, however, *Nature* were not inclined to be helpful.

> In your message, you also draw our attention to some points concerning the Corrigendum published on 1 July by Mann et al. First, the published phrase 'Mann et al., in review' does not constitute a break of *Nature*'s embargo policy because it does not specify the journal involved; neither does it pre-empt your Communication Arising which, in the event that it is accepted for publication, will be published alongside a reply from Mann et al.[72]

This was a very peculiar statement because it clearly did pre-empt the publication of McIntyre's paper (and as we have seen, *Nature* managed to avoid publishing that document anyway). The policy at issue, which McIntyre had quoted in his email, didn't mention the specification of journals at all. It merely stated that the comments should not be used for any purpose other than making a response:

> The responders [i.e. Mann] must keep the comment confidential and must not use it for their own research or for any other purpose apart from replying to the comment, nor can they distribute it without first obtaining *Nature*'s permission.

Yet here was Mann using the comment not only in the Corrigendum but also apparently in the submission to *Climatic Change*. The only suggestion from *Nature*'s editors was that McIntyre should add discussion of Mann's statement to his own submission, an idea which must have provided little comfort since that paper had been in publishing limbo for over three months at the time. McIntyre made some last despairing attempts to get *Nature* to withdraw the critical sentence but his emails went unanswered. It was pretty clear that once again, the journal stood between McIntyre and his search for the truth. This battle had been lost.

Mann's bulldog – an interlude

In 2008, several years after the rejection of McIntyre's *Nature* submission, there was another equally strange attempt to defend short-centred PCs. A scientist supporter of Mann, known only as 'Tamino' (although he also styled himself 'Mann's bulldog'), had been trying to knock down the idea that full-period centring was a critical part of PC analysis and to promote the idea that short-centred standardisation was a valid alternative. He claimed that the Mannian approach was supported in the statistical literature and he invoked in his support none other than Ian Jolliffe.

> Centering is the usual custom, but other choices are still valid; we can perfectly well define PCs based on variation from any 'origin' rather than from the average. It fact it has distinct advantages *if* the origin has particular relevance to the issue at hand. You shouldn't just take my word for it, but you *should* take the word of Ian Jolliffe, one of the worlds foremost experts on [PC analysis], author of a seminal book on the subject. He takes an interesting look at the centering issue in this presentation . . .[73]

Tamino must therefore have been completely mortified when the following appeared on his website:

> IAN JOLLIFFE: It has recently come to my notice that . . . my views have been misrepresented, and I would therefore like to correct any wrong impression that has been given. . . . An apology from the person who wrote the page would be nice.[74]

Jolliffe went to to explain that his presentation in no way supported the idea of short centring,* and also rather surprisingly said that

---

*   Jolliffe referred to it as 'decentred'.

until the second half of 2008 he had had no idea of exactly what Mann had done in MBH98. Up until then he had been labouring under the misapprehension that Mann had used a technique called uncentred PC analysis, which is essentially PC analysis with no centring at all. Despite having reviewed McIntyre's *Nature* submission, he had apparently still not realised that the technique used by Mann was short centring. He even went so far as to say he had doubts whether even *standard* PC analysis was a suitable technique for temperature reconstructions.

Jolliffe had killed off the idea of short centring, but this exchange didn't take place until 2008. Back in 2004, McIntyre was to have a long fight ahead of him to win an argument that someone with Jolliffe's authority could have settled in minutes, if only he had noticed what Mann had done when he reviewed McIntyre's *Nature* submission.

6 Fighting Back

Universities incline wits to sophistry and affectation.

Francis Bacon

Although *Nature* had declined to publish McIntyre and McKitrick's critique of MBH98, it was clear to both sides that the debate was not over and that they would attempt to have their work published elsewhere. There would have been a great deal of cachet in appearing in *Nature*, who, as the publishers of the original study, should also have published the correction. But so long as his arguments got into print somewhere, McIntyre was reasonably content.

Two papers and twelve hockey sticks

As he and McKitrick pondered how best to proceed, the idea came to them to publish not one but two revised papers. The issues around MBH98 were so complex that there was no shortage of material and the extra space would allow them to develop their arguments as fully as was necessary. Eventually, it was decided that they would submit one paper to *Geophysical Research Letters* (GRL) examining Michael Mann's incorrect use of PC analysis and the question of the verification statistics, while a longer paper, looking at the sensitivity of Mann's reconstructions to various changes in the data and methods, would be submitted to *Energy and Environment*. To outsiders, the outlet for the second paper was something of a surprise, as many critics had tried to dismiss McIntyre's work in MM03 on the fallacious grounds that it had been published in such an obscure journal. But McIntyre and McKitrick felt that they owed Sonia Boehmer-Christiansen a favour – it was she who had taken a risk on publishing them in the first place and now they were attracting so much attention, it only seemed fair to give her a small payback.

Work on the new papers continued throughout the summer of 2004, McIntyre and McKitrick developing and extending the arguments they had presented in their *Nature* submission in order to cover all of the flaws in Mann's papers. The new critique had to be watertight and McIntyre was kept busy developing a whole new series of simulations of the Hockey Stick – firing red-noise at the short-centred standardisation routine and analysing the results. His hope was to develop a much more sophisticated analysis of the effect of short centring on the data. With the expectation that more space would be available to him in GRL, McIntyre wanted to explain and quantify exactly what was going on. By the time he had performed 10,000 simulations he had some very damning evidence indeed. In fact, the Mann algorithm managed to deliver a hockey stick from these random data series *over 99% of the time*.

Towards the end of the year, McIntyre was invited to present some of his results in a poster at the Fall Meeting of the American Geophysical Union (AGU) in San Francisco, and part of his presentation was a graphic showing some of these simulations. This is shown in Figure 6.1. The chart shows 12 hockey stick plots, all of which were generated using short centring. Only one of them, however, is based on real proxy data – it was the Hockey Stick itself – while the other eleven are generated from red noise. The game was to guess which hockey stick was the real one, and of course none of the visitors to McIntyre's poster at the AGU could do it.

The GRL paper

The new papers were ready by the October 2004 and spent the following winter tied up in the usual to and fro as reviewers demanded revisions and clarifications. GRL had decided not to appoint Mann as a reviewer, so there was to be a much easier time ahead for McIntyre and McKitrick.

As we have seen, the GRL paper (which will henceforward be referred to as ' MM05(GRL)') focussed on PCs and the verification

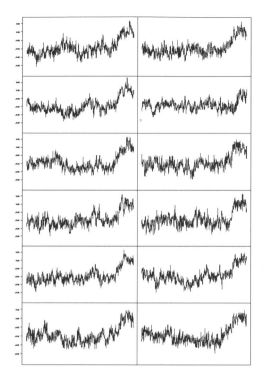

FIGURE 6.1: Twelve hockey sticks

statistics.[75] The central theme of the PC argument was unchanged – Mann's novel short-centring procedure was biased and therefore produced hockey stick shaped graphs. However, the arguments could now be made much more persuasively. Armed with the battery of new simulations, McIntyre had categorised each of the 10,000 dummy hockey sticks according to a new 'hockey stick index'. In essence the more the twentieth century portion of the graph deviated from the long-term average, the higher its hockey stick score. If the twentieth century excursion of the first PC was less than the standard deviation (which, you may remember, can be thought of as 'how far the line normally deviates from the mean'), then it was allocated a hockey stick score of 0. If the excursion was more than one standard deviation it scored 1, two standard deviations scored 2 and so on.

Using this simple methodology, McIntyre was able to demonstrate that Mannian short centring would almost never produce anything other than a hockey stick. In fact, three-quarters of the time it would produce a hockey stick with a score of 1.5 – in other words a stick with a big blade.

Having demonstrated the effect of short centring on random data, McIntyre went on to discuss its effect on the NOAMER PC1. With short centring, the bristlecone pines completely dominated the PC1 – in fact 93% of the variance could be shown to be due to 15 bristlecone pine series. But with standard centring, the bristlecones didn't appear in the PC1 at all, but were relegated to the PC4 – the fourth most significant pattern in the data. In other words short centring made the hockey stick shape of the bristlecones appear to be the dominant pattern in North American tree rings. Correct centring relegated them to what they were – the growth pattern of an obscure species of tree from one small area of the western USA.

Preisendorfer

The theme was taken up again in the *Energy and Environment* paper (henceforward 'MM05(EE)'). In his final reply to *Nature*, to which McIntyre and McKitrick had been unable to respond, it being the final round of submissions, Mann had invoked a new argument to defend his short-centring methodology. In essence he argued that whether the methodology was biased or not was irrelevant: the final reconstruction you got with short centring was the same as the one you got with correctly centred data. He agreed that when correctly centred, the bristlecones remained stuck in the PC4. In the original Hockey Stick paper, he had only used the PC1 and the PC2 of the NOAMER network in the reconstruction of the early fifteenth century. The PC4 was not carried forward to the calibration and hence did not influence the final reconstruction. Now, however, he was claiming that the first *five* PCs should be used and hence the PC4 would still be carried forward. Once there, it didn't matter that it was only a PC4 that explained only a small fraction of the total variance in the dataset. This is because, as we have seen, the weighting a series receives in the final reconstruction depends only on how well it correlates to the temperature PC during calibration. The hockey stick shape of the bristlecones correlated well with temperatures, so it didn't matter to Mann *which* PC the hockey stick signal was in, so long as it was one of the ones retained and carried forward to calibration. Once in the calibration, the hockey stick shape would imprint itself on the reconstruction.

The argument implicit in Mann's claims was that bristlecones in the White Mountains of California were somehow picking up a temperature signal of the whole Northern Hemisphere that was absent from the rest of his dataset. An objective observer would surely question if the reconstruction could be considered robust and reliable if its findings stood or fell on the inclusion of a PC4, representing one type of tree in one small area of the USA.

The number of PCs that is retained from a PC calculation is not something that is set in stone. As we saw in Chapter 2* it depends very much on the particular data set being examined. Mann, however, had said he had used 'the standard rule' for determining such things. The rule to which he referred, which goes by the slightly forbidding name of 'Preisendorfer's Rule N', is actually one of a number of possible approaches that statisticians use for working out how many PCs are significant and should be carried forward to subsequent calculations. McIntyre later cited one statistical authority who had listed a whole host of exotic-sounding alternatives, with names like the Bootstrapped Kaiser–Guttman Criterion, the Broken Stick, and Bartlett's Test of the Equality of Variance.[76] So there are no hard and fast rules in this area, and Preisendorfer's Rule N is no more than one rule of thumb among many, despite Mann's claims that it was the 'standard rule'.

---

*   See page 49.

That said, the original MBH98 paper *did* refer to the use of Preisendorfer's Rule N, but unfortunately for Mann's case, this was only in the context of the retention of *temperature* PCs. There was no mention at all of any retention policy for *tree ring* PCs. Of course, Mann might well argue that Rule N was used for tree rings too, but his problem here was that there was no sign that he had. Wildly different numbers of PCs were being retained in the different PC calculations. For example, Mann had retained two PCs in one of the Vaganov network calculations and nine in one of the Stahle ones. Whatever policy he was using it appeared to be neither rational nor consistent and, to this day, it remains one of the unsolved mysteries of MBH98.

Verification statistics

The story of the verification statistics plays an important part in the rest of the Hockey Stick story, and it is probably worthwhile recapping how they fit into the overall scheme of things. In Chapter 2 we showed how, once the paleoclimatologist has developed his mathematical model of how tree ring widths relate to temperature, he has to assess whether he got the answer by chance or whether he has in fact come up with something he can rely on for a full-scale temperature reconstruction. To do this, he keeps back a portion of instrumental temperature data (the verification period) and then tries to recreate those temperatures using only the ring widths. The verification statistics simply measure how well these reconstructed temperatures match up against the instrumental ones. If they are close enough ('significant', in the jargon) then he will go ahead and perform the rest of the reconstruction.

There are a number of approaches to measuring how well two sets of numbers match up against each other. As we've seen, Mann had concentrated on the Reduction of Error (RE) statistic, citing a paper by one of his associates, Ed Cook, in his support. Although Cook was a paleoclimatologist rather than a statistician, he was regarded as something of an authority on statistical matters in climatological circles. However, unfortunately for Mann, the Cook paper he cited in favour of the RE statistic actually stated that a *suite* of verification statistics should be used. Cook's list of suggestions was headed by McIntyre's own preferred measure, the R^2, alongside the RE and such exotica as the CE statistic, the sign test and the product mean test.

We also saw in Chapter 3* that for many sciences the R^2 is the default choice for measuring the correlation between two sets of numbers. The RE statistic, on the other hand, is little used outside climatology and, as a result, has not been studied by theoretical statisticians. This means that the way it behaves in different situations and the circumstances in which it is safe to rely on it are not fully understood. However, despite these drawbacks, the RE *is* widely used in climatology and it was not going to help McIntyre's case to dispute Mann's application of it in MBH98. The R^2 was not without its difficulties either and Mann had made much of these. When responding to McIntyre's *Nature* submission, Mann had positively raged about its unsuitability for climate reconstructions, calling R^2 scores 'inappropriate measures of forecasting or reconstructive skill'.[61] These remarks had been picked up by Ian Jolliffe in his review comments:

> The advocacy of RE in preference to [R^2] by MBH is a bit extreme.** [R^2] certainly has drawbacks, but no verification measure is perfect, and I see no evidence in the verification literature . . . that RE is the standard preferred measure. Indeed the only one of the 3 references . . . cited in the revised response that was available to me is somewhat critical of RE. My preference would be not to rely on a single measure . . .[60]

---

*   See page 66.

**   Jolliffe actually referred to r, which is a slightly different measure of correlation to R^2, but they are closely related, R^2 being the square of r. I've changed it to R^2 for the sake of ease of narrative.

... which, as we saw above, is precisely the position of Mann's own quoted authority, Cook. The whole argument was somewhat strange anyway because, despite what he was now saying about R^2 being 'inappropriate', according to what he had reported in MBH98, Mann *had* calculated the R^2, as well as its close variant, r:

> [RE] is a quite rigorous measure of the similarity between two variables, measuring their correspondence not only in terms of the relative departures from mean values (as does the correlation coefficient r) but also in terms of the means and absolute variance of the two series. For comparison, correlation (r) and squared-correlation [R^2] statistics are also determined.[14]

In fact, MBH98 included a colour-coded map of the verification R^2 for the 1854–1901 step of the reconstruction, so there can be little doubt that the figures had been calculated. Tellingly, the equivalent figures for earlier periods, including the critical AD 1400 step, were nowhere to be seen.

McIntyre's *Nature* submission had been almost silent on verification statistics; he and McKitrick were quite clear that they were not creating their own reconstruction of past temperatures, merely demonstrating that Mann's was not reliable. Any discussion of verification statistics had therefore seemed pointless. However, when he read the reviewers' comments, the inordinate length at which they considered the issue surprised McIntyre. It was out of all proportion to the weight he and McKitrick had given it in the paper itself; it all seemed rather odd. But when he put this fact together with Mann's fulminations against the R^2 statistic, the realisation dawned that he might have unearthed a whole new problem with Mann's paper. Could it be that the MBH98 reconstruction had actually failed the R^2 completely? When the reviewers had asked McIntyre to produce his own verification statistics, he had noticed that the R^2 was extremely low, but he just hadn't seen what a can of worms he, Jolliffe and Zorita had stumbled across. If the RE was high, but the R^2 was low, there was a real possibility that the MBH98 result was entirely spurious.

Spurious significance

Statisticians have understood for many years that just because your chosen statistical measure indicates that the result is significant, it doesn't mean that it actually *is* significant. In fact, it is entirely possible that the result is entirely *in*significant. This is just as true of the R^2 as it is of the RE or any of the other measures.

Spurious correlations, as these statistical foul-ups are known, have been written about since the start of the twentieth century, the classic case being documented in the 1920s by the Scottish statistician Udny Yule. Yule revealed the remarkable fact that there was an amazingly close correlation between the proportion of marriages that took place within the Church of England and the mortality rate, the apparent implication being that getting married in the Anglican rite would kill you. Of course, this was patently absurd but it is by no means an isolated case. Other examples include the correlation between the number of ordained ministers and the rate of alcoholism, and that between the salaries of Presbyterian ministers in Massachusetts and the price of rum in Havana. The correlation scores on some of these classic studies can be extraordinarily high, and statisticians have learned to take great care when they find a high correlation to ensure that it is not in fact spurious – this was what Jolliffe was alluding to when he recommended looking at more than one statistical measure. McIntyre noted subsequently that Yule's 'nonsense' correlation between Church of England marriages and mortality would certainly have passed the RE test with flying colours.

If he was going to confirm the possibility of spurious significance, McIntyre had to show that the R^2 of the Hockey Stick was as low as he thought. In order to do this he had to persuade Mann to release either his R^2 figures, particularly those from the critical AD 1400 step, or alternatively the 'residuals' from which he had calculated them. These latter figures were simply the differences between the reconstructed and the actual temperatures in the verification period, which would allow McIntyre to recalculate the R^2 himself.

McIntyre had been asking Mann for the residuals since way back in 2003, together with the source code for all of the MBH98 calculations. After Mann's initial refusal, he had ended up copying the request to the National Science Foundation (NSF), the US government body that had originally funded Mann's research. This, he hoped might concentrate Mann's mind sufficiently to elicit some action. Remarkably though, the NSF replied that Mann was under no obligation to deliver up the code, which they said was his personal property(!), a view which might have troubled any American taxpayers within earshot.

In 2004, McIntyre had made another attempt to force the data from Mann's grasp. You may remember that McIntyre had been invited to peer review a Mann submission to the journal *Climatic Change* that was highly critical of MM03.* McIntyre had taken advantage of this invitation to request the source code and residuals again, as part of his peer review. The response from Stephen Schneider, the journal's editor, was very instructive. He explained that in 28 years of editing journals, he had *never* had a request for this kind of information as part of a peer review. This admission demonstrates something very profound about the nature of the peer review process, which is a subject we will return to later in this story.

The request to *Climatic Change* for the residuals dragged on for a short time, before eventually Schneider decided to cut short the argument. He stated that reviewers were not expected to run code, but he did indicate that the journal would be adopting a new policy on data and materials, which would require authors to deliver up data on request. Unwilling to be thwarted this way, McIntyre again requested that *Climatic Change* obtain the residuals on his behalf, pointing out that these figures were undoubtedly 'data' within the

---

*   See page 146.

meaning of the policy, but again there was a point blank refusal from Mann.

> It is not our responsibility to provide [the residual series], we have neither the time nor the inclination to do so. These can be readily produced by anyone seeking to reproduce our analysis, based on the data we have made available, and our method which we have described in detail . . .[77]

Another dead end. McIntyre wrote his review, pointing out that Mann was thumbing his nose at the journal's policies, and should therefore be automatically disqualified from publication. In the event, nothing more was heard of Mann's paper, which was quietly withdrawn, although not before it had been cited in other papers attacking McIntyre's work.[58] It did rather look as if Mann had withdrawn the paper rather than release the residuals. Even then, McIntyre hadn't quite cottoned on to the possible implications.

The *Nature* materials complaint, which we looked at in the last chapter, offered another opportunity for McIntyre to force the issue. However, as we have seen, *Nature* was not inclined to make Mann and his colleagues do anything very much and the draft Corrigendum did not include any reference to the residuals. Amazingly, McIntyre refused to give up. After the issuing of the Corrigendum, he made a second request to *Nature* (which received an extemporising reply) and then a third one, this time directly to the physical sciences editor, who refused outright:

> And with regard to the additional experimental results that you request, our view is that this too goes beyond an obligation on the part of the authors, given that the full listing of the source data and documentation of the procedures used to generate the final findings are provided in the corrected Supplementary Information. (This is the most that we would normally require of any author.)[78]

Benchmarking

Without the residuals, it was difficult for McIntyre to prove his point, but there was another aspect to the verification statistics that was rather more fruitful and this was the area of benchmarking.

As we have seen, all parties agreed that R^2 is not perfect. It does, however, have some great advantages. Because of its ubiquity in other sciences, its behaviour in different circumstances is well understood, in direct contrast to RE. It is also possible to refer to look-up tables that will show you how significant your R^2 is for any given set of numbers. This is because, in the jargon, it has a theoretical distribution. RE, on the other hand, has no theoretical distribution and you therefore have to assess whether your result is significant by other means, usually by creating a benchmark.

There is a rule of thumb for benchmarking RE, which says that any positive number is significant, although this only applies to simple linear situations, rather than the more complex ('multivariate') model of the Hockey Stick. Because of this, Mann had not simply used the rule of thumb but had gone to the trouble of constructing some justification for his benchmark. In order to do this, he used what is called a Monte Carlo method.

The principle of a Monte Carlo method is to see what happens to lots of random number series when you put them through your process. If the score your actual data achieves comes out at the top end of what any of the random number series scored, you can be pretty sure that what you got was not just chance, but represents a real, meaningful result. To elaborate slightly: you have to create simulations of the reconstruction process, but using random 'red noise' data rather than real proxies. These dummy data series are known as pseudoproxies and are rather like the ones McIntyre used to test the effect of short centring on the PC calculation.* As before, the pseudoproxies are carefully designed so that their statistical characteristics are of the same type as the

---

*   See page 113.

actual data – in our case, the tree rings. You then crunch the pseudoproxies through the PC calculation and the calibration and then you see what RE score you get against the real temperature data in the verification period. This is your first simulation. However, you need to have lots of simulations – let's say 10,000. So you repeat the process 10,000 times and then you list the 10,000 separate RE scores, highest at the top, lowest at the bottom. The scientist doing the study then has to decide how confident he wants to be that his result is significant. Mann went for a high confidence level of 99% This would mean that there would be only a 1% probability that he had got his results by chance. Having selected this level, all he had to do was to count down to the 100th simulation in his list (1% of 10,000 is 100) and read off its RE score. This RE score was the new benchmark. Then, any reconstruction he performed on real data that had an RE in excess of this score was deemed significant.

When Mann performed the Monte Carlo benchmarking, he came up with a score of zero. In other words any verification routine with an RE score greater than zero would be accepted as having 'statistical skill', which is to say that the reconstruction could be considered significant. In MBH98 he had reported that the RE score in the AD 1400 step was 0.51, well in excess of the benchmark, and was therefore able to claim a high degree of skill.

Mann's benchmark

While Mann had given a brief summary of his benchmarking routine in MBH98, the paper contained little of the detail necessary to fully replicate what he had done. He had disclosed the particular statistical model used to simulate the tree ring series; this was a standard statistical model called AR1. There were also some details of the parameters used for the model.

When you are creating this type of simulation, it is very important that your simulated data has the same statistical properties as the real data and also that you treat the simulations in exactly the same way as the real data. Whether this had happened in practice was hard to tell from the text of MBH98. Was AR1 an appropriate model? Had the simulated data been through all of the same steps as the real data? This latter point was potentially crucial. In the MBH98 methodology, the short centring routine was doing some very strange things to the data.

This insight prompted McIntyre to try to create new RE benchmarks from scratch using the 10,000 simulations he'd created to demonstrate the effect of short centring. Instead of feeding random data into the correlation and the RE calculation, as per Mann, the idea was to feed it random data that had been processed through the short centring routine instead. Getting the calculations done was straightforward and the results were immensely gratifying. The correct benchmark level for the RE statistic turned out to be 0.59, as compared to Mann's zero. In other words, any temperature reconstruction that came out of Mann's data needed to score over 0.59 to be considered significant. As we have seen, the Hockey Stick itself had only scored 0.51 on the AD 1400 step, so it could no longer be considered statistically significant, even on the somewhat dubious RE measure.

For McIntyre, what was even better about the simulations was that they appeared to replicate very faithfully what was seen in the real data, namely high RE scores and low R^2. This was extremely strong evidence that the RE was in fact entirely spurious. It looked very much like game, set and match to McIntyre.

The *Energy and Environment* paper

The submission to *Energy and Environment*, MM05(EE), was to be an altogether broader paper, summarising much of McIntyre's work to date, looking at the differences between MBH98 and his attempts to recreate it, and examining the sensitivity of the reconstruction to various changes in the data network, particularly Gaspé and NOAMER.[79] McIntyre and McKitrick decided that they would also examine these last two series in more detail, including a lengthy discussion of their validity as proxies. To round things off, there was also to be a section rebutting each of the main counter-arguments put forward by Mann and another on the broader implications of the whole Hockey Stick affair.

Reconciling MBH98

As his understanding of exactly what Mann had done had improved, McIntyre had been able to pin down the differences between MBH98 and his emulation of it to just two main factors – Gaspé and NOAMER. In order to demonstrate how the reconstruction was influenced by these two series, just a fraction of the total data, he and McKitrick had decided to present a sensitivity analysis. This would demonstrate how removing the series, or making apparently insignificant changes to them, would dramatically change the final reconstruction.

As we have seen, Gaspé was included twice in the proxy network and, uniquely among the MBH98 proxies, was artificially extrapolated back to the year 1400, allowing it to be included in the critical AD 1400 roster, where it had the effect of depressing the reconstructed temperatures in the Medieval Warm Period. The extrapolation was not disclosed in the original paper, but Mann had been forced to acknowledge it in the Corrigendum. Of course, the Corrigendum also said that this didn't materially affect the MBH98 results, so to the reader, the extrapolation appeared, as Mann had put it, a mere 'technicality'.

McIntyre's arguments about Gaspé had been relegated to a footnote in the original *Nature* submission in an attempt to reduce the word count, but now with room to explain it in full, he could afford to set down all of the idiosyncrasies of the series and Mann's treatment of it. For example, from the start of the series in 1404, through to 1421, Gaspé was based on a single tree, and there were only two in the subsequent period up to 1447. This obviously raised enormous doubts over the reliability of this section of the data. In fact, the original authors of the Gaspé study, Jacoby and D'Arrigo, had not used the early portion of the series at all in their own reconstruction, deeming the data too sparse. Their lead had been followed by later scientists: only Mann, it seemed, had ever used the dubious early part.

So there were a number of Mann's decisions regarding Gaspé that were open to question. Some, like extrapolation, might have appeared insignificant, while others, like the small number of tree cores in the earlier centuries, might have looked a little more troubling. McIntyre was able to show, however, that all of the decisions Mann had taken had a huge influence on the final shape of the reconstruction. For example, assuming you had first fixed the short centring routine, if you removed the early portion of the Gaspé record (based on one or two trees) you kept the Medieval Warm Period. If you removed the individual proxy version of Gaspé (but not the NOAMER version), you kept it too. In fact, you could only get rid of the Medieval Warm Period by using the Gaspé series twice and by including the unreliable early portion, *and* by extending this highly dubious data back to the start of the fifteenth century.

Mann had claimed that the extrapolation was justified by the need to keep a representative of the northern treeline series in the reconstruction of the fifteenth century. This claim was also shown to be wrong on a number of scores. For a start, Gaspé, which is located south of Québec, is nowhere near the northern treeline. But for the purposes of the sensitivity analysis, McIntyre merely replaced Gaspé with another northern treeline series, Sheenjek River. This series was based on a higher number of trees and should therefore have been more reliable, and McIntyre was able to show that when this was done, the Medieval Warm Period appeared once more.

The other side to the sensitivity analysis was to look at various configurations of the NOAMER PC series. Assuming you had first removed the Gaspé extrapolation, you could get wildly different results depending on how the NOAMER PC calculation was performed. It turned out that the Medieval Warm Period was eliminated from the reconstruction *only* if you had bristlecone pines in your proxy database *and* a short-centred PC algorithm (assuming you retained the same number of PCs as Mann had used in MBH98). If the PCs were centred you lost the influence of the bristlecones, which lingered down in the PC4. Now, of course, Mann had pointed out that if you retained the PC4 too, you could get the bristlecones back into the reconstruction and 'get rid of the Medieval Warm Period', but this was hard to justify, as we saw above, because the PC4 represented such an obscure pattern in the data.

Even if Mann could have justified the use of the PC4, there was another problem with NOAMER. Because the bristlecones were not reliable proxies in the first place – as we have seen the growth pattern was thought to be contaminated with a non-climatic signal – Mann shouldn't have been using them in his dataset anyway. McIntyre was able to show that if you eliminated the bristlecones from the proxy roster, it didn't matter *which* centring convention you used, the Medieval Warm Period remained in the record.

Flipping proxies

The sensitivity analysis gave some stark results, but there was to be one last amazing demonstration of the perverse results that had been caused by Mann's short-centred PC convention. A reader of McIntyre's website had written in to ask what happened if the ring widths of the non-bristlecone sites were artificially inflated in the early fifteenth century. The idea was to see how this new dummy information would show up in the final result. Intrigued by the suggestion, McIntyre ran another simulation, using his best approximation of Mann's methodology and data, but adjusted in the way that the reader had suggested. The results were simply extraordinary. Adding extra ring width in the early fifteenth century caused the reconstructed temperature for that period to go *down*. In other words the algorithm was reading *wider* ring widths as evidence of *lower* temperatures. This simply could not be the case. If it were, it would be the end of dendroclimatology as a science.

McIntyre observed, 'these results are initially very counterintuitive and have provoked some disbelief'[79] and a section was added to the *Energy and Environment* paper detailing the exact computer code used in the calculation so that doubters could test the result for themselves. This was a stark contrast with Mann, who still firmly refused to release any of his scripts. McIntyre went on to explain exactly what was happening.

> This rather perverse result nicely illustrates a problem of mechanically applying a numerical algorithm like PC analysis without regard to whether it makes sense for the underlying physical process. PC methods are indifferent to the orientation (up or down) of a series – the difference is merely the presence or absence of a negative sign . . .
>
> Under the MBH98 algorithm, the addition of the extra values in the first half of the 15th century causes the algorithm to flip the series upside-down so that they match as well as possible to the bristlecone pines, whose hockey stick pattern is imprinted on the PC1.[79]

The point he was making is that PC methods see upticks and downticks as the same thing – to a PC calculation they're all just 'deviations from the mean'. Presented with a twentieth century uptick and a twentieth century downtick, the PC algorithm would read these as 'big twentieth century deviations'. It would then flip one of the series over when providing the answer. Which one it flipped would depend on their relative sizes and depending on which one this was, the final result could look like a downtick rather than an uptick (or vice versa).

If you think back to the metaphor of the shadow cast by a comb, with which we explained PC analysis back in Chapter 2, the point may become clearer: it doesn't matter *which* way up the teeth are pointing, the shadow is in essence the same. So if we have a set of combs rather than the single one in our previous example, and we shine a light at them, PC analysis will combine all the shadows from all the different combs, some of which have their teeth pointing upwards and some downwards, and return the answer 'comb'.

So when McIntyre had added some extra dummy ring width to proxy records in the fifteenth century, it turned out that the PC algorithm had found the best way to summarise it was to flip the whole series over to match fifteenth century downticks in other series. Extra fifteenth century growth led to reconstructed temperatures that were apparently lower.

In fact there was another remarkable aspect to this particular statistical oddity. It turned out that when Mann had archived results from his second hockey stick paper (MBH99) the PC algorithm had been flipping again: the final PC1 had ended up with a twentieth century downtick, so it was actually *upside-down*. Mann had therefore had to flip it back again for presentation purposes.

Robustness

Apart from the sensitivity analysis, McIntyre also wanted to present a demonstration of just how it was that Gaspé and NOAMER could make such a huge difference to the AD 1400 reconstruction. If you remember, once the PC calculations are complete, the 112 (or perhaps 159) PCs and individual proxies are lined up against the temperature records and a weight is given to each, depending on how good the correlation is. It was all very well having a hockey stick shaped series or two, but McIntyre needed to make it clear how these came to dominate the temperature reconstruction.

In order to do this he constructed a plot of the hockey stick index against the calibration correlation. For each series, the hockey stick index simply measured how big the blade of the stick was compared to the handle. The calibration correlation was, as the name suggests, a measure of how well the series matched up against the temperature records in the calibration period. Those that had high correlations would be heavily weighted and their shapes would therefore dominate the final reconstruction.

The plot is shown in Figure 6.2, with Gaspé and NOAMER marked in the top right hand corner. The implications were clear, as McIntyre and McKitrick explained:

> Except for [Gaspé NOAMER] there is an overall negative relationship between [the hockey stick index] and the

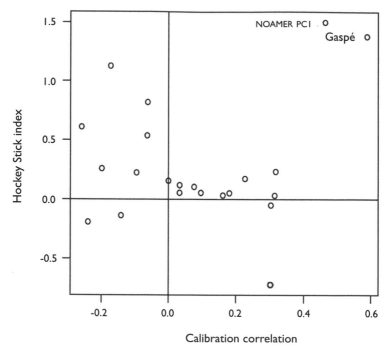

FIGURE 6.2: Outliers in the dataset

> correlation with temperature: i.e. hockey stick series fit the temperature data relatively poorly in the calibration interval. But the [NOAMER] and Gaspé series are such influential outliers that they reverse this pattern for the model as a whole.

In other words, the group of points on the left have profiles that are hockey stick shaped to various degrees. However, the more hockey stick shaped they are, the less well they correlate with temperature. This is in stark contrast to Gaspé and NOAMER, the two outliers, which correlate better than any of the other series in the database. Because of this their hockey stick shape is imprinted on the final reconstruction.

Credibility of the critical series

The final piece in the jigsaw was a look at the credibility of Gaspé and the NOAMER bristlecones as proxies. They may have been outliers, but it was conceivable that there might have been an unusual temperature signal affecting their growth. Although this wouldn't justify the extrapolation of Gaspé or the short-centred PC calculation, it would at least justify their inclusion in the network in the first place.

As we've seen, McIntyre knew from his own experience that there was nothing going on in the climate of Québec in the late twentieth century that would have caused such a dramatic growth spurt in Gaspé. We've also seen that it was widely agreed that the growth spurt in the bristlecones was non-climatic too. It was necessary, however, to bring all this information together in one place so that any arguments that these two series were valid could finally be laid to rest. This was even more important in the 1000–1399 extension of the Hockey Stick in Mann's second paper, MBH99, where Gaspé and the bristlecones were even more dominant in the final network because of the reduced number of proxy series that were available for earlier centuries.

McIntyre's paper therefore laid out in copious detail the problems with the two datasets. He showed that the bristlecone growth was anomalous by comparing NOAMER to an earlier Northern Hemisphere reconstruction by Keith Briffa, which was based on a *mixture* of tree species.[80] Briffa's reconstruction showed nothing like the same shape as MBH98, which was dominated by the bristlecones. In fact, the twentieth century correlation between the two reconstructions was precisely nil. As McIntyre, deadpan, observed, whatever major temperature signal Mann had captured had apparently escaped detection in Briffa's paper.

The scientific literature appeared to hold the explanation for this discrepancy. As we saw in the last chapter, Donald Graybill, the original author of the bristlecone studies, had stated that the twentieth century growth spurt in these trees could not be explained by temperature changes. He believed growth was being affected directly by carbon dioxide levels, an effect known as 'carbon dioxide fertilisation'. So when it came to the bristlecones, the growth spurt in the twentieth century was probably due to changes in both carbon dioxide levels *and* temperature, which inevitably made it very hard to isolate the temperature signal. There were also any number of other possibilities that might have been at the root of the growth spurt: nitrogen fertilisation, a change in the ecosystem, the influence of sheep and of rainfall.

In fact, the equation was even more complex than this. Bristlecones often exist in a strange 'stripbark' form, where the bark on one side of the tree dies back. It turned out that Graybill had actively *sought out* stripbark trees when he collected the samples that ended up in the NOAMER PC1, believing that these would be more susceptible to carbon dioxide fertilisation. Although some authors had claimed that bristlecones could still be used for temperature reconstructions because *only* the stripbark form was affected in this way, it was simple enough to demonstrate that there were only stripbark trees in Graybill's samples. The NOAMER PC1 and Mann's results were therefore inherently unreliable.

Each and any of these factors should have been considered and eliminated in the original MBH98 paper before it was published. That they weren't was something of an indictment of the system of peer review. That a paper that relied on these unreliable results passed the subsequent IPCC review and reached a position of such importance in the case for manmade global warming was remarkable.

It was fairly clear then that there were major problems with the credibility of the bristlecones in NOAMER. What then of Gaspé? McIntyre explained in the *Energy and Environment* paper that the data was derived from cedars. These trees, like the bristlecones, exist in strip bark and whole bark forms, and are very slow growing. Cedar chronologies are not unique in tree ring studies, but they are few and far between. Apart from Gaspé, those that do exist show either no twentieth century growth spurt or a negative relationship between growth and temperature. Gaspé seems to be unique in having a hockey stick shape.

Even if this fact had not been enough to raise suspicions over the series' reliability, there was another feature of cedar chronologies noted in the literature that was potentially fatal to Mann's thesis: the response of cedars to changes in temperature was non-linear. In other words, it appeared that the trees grew fastest in cool and wet conditions but in very hot or very cold weather they slowed down. A chart of growth versus temperature for cedars

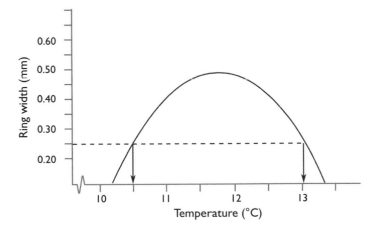

FIGURE 6.3: Upside-down U-shaped temperature response

would therefore be an upside-down U shape. The possibility of a non-linear growth response had been observed in other species too,[81,82] and as McIntyre pointed out, if true it would completely undermine the whole basis of paleoclimatology, which relies on the relationship between temperature and ring width being linear. If the curve was an upside-down U, as shown in Figure 6.3, any given ring width could imply one of two different temperatures, with no way of knowing which one was right. In the example shown, a ring width of 0.25 mm could imply a temperature of either 10.5°C or 13.0°C. McIntyre performed a test of the correlation between the Gaspé ring widths and the local temperature and there was no

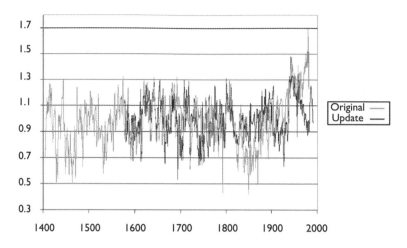

FIGURE 6.4: Gaspé – original and updated series

The original series has a sharp twentieth century uptick, while the update shows a sharp decline in ring widths starting in the 1960s.

relationship apparent, suggesting that this was indeed a major problem.

McIntyre had even more startling revelations to make about Gaspé though. During the course of his research, an update of the Gaspé series had come into his possession that threw further doubt on its reliability. The original study was based on samples taken in 1982, but it turned out that new ring cores had been taken in 1991. The new data were astonishing – they showed nothing like the same growth spurt seen in the earlier figures (see Figure 6.4).

Extraordinarily, Jacoby, the author of the new study, had refused to publish these revised figures or to archive the data, on the grounds that the older data showed the temperature better and because their research was 'mission-oriented'. [83] In fact he was not even willing or able to identify the location of the site to allow re-sampling, repeatedly turning down McIntyre's requests for the map co-ordinates. Eventually, once the request for the location was put to him via the original journal, Jacoby claimed that the location had been lost, a remarkable fact for a site that had been sampled twice. The team who had made the update had apparently never found the original location. The exact whereabouts of the site remain a mystery to this day.[84]

Announcement

By January, with the papers accepted for publication and nearing print, McIntyre was ready to tell the world that he and McKitrick were about to start their fightback. He posted up an article on his Climate2003 website announcing the publication of no less than three papers about the Hockey Stick – in GRL, in *Energy and Environment* and a third article in a Dutch popular science magazine.

The trouble was only just beginning.

7   Commentary

*For a successful technology, reality must take precedence over
public relations, for Nature cannot be fooled.*

Richard Feynman

RealClimate

On 19 November 2004 a new Internet domain, 'realclimate.org',
was registered by Betsy Ensley, an employee of Environmental
Media Services (EMS), a PR firm based in Washington DC. EMS was
a pivotal organisation in the green movement and Ensley was a
committed environmental activist, having previously been
involved in setting up such campaigning organisations as
bushgreenwatch.org and womenagainstbush.org.

EMS itself was run by David Fenton, a powerful PR executive,
as part of his lobbying organisation, Fenton Communications.
Fenton has been called 'one of the most influential PR people of
the twentieth century', a claim that was based in part on the
leading role he played in promoting the notorious Alar scare in the
1980s, when apple growers across the USA were ruined by an
unsubstantiated claim that a pesticide they used caused cancer.

The setting up of RealClimate in November 2004 signalled
the start of a new phase in the war of the Hockey Stick – the blog
campaign. Blogs – Internet diaries and journals – were big news at
the end of 2004, promising to revolutionise the Internet by making
web publishing simple and accessible to everyone. There had been
an explosion of online activity, with all sorts of people piling onto
the bandwagon and setting down their opinions and their most
intimate thoughts for all to read online. Public relations people like
Fenton had rapidly cottoned on to the significance of blogging as

a campaigning tool; it was clear from the start that RealClimate was to be a weapon in the environmentalists' fight and that much of its campaigning was to be directed at McIntyre and McKitrick.

New media war

Ever since the publication of MM03, and even before then, supporters from both sides had been hurling brickbats at each other on the Internet. As McIntyre and McKitrick's new papers in *Energy and Environment* and *Geophysical Research Letters* neared publication, the war of words was beginning to be stepped up, and looked as though it would explode in the fairly near future.

With experience suggesting that a torrent of criticism would shortly be heading their way, McIntyre and McKitrick were very grateful for any support they could get, and, in the event, some of it appeared from surprising directions. A highly supportive article was published in the MIT *Technology Review* by an eminent physicist from the University of California, called Richard Muller.[85] The headline 'Global Warming Bombshell – A prime piece of evidence linking human activity to climate change turns out to be an artefact of poor mathematics' gives something of the flavour of the piece. Muller was highly critical of the way the scientific community had treated McIntyre and McKitrick, and in particular the refusal of *Nature* to publish their critique of MBH98. It was, he said, 'more dangerous to have a phony hockey stick than a broken one'.

Mann's supporters saw Muller's article as a gauntlet thrown down, and within days of its publication, it was coming in for sustained criticism around the Internet. Mann later accused Muller of a 'scurrilous parroting' of McIntyre and McKitrick's arguments,[86] but the initial reaction came from one of Mann's supporters. William Connolley, a climate modeller and sometime green politician, commented in a web forum called sci.environment that he thought McIntyre and McKitrick's claims about the effect of short centring were 'probably wrong'.[87] However, when he examined Mann's code, the best he could come up with was a less forthright statement that the short centring looked 'odd', although he also opined that it wasn't obviously harmful. Connolley also thought that the short centring would reduce the impact of series with a twentieth century uptick rather than exaggerate them, suggesting that his conclusions were only based on a brief review of McIntyre's work. Another climate modeller, James Annan, commented that McIntyre and McKitrick didn't know the difference between multiplication and division and said that he intended to start looking into the issue in more detail.[88] In the event, though, it was in fact Connolley who started the counterattack, setting up a web page to 'audit the auditors', but whether he eventually worked out that McIntyre was in fact correct or he lost interest in the subject, he soon moved on to other things.[89] Annan also suddenly seemed to lose interest in McIntyre's work.

Annan and Connolley may have moved on, but their earlier thoughts had been picked up and propagated across the left-wing blogosphere, where prominent voices such as those of Brad DeLong and Tim Lambert spread the idea that McIntyre and McKitrick's new papers were fatally flawed,[90,91] a remarkable thing considering that they had not even been published at that time.

It may have been the wish to undermine a prominent supporter of McIntyre's work that led to all the criticism of Richard Muller, but his position as one of America's most eminent scientists also attracted the interest of others with a more neutral standpoint; people who were just interested and came with no preconceptions of McIntyre's work or of Mann's. One of these was Marcel Crok. Crok was a Dutch science journalist, who worked for a popular science magazine called *Natuurwetenschap & Techniek* (which translates into something like 'Natural Science and Technology'). Crok had read Muller's article and was intrigued that such an important scientist should have stuck his neck out so far in order to support two outsiders such as McIntyre and McKitrick. Perhaps, Crok wondered, there was more to their claims than one might think from the barrage of criticism and derision that was emanating from Mann's supporters. He resolved to take a closer look.

Shortly afterwards, in November, the RealClimate website went live in something of a blaze of publicity, the cheerleading led by none other than McIntyre's old friends at *Nature*, who welcomed the new venture with a supportive editorial.[92] Billing itself as 'a commentary site on climate science by working climate scientists for the interested public and journalists', RealClimate boasted nine prominent climatologist/writers led by Mann himself and a climate modeller called Gavin Schmidt, who worked at NASA's Goddard Institute. Also in the RealClimate lineup were Ray Bradley, one of Mann's co-authors on MBH98, and William Connolley, the environmentalist-cum-climate modeller who had been criticising McIntyre on sci.environment.

Posting on Rutherford

Within days of the launch, Mann was using this new platform to carry on the fight against McIntyre. One of the first postings on the site was a summary of a forthcoming paper by none other than Mann's assistant, Scott Rutherford, whom we met at the start of Chapter 3, sending the data files to McIntyre. The new paper, known as Rutherford et al 2005 (it was to be published in the new year), included several prominent members of what was to become known as the Hockey Team – Mann, Bradley and Hughes, plus three British scientists – Keith Briffa, Phil Jones and Tim Osborn[93] – all familar names from earlier in the story.

Rutherford et al 2005 completely discredited the claims of McIntyre and McKitrick, at least according to Mann, who authored the blog posting.[94] He was certainly not mincing his words in the text either, talking of MM03's 'falsely reported putative errors' and McIntyre's 'misunderstanding of the methodology used' in MBH98. To anyone familiar with what had gone before, it was clear that Mann was largely reiterating the arguments he'd used earlier in the year to attack the *Nature* submission – that McIntyre had used the wrong dataset, that he hadn't used a stepwise procedure and so on. It was technically relatively simple for McIntyre to rebut Mann's claims of course, he had done this before, but it quickly became clear that as the new papers neared publication, RealClimate and the rest of Mann's web-based supporters were going to step up the level of attacks. He was going to be hard pressed to keep on top of the task of defending himself from all of Mann's supporters.

The next offensive was not long in coming. Two new Mann postings appeared on RealClimate on the same day in December. The first was an excoriating rant with the catchy title of 'False claims by McIntyre and McKitrick regarding the Mann et al. (1998) reconstruction'.[95] This piece was an attempt to rebut the new criticisms of the Hockey Stick ('spurious criticisms', in Mann's words) before they were published, and largely revisited the arguments of his replies to *Nature* – Preisendorfer's Rule N, the argument that if you used the NOAMER proxies without processing them through the PC calculation you still got a hockey stick and so on. It was no doubt very effective in persuading his readers.

The second posting was called 'Myth vs fact regarding the hockey stick' and it included another sustained attack on McIntyre and McKitrick (the brackets are in the original):

> False claims of the existence of errors in the Mann et al (1998) reconstruction can also be traced to spurious allegations made by two individuals, McIntyre and McKitrick (McIntyre works in the mining industry, while McKitrick is an economist). The false claims were first made in an article . . . published in a non-scientific (social science) journal '*Energy and Environment*' and later, in a separate 'Communications Arising' comment that was rejected by *Nature* based on negative appraisals by reviewers and editor [as a side note, we find it peculiar that the authors have argued elsewhere that their submission was rejected due to 'lack of space'].[86]

At this point, readers may care to refer to the final letter from *Nature*, which can be found at page 130.

Climate Audit

As the attacks grew more numerous and more ferocious, and his work drew more and more attention, McIntyre found himself having to spend an increasing amount of his time explaining what he was doing. He therefore started to post regular updates on his Climate2003 website in order to try to fend off some of the more frequent critiques and questions. This site, now defunct,* predated blogging and was an old-style flat webpage with a few screens of information. Because of this, it had always been difficult for readers to use and newcomers would have found it very hard to find their way around the site. As the volume of information on the site started to increase, this situation was becoming more confusing and it soon became clear that Climate2003 was no longer up to the job. Maintaining the site was becoming a daily struggle which was occupying too much of McIntyre's time and distracting him from his research. As supporters were pointing out to him, the Internet world was gravitating towards blogs and he was going to have to tag along too or get left behind, his voice drowned out in the climate cacophony. The decision was taken to move to a blog format.

McIntyre didn't have a PR adviser to help him get his new venture off the ground, but with the help of a supporter, a blog was set up to McIntyre's specifications and he was able to start posting articles towards the end of 2004. Its name was Climate Audit and its inception was to be a turning point in McIntyre's fortunes.

The first few posts on Climate Audit merely reproduced earlier articles from Climate2003, but with years of research behind him, stretching right back to the Climate Skeptics days, there was no shortage of new material and McIntyre was soon into the swing of things. Postings, visitors and comments were slow at first, but McIntyre managed to keep a relatively civilised tone to proceedings and even gave critics a friendly welcome. As the months passed, and visitor numbers and comments rose, a regular readership

---

* It is, however, well indexed on the Wayback Machine for any readers who might want to see its contents.[96]

developed, with a remarkable array of scientific expertise, particularly in areas that were directly relevant to McIntyre's research, such as statistics and signal processing. The Hockey Team were soon confronted by a very well-equipped opposition – the Climate Auditors. While some of the blog's readers were not always welcoming to those opponents who left comments on the site, those opponents were at least allowed to do so without being censored, in stark contrast to RealClimate, which had quickly developed a reputation for deleting or editing comments.

Publication of the new papers

As the final days counted down towards the official appearance of McIntyre's new papers, Mann started a new set of attacks. A week before the publication date, he and Schmidt posted up a long two-part article on RealClimate on the subject of peer review, outlining a litany of scientific articles that had passed the peer review process but had subsequently proved controversial or just plain wrong. In the words of the article's title, peer review was 'a necessary, but not sufficient condition'.[97] It was obvious where this was leading, and sure enough, the second part of the article, timed to coincide with the publication of McIntyre's new papers, was an aggressive denunciation of McIntyre and McKitrick's work. Their studies were 'flawed', 'deeply flawed', 'botched', 'bizarre', and their claims were 'false and specious'. The paper had 'managed to slip through the imperfect peer review filter of [*Geophysical Research Letters*]'.[28]

This was strong stuff, but Mann and Schmidt didn't just throw mud. They went through the whole catechism of their earlier arguments against McIntyre and McKitrick for the benefit of new readers, in a vain attempt to try to stop the Climate Auditors's story gaining any media momentum. However, by now McIntyre had his own blog, from which he could shoot back, and now there was mainstream media interest in his work as well.

Natuurwetenschap & Techniek

As we have seen, the Richard Muller piece in MIT *Technology*

Review had been picked up by the Dutch science journalist, Marcel Crok. Crok had been unsure what to make of the story at first, but he had arranged to interview McIntyre in December 2004, and since then had done a great deal of investigative work, digging into the background of the story and also seeking independent opinions from Dutch scientists on the statistical issues. The more he continued his investigations, the more convinced he became that there was a big story to tell.

As his interest grew, Crok decided to contact Mann in order to get the other side of the story and to try to draw out responses to the allegations that McIntyre had just made in the two new papers. Some of the questions were also prompted by McIntyre himself as a way of pointing Crok to the pertinent issues – the differences in their positions and where he felt the weaknesses in Mann's claims lay.

Mann's reply, when it arrived, was a prime example of his unmistakeable style. In a long email, he went to considerable pains to point out his low opinion of the Canadians and made it abundantly clear that he wanted nothing favourable about McIntyre or McKitrick to appear in the Dutch press:

> I hope you are not fooled by any of the 'myths' about the hockey stick that are perpetuated by contrarians, right-wing think tanks, and fossil fuel industry disinformation . . . I must begin by emphasizing that McIntyre and McKitrick are not taken seriously in the scientific community. Neither are scientists, and one (McKitrick) is prone to publishing entirely invalid results apparently without apology. 'New Scientist' considered running an article . . . on [McIntyre and McKitrick's] claims. The editor decided not to run an article, concluding that their claims were suspicious and spurious after interviews with numerous experts and after it was revealed that they had suspiciously close ties with the fossil fuel/energy industry.[98]

When he accused McKitrick of publishing invalid results 'without apology', Mann was referring to a paper that McKitrick had published earlier in 2004.[99] Shortly after publication, the paper had been found to contain an error* and a correction was issued shortly afterwards. It is probably fair to say, though, that a correction is not normally accompanied by an 'apology', so Mann's complaint seems somewhat overstated. The citing of this error has been a regular form of attack on McIntyre and McKitrick over the years,[100,101] despite the fact that McKitrick's correction showed that error didn't affect the paper's conclusions.

Claims of McIntyre's alleged closeness to big oil have also been made regularly by the Hockey Team and their supporters, the 'evidence' usually consisting only of dark mutterings about McIntyre's background in the extractive industries. McIntyre had made clear statements in all of his papers that he had received no financing for his work – indeed he had spent thousands of dollars of his own money pursuing it up to that point. The most definitive accusation made by the doubters, which was the one referred to by Mann, was a claim by the Environmental Defense Fund that McIntyre was being funded by ExxonMobil. However, the evidence appeared to consist only of the fact that McIntyre had once written an article for a think tank that had received funding from Exxon.

Mann clearly felt, however, that this was persuasive evidence and advised Crok to treat McIntyre with 'appropriate suspicion'. Instead of talking to McIntyre and McKitrick, he said, Crok should speak to people like Jones, Briffa, or Jonathan Overpeck – all core members of the Hockey Team (although that is not the way Mann described them).

Much of the substantive content of Mann's email – where he discussed the scientific controversies – merely went over old ground. However, he did manage to come up with some new information on the verification statistics. We have seen above that one of the chief criticisms of the Hockey Stick was the fact that

---

* Practicing what he preached, McKitrick had posted his data and code online, making it much simpler to check his work.

Mann had not published his verification R^2 so that it was impossible for anyone to gauge the reliability of the reconstruction.* In his replies to McIntyre's *Nature* submission, he had argued forcefully that the R^2 was a flawed statistic and that is was inappropriate to the particular circumstances of the Hockey Stick. Since that time, however, McIntyre had got hold of some useful new intelligence on the subject. Crok had posted his correspondence with Mann on the Internet and this revealed an interesting exchange on the subject of verification statistics.

> CROK: There is a severe debate between you and [McIntyre and McKitrick] about the skill of the calculation. You claim a high RE-statistic. [McIntyre and McKitrick] show that their simulated hockey sticks also give a high RE-statistic but a very low R^2 statistic.

> MANN: . . . Our reconstruction passes both RE and [R^2] verification statistics if calculated correctly. Wahl and Ammann (in press) reproduce our RE results (which are twice as high as those estimated by [McIntyre and McKitrick]), and cannot reproduce [their] results. There is little, if anything correct, in what [McIntyre and McKitrick] have published or claimed. Again, none of their claims have passed a legitimate scientific peer review process![98]

So Mann was admitting that he *had* calculated both RE and R^2, and also claiming that the Hockey Stick *passed* the R^2 test. This made something of a nonsense of his claim that R^2 was a flawed statistic.

By the end of Crok's investigations, he was in no doubt where the truth lay, and his final article carried the title 'Kyoto protocol based on flawed statistics' and the subtitle 'Proof that mankind

---

* See page 157.

causes climate change refuted'.[102] In the ten pages of the article, he told the full story of McIntyre's work up to that time in a way that was hard to ignore. In particular, Crok had discussed McIntyre's work with a number of eminent scientists, all of who supported the criticisms of MBH98. For example, Eduardo Zorita confirmed that he had never heard the number of 159 proxy series that Mann now claimed were used in MBH98. In Zorita's work on MBH98 the number was 112, just as it was in McIntyre's original study, and just as it was in the text of MBH98. Crok was also able to report the comments of two important scientists both of whom confirmed that the Mann algorithm 'mined' the data for hockey sticks. One of these was Mia Hubert, an expert in robust statistics. The other was Hans von Storch, an eminent climatologist and a colleague of Zorita's at the Institute of Coastal Research.* Von Storch is one of the big names in climatology and had been one of the editors who had resigned from the board of Climate Research over its publication of the Soon and Baliunas paper,** but he was not a member of the Hockey Team either. Later though, he was seen as being in competition with McIntyre to be the man who broke the Hockey Stick. It is fair to say that he was, and remains, respected by both sides of the debate and we will meet him again later in the story.***

Regalado in the *Wall Street Journal*

Marcel Crok's article was just the start of it. In mid-February, and with the papers in print, McIntyre made the front page of the *Wall Street Journal*, in a long article which was very supportive of his work.

---

* By strange coincidence, Heike Langenburg, the *Nature* editor we met on page 135, used to be a postdoctoral researcher in von Storch's laboratory. I don't imply any great significance to this fact beyond an observation that climatology is a very small world.

** See page 56.

*** Von Storch is a colourful character who once founded a club to defend Donald Duck against accusations of indecent behaviour, and for some years was the editor of a Donald Duck magazine, *Der Hamburger Donaldist*.

Since it was published four years ago in a United Nations report, hundreds of environmentalists, scientists and policy makers have used the hockey stick in presentations and brochures to make the case that human activity in the industrial era is causing dangerous global warming.

But is the hockey stick true?

According to a semiretired Toronto minerals consultant, it's not.[103]

The article contained some interesting revelations. Its author, Antonio Regalado, had asked Mann about his refusal to release his code to McIntyre. As Regalado explained:

Mr. McIntyre thinks there are more errors [in Mann's work] but says his audit is limited because he still doesn't know the exact computer code Dr. Mann used to generate the graph. Dr. Mann refuses to release it. 'Giving them the algorithm would be giving in to the intimidation tactics that these people are engaged in', he says . . .

. . . which was a remarkable statement, given that McIntyre had already published all of his correspondence with Mann on the old Climate2003 website, and *none* of it could even remotely be construed as 'intimidation'.[31]

Regalado's other scoop was the news that Mann had contacted the editor of *Geophysical Research Letters* (GRL) in order to denigrate McIntyre's new paper.

The editor [of GRL], Steve Mackwell, says Dr. Mann contacted him to argue that the Canadians' work was deeply flawed. Dr. Mann then put a critique on his blog, 'Realclimate.org', calling the Canadians' new paper 'demonstrably specious'.

A few days later, McIntyre was in the *Wall Street Journal* again, this time as the subject of an editorial.[104] He was suddenly hitting the big time.

Media blitz

The coverage rapidly turned into a media blitz, although it was a long way short of uniformly favourable. David Appell, who had been one of the earliest critics of McIntyre,* wrote a hagiography of Mann for *Scientific American*,[105] while Mann himself was treated to a primetime spot on BBC radio, in which he was allowed to promote his side of the argument largely unchallenged.[106]

McIntyre was picking up plenty of support of his own. Hendrik Tennekes, a former head of research at the Royal Netherlands Meteorological Institute emailed McIntyre to say that he thought Mann was 'a disgrace to the profession',[107] and climatologist Kevin Vranes had this to say about Mann's withholding of data and code:

> The [*Wall Street Journal* (WSJ)] highlights what Regaldo and McIntyre say is Mann's resistance or outright refusal to provide to inquiring minds his data, all details of his statistical analysis, and his code. The WSJ's anecdotal treatment of the subject goes toward confirming what I've been hearing for years in climatology circles about not just Mann, but others collecting original climate data . . .
>
> As concerns Mann himself, this is especially curious in light of the recent RealClimate posts . . . in which Mann and Gavin Schmidt warn us about peer review and the limits therein. Their point is essentially that peer review is limited and can be much less than thorough. One assumes that they are talking about their own work as well as McIntyre's, although they never state this. . .
>
> Of their take on peer review, I couldn't agree more. In my experience, peer review is often cursory at best. So this

---

*   See page 94.

is what I say to Dr. Mann and others expressing deep concern over peer review: give up your data, methods and code freely and with a smile on your face. That is real peer review . . .

Your job is not to prevent your critics from checking your work and potentially distorting it; your job is to continue to publish insightful, detailed analyses of the data and let the community decide. You can be part of the debate without seeming to hinder access to it.[108]

The trickle of support started to grow. It seemed as though McIntyre's paper was emboldening those scientists who doubted Mann's findings. It was as if the whole climatological community had been fearful of speaking out until an outsider had pointed out the flaws in MBH98. Suddenly, the logjam burst open and climatologists outside the Hockey Team began to air their concerns in public for the first time. Ulrich Cubasch, an eminent German researcher and IPCC lead author, announced that his team were also examining the Hockey Stick, and that they could not reproduce Mann's results. Moreover he said that they had found 'a can of worms'.[109] Even von Storch was getting in on the party, referring to 'Mann's shoddiness' in the same article.[110] When even mainstream scientists felt they could speak out against the Hockey Stick, it was clear that something had changed. The tide had started to turn.

8   Big Mac and the Two Whoppers

Criticism is prejudice made plausible.

HL Mencken

A man's a man for a' that.

Robert Burns

McIntyre and McKitrick's GRL paper attracted no less than four formal responses, a surprisingly large number. As we have seen, when formal comments on a paper have been received, the journal will normally invite the paper's authors to review these submissions and to formulate a written response. With four comments in play, McIntyre might have been concerned at the amount of work he was going to have to do to fend them all off, but as he looked over the manuscripts, he knew he could relax. There was little in any of them that struck serious blows at his work.

Von Storch and Zorita

Since the publication of MM03, Hans von Storch and Eduardo Zorita had got in on the act of investigating the Hockey Stick. In 2004 they had published their own critique of MBH98, in which they had concluded that Michael Mann's methodology was artificially reducing the size of the wiggles in the Hockey Stick's long 'handle'.[111]

Von Storch now picked up on this approach in the comment he and Zorita submitted on MM05(GRL). He argued that, while the effect of short-centred PCs was just as McIntyre and McKitrick had described, it was not significant in the final MBH98 result.[112] Von Storch had reached this conclusion by using artificial tree ring series ('pseudoproxies'), which were created by taking his climate model's temperature output for a particular point on the Earth's surface (a 'gridcell') and adding noise to it to make it look more like a real tree ring series. Because the pseudoproxies had been created under controlled circumstances, they were a kind of idealised tree ring record, whose properties were understood exactly. The pseudoproxies could then be fed into Mann's algorithm in place of the real proxy data, theoretically allowing von Storch to discover exactly what the effect of Mann's procedures was.

Von Storch and Zorita's approach was plausible, but in McIntyre's view, their implementation of it was problematic on at least two counts. Firstly, von Storch and Zorita assumed for the sake of their calculations that there was a reasonable correlation between temperature and pseudoproxy in the gridcell, in other words that the trees were responding to their local temperature, and therefore the tree ring widths would be a simple mathematical function of that temperature. This was not an unreasonable assumption because, as we saw in Chapter 2, paleoclimatologists pick trees that are at the upper treelines in the belief that these trees are responding to temperature and not to any of the other factors which can affect tree growth. The problem with this approach, however, was that in the real MBH98 data there was *no* correlation at all between gridcell temperature and the tree ring widths. Mann, you may remember, created his reconstruction by correlating temperature directly against temperature patterns covering much larger areas rather than the immediate locale. He had argued that these trees were not responding to their local temperature, but rather to temperatures over these wider areas, by means of the 'teleconnections' that we met in Chapter 2.*

The other major problem with von Storch's approach was that he assumed the pseudoproxy data should consist only of a climatic trend plus some noise. This missed a fundamental point about

---

*   See page 47.

McIntyre and McKitrick's claims, and one that the two Canadians had made again and again, without anyone seeming to quite get the message. Their argument was not simply that the short centring would produce hockey sticks; it was that it would pick out hockey sticks to the exclusion of everything else. The point was subtly different, and emphasised the interaction between the short centring and poor quality data – 'a few bad apples' as McIntyre was wont to put it. Mann's algorithm could be imagined as 'scanning' the proxy database for hockey stick shaped series – it was in essence an automated method of 'cherrypicking' hockey stick shaped series.* If there was a bad apple – a series whose hockey stick shape was of non-climatic origin (like the bristlecones) – the algorithm would be likely to declare this the dominant pattern in the data, to the exclusion of anything else.

In their reply to von Storch and Zorita, McIntyre and McKitrick created some new simulations that powerfully demonstrated this point.[113] You will remember that there were 16 bristlecone pine series in the NOAMER network. Starting with the original MBH network of 112 series (i.e. including all of the non-bristlecone data too), McIntyre removed one bristlecone series at a time, measuring how much of a hockey stick shape the resulting PC1 had after each step (using the hockey stick index).** For effect, he kept the most heavily weighted series, Sheep Mountain, until last. The effect was extraordinary. The hockey stick index of the PC1 started at something over 1.4 when all the bristlecone series were present. As each of the 16 series was removed, the index remained to all intents and purposes entirely unchanged. Even when there were only three left, the algorithm *still* produced a virtually unchanged hockey stick. With two bristlecone series in the network, the shape of the PC1 was only slightly attenuated, with a hockey stick index of 1.1. It wasn't until bristlecone representation in the network was pruned to just the Sheep Mountain series that the

---

*   The subject of cherrypicking is considered in more detail on page 236.
**   See page 153.

hockey stick shape disappeared from the PC1. Even then it reappeared in the PC2, which Mann would presumably argue could be carried forward to the calibration under his supposed application of Preisendorfer's Rule N.

This then, was the crux of McIntyre's argument. If you had a hockey stick shaped series the short centring would put it in the PC1. Then, because hockey stick shaped series correlated well with the temperature PCs, the final temperature reconstruction would have a hockey stick shape too. It was, as he said, a complete answer to von Storch's comment.

Huybers

Peter Huybers' was one of the more challenging responses to McIntyre's new paper. Huybers was a post-doctoral fellow at the Woods Hole Oceanographic Institution in California and he had contacted McIntyre soon after the publication of the new papers. There had been a regular exchange of emails since, which McIntyre describes as 'mostly cordial'.*

Huybers' correspondence with McIntyre seemed to suggest that there were many areas of agreement between them. For a start, he seemed quite convinced that short centring would indeed bias the PC calculation to find hockey sticks, although when McIntyre tried to make this agreement clear in his reply to Huybers' comment,[117] he was at first shot down by the GRL editor, who felt that this was an attempt to divert attention from differences of opinion. This prompted McIntyre to write again to Huybers. It was important, he said, that the wider climatology community understood the points of agreement as well as any bones of contention. When Huybers consented, they were able to put on a united front, and the journal somewhat reluctantly agreed that McIntyre could state in his reply that Huybers 'concurred' that Mann's algorithm was biased.

---

* McIntyre's side of the story of Huyber's comment and his response to it is told over a series of Climate Audit postings.[114–116]

From his correspondence, Huybers also seemed to agree with McIntyre that bristlecones were unsuitable proxies. However, in his comment Huybers made a complete *volte-face* and wrote that their suitability as proxies should be assessed in later studies, a most peculiar position when there was already a considerable body of literature suggesting that there were unidentified non-climatological effects distorting the temperature signal.

Covariance matrix, correlation matrix

Huybers' next criticism addressed a fairly obscure corner of the Mannian algorithm. When the proxy series are summarised down to the PCs, the first step, as we've seen, is to centre the data by taking away the series average. In the next step, there are different ways of performing the calculation: Mann had used an approach called SVD,* which was generally agreed to have made his methodology even more biased than it already was. Huybers, however, was taking issue with McIntyre's use of another method called the covariance matrix, rather than a third method, the correlation matrix, saying that it exaggerated how bad Mann's method was. The two methods – covariance and correlation – are actually very similar; the only difference is that in the correlation matrix the data is adjusted to put everything on the same scale – simply by dividing by the standard deviation. At this point you might well be saying to yourself 'but wait a minute, isn't everything directly comparable already?' and you'd be quite right. For paleoclimate reconstructions, standardisation is not necessary within the matrix because the data has already been standardised prior to processing. If a correlation matrix is used, it would mean that the data would in effect be re-standardised. Because of this, McIntyre had used the covariance matrix instead, an approach that he found was endorsed by most of the statistical authorities he could find in the literature. However he didn't rule out using the alternative approach.

---

*   Singular value decomposition.

The use of a correlation matrix (i.e. re-normalizing) is certainly an option, but climate history should not stand or fall on this choice. The bristlecones do get promoted higher with a correlation matrix than with a covariance matrix.[114]

This was a position with which Huybers seemed inclined to agree. In fact, he explained to McIntyre:

It is . . . rather unsatisfying that answers are so sensitive to seemingly small changes in technique.[114]

So when McIntyre read the submitted comment, he was surprised to see that Huybers was now in effect arguing that the choice he and McKitrick had made to use the covariance matrix was exaggerating the apparent bias of the MBH98 methodology.

In summary, [McIntyre shows] that the normalization employed by MBH98 tends to bias results toward having a hockey-stick-like shape, but the scope of this bias is exaggerated by the choice of normalization and errors in the RE critical value estimate.[117]

In order to demonstrate this alleged exaggeration Huybers presented a graph that showed how each of the three methods – MBH98 (short-centred), McIntyre's (with a covariance matrix) and Huybers (with a correlation matrix) – when applied to the AD 1400 proxy roster, compared to the simple average of the underlying data. As he presented it, there was a gap between the result you got from McIntyre's covariance PC1 and the simple average for most of the length of the record, the implication being that the way McIntyre was standardising the data was making MBH98 look worse than it really was. This is shown in Figure 8.1. McIntyre's version, using the covariance matrix, is shown at the top and for most of the series there is a gap between the PC1 and the average. Huybers' correlation matrix is at the bottom, matching the average much better. Huybers' point was somewhat moot, however, because, as McIntyre observed, if your intention was to recreate the average of the data, it might be simpler to actually *use* the average. As we saw in the simple examples in Chapter 2, the whole point of PC analysis is to capture features that are hidden by simple averages.

McIntyre also noticed a neat device that Huybers had used to make his claims more credible. When you present two graphs on the same axes in this way, it is normal to try to make the whole length of the two lines match. But when Huybers had put the series average up against the PC1s in his paper, he had not done this; he had only tried to match the twentieth century portion of the graph. It was this change that had opened up the gap between McIntyre's covariance PC1 and the series mean, apparently supporting Huybers' thesis that the covariance matrix was exaggerating things. However, McIntyre was able to show that if you centred the two lines properly, the covariance PC1 would have been pushed down so that it matched the average for most of its length, but opening up a small divergence in the twentieth century. This supported McIntyre's position, which was originally Huybers' too, at least in their correspondence, that the issue was relatively trivial.

McIntyre also pointed out that when you did the same comparison on the *full* network, rather than just the AD 1400 network, with its very small number of series, the average was most like the result from the covariance method. This suggested that it was Huybers' correlation matrix that was biasing the results, not the covariance matrix.

Huybers also tried to find support for the use of the correlation matrix in the scientific literature, and cited texts by two eminent statisticians, Preisendorfer (of Rule N fame) and Rencher. Strangely, though, Huybers did not include any page numbers with his references to these experts' work, as is normal when citing a specific part of a book. When McIntyre referred to the texts himself, both authors turned out to be unequivocal in their support

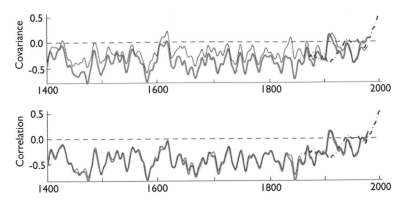

FIGURE 8.1: Correlation matrix versus covariance matrix

Thin black line: PC1; Thick grey line, series mean.

for the use of the *covariance* matrix and not the correlation.* In the face of what appeared to be a misrepresentation of what Preisendorfer and Rencher had said, McIntyre decided to take the point up with GRL's editor, Jay Famiglietti.**

> Both citations are texts, yet no chapter or page citations are given. The reason, I suggest, is that it would be impossible for Huybers to provide a direct quotation from either authority on this point because they do not support the procedure Huybers proposes . . . In fact, their explicit statements run in the opposite direction. As a general matter, it is simply false that scaling to unit variance is a 'standard practice in [PC analysis]' for data sets already standardized to dimensionless units with a common mean. The onus is on Huybers to back up this claim. Neither of

---

*   Relevant quotations from Preisendorfer and Rencher were provided by McIntyre on one of his postings on Huybers'comment.[116]

**   The story of how Dr Famiglietti came to be handling their paper is revealing and will be covered a little later in this chapter.

his authorities do so. He should either produce explicit
support, such as a chapter and page number or remove the
claim. The onus should not be on us to use up our word
limit providing extensive quotations from his own sources
to show that they do not support him.[116]

In their draft reply to Huybers' comment, McIntyre and McKitrick
had tried to address these issues, but were advised by Famiglietti
that they should remove these sections since they would be dealt
with editorially – in other words that the journal editors would
speak to Huybers and have him amend his submission. McIntyre
moved on to the more substantive issues.

Huybers on standardisation
Huybers had also spoken of McIntyre and McKitrick having
attempted to 'remove the bias' in MBH98 by applying
standardisation based on the full length of the series, and described
their standardisation procedure as 'questionable', referring to it as
'MM [i.e. McIntyre and McKitrick] normalization'. This seemed to
be an unnecessary personalisation of what was actually the text
book approach to the issue. McIntyre was also irritated because
these statements appeared to imply, as so many critics of his work
had done, that he and McKitrick were trying to create an
alternative temperature reconstruction to Mann's and that they
were advocating the use of the covariance matrix to do this
correctly. This was something they had consistently stated was *not*
their intention, since they did not believe that it was possible to
extract a meaningful temperature signal from tree rings by *any*
method. McIntyre was understandably irritated with Huybers and
asked Famiglietti to intervene but to no avail.

Huybers on verification statistics
Huybers' points on PC methodologies may have been easily dealt
with, but he had some much more challenging points to make
about the verification statistics. Privately, Huybers conceded that

MBH98 catastrophically failed the R^2, although it took a little persuasion to get him and the journal to allow McIntyre and McKitrick to say so in their reply. He was not convinced that it failed the RE test however. His issue was with the benchmarking of the RE. We've seen how Mann and McIntyre had used Monte Carlo methods – generating red noise pseudoproxies and processing them through the MBH98 algorithms – to assess how high the RE needed to be to indicate a significant result. Huybers said in his comment that McIntyre had missed out an important part of the MBH98 procedure and that if he replicated the exact steps of the paper, he would get a benchmark level of zero, just as Mann had. The missing step was part of the calibration process, where the proxy records are matched up against the temperature records in order to define the mathematical relationship between them. Huybers had noticed that the pseudoproxy variances (the size of the wiggles in the line) was less than the variance of the target – the temperature data – and argued that the pseudoproxies should be adjusted by 'rescaling' them. This, he said, was what Mann had done with the real data and Huybers claimed that it was a critical step in the process, although he didn't explain why.

There was no mention of this rescaling step in the original MBH98 paper, but Mann's recently released code (see Chapter 9) showed that he and Bradley and Hughes *had* rescaled their variances, although in a rather different way to that described by Huybers. McIntyre was able to recreate the rescaling the way Huybers had performed it and agreed that, if the calculation was done as Huybers described, then the benchmark should be zero. However, the difference between Huybers' method and Mann's turned out to be important.

In the AD 1400 step, Mann had taken the NOAMER PC1 and combined it with the 21 other proxy series. Then he'd gone through the steps (calibration, verification, reconstruction) to take him to the reconstructed PC1. Then he had rescaled so as to make the variances of the reconstructed PC1 and the temperature PC1 the same. Huybers, on the other hand, simply took his pseudoproxy

PC1s and inflated them to match the variance of the observed temperature record. So the difference between the two methods was that in Mann's procedure, the PC was mixed with other data before rescaling, whereas in Huybers' it wasn't. Both procedures inflated the variance by the required amount, but with rather different side effects.

In order to reveal the effects of this difference in methodology, McIntyre created some new simulations where the pseudoproxy PC1s were combined with 21 white-noise series representing the 21 other proxy series used in the original AD 1400 step. These were processed through the whole reconstruction process, including the newly revealed rescaling, right through to the reconstructed temperature PC. This replicated as closely as possible the exact Mann algorithm as it was currently understood. When this was done for all 1000 simulations, the revised RE benchmark was calculated to be 0.54, rather than the 0.59 calculated in the original MBH98 paper. While this was a slight reduction, it was well in excess of the zero score argued for by Mann and Huybers and still above the score achieved by the Hockey Stick itself.

All of Huybers' criticisms had therefore been dealt with. Huybers'comment and the Climate Auditors' response[118] would go forward for publication, but in reality not a finger had been laid on McIntyre's critique of MBH98.

The Hockey Team comments

The two remaining comments on MM05(GRL) were both from Hockey Team members.

The first was from from David Ritson, a physicist from Stanford and sometime guest author at RealClimate. Ritson's comment was rather strange. McIntyre described it as 'goofy'[119] and when he pointed out some of the problems to the editors they chose to drop it in its entirety rather than ask him to reply formally. Strangely though, this was not the last that was heard of Ritson's comment.

Wahl and Ammann

The other, more important, Hockey Team contribution came from Caspar Ammann and Eugene Wahl. Ammann, who was the lead author of the comment on McIntyre's paper, had been a PhD student of Ray Bradley's and had since published several papers with Mann. In fact, so close was Ammann's association with Mann that Climate Audit readers unkindly referred to him as 'mini-Mann'. He now worked at University Corporation for Atmospheric Research (UCAR), a major US centre for climate change research in Boulder, Colorado.*

McIntyre had first come across Ammann back in 2004 at the AGU Fall Meeting, where he had been presenting the poster on his work and inviting visitors to 'guess which was the real Hockey Stick'.** When not speaking to visitors to his poster, McIntyre had taken the chance to listen to some of the main presentations, and one of these had featured Ammann explaining that he was in the process of replicating the Hockey Stick. In fact, Ammann had gone so far as to tell the newspapers that he could recreate Mann's work precisely. This had sounded rather surprising to McIntyre, who was still unable to replicate Mann's results exactly. He had therefore written to Ammann to enquire when his results would be published and whether it would be possible see a draft of the paper. However, despite writing twice, he never received a reply.

Press release

Nothing more was heard of Ammann's replication of the Hockey Stick for six months. Then, in May 2005, with the controversy over the papers raging on all sides, McIntyre and McKitrick were invited to give a lecture on their findings at a meeting of the Heritage Foundation, a political think tank in Washington DC. On the morning before McIntyre was due to speak, Ammann and Wahl

---

* Strictly speaking, Ammann works for the National Center for Atmospheric Research (NCAR), one of the laboratories run by UCAR. For ease of narrative, I refer throughout to UCAR.

** See page 152.

suddenly issued a press release via the UCAR website. In it they claimed that they had submitted two manuscripts for publication, which together showed that the Hockey Stick could be exactly replicated, confirmed its statistical underpinnings and demonstrated that McIntyre and McKitrick's criticisms were baseless.

> [Caspar] Ammann and Eugene Wahl of Alfred University have analyzed the Mann-Bradley-Hughes (MBH) climate field reconstruction and reproduced the MBH results using their own computer code. They found the MBH method is robust even when numerous modifications are employed. Their results appear in two new research papers submitted for review to the journals *Geophysical Research Letters* and *Climatic Change*. The authors invite researchers and others to use the code for their own evaluation of the method.[120]

Ammann's submission to GRL then, was the last of the comments on McIntyre's paper.[121] The circumstances of the press release caused several observers to raise eyebrows: the practice of using a press release to announce scientific findings was dubious in itself, but some people also noted that when announcements of this kind are made, they tend to be about papers that have just been published, or at least accepted for publication. To make such a dramatic statement about the *submission* of a paper was unusual in the extreme. McIntyre was certainly unable to find any other UCAR press releases announcing papers in similar circumstances. Unfortunately, this sort of subtlety probably went unnoticed by the majority of readers, and if any of the journalists who wrote about Ammann and Wahl's work did spot it, they failed to point it out to their readers.

The comment

Ammann and Wahl made two main points of attack in the comment they submitted to GRL ('the GRL comment'), covering much of the same ground as Mann was making in his comments at

RealClimate. Firstly, they took issue with McIntyre's arguments over PC analysis, which they attempted to do by means of a subtle (or not-so-subtle, depending on how close the reader was to the debate) misrepresentation of what McIntyre was saying. While McIntyre and McKitrick had always said that short centring of the PCs biased the algorithm so as to overweight the bristlecones, Ammann and Wahl chose to paraphrase this position as follows:

> [McIntyre and McKitrick] claim that the standardization approach chosen by MBH biases the [tree ring] information towards a 'hockey stick' shape . . .[121]

which could be construed as suggesting that the Canadians believed that the hockey stick shape was introduced by the algorithm (rather than being in the mix already in the shape of the bristlecones). They made this insinuation more plain a little later in the article:

> [McIntyre and McKitrick] emphasize that the 'hockey stick' shape is introduced because the standardization is performed relative to a subsection rather than the full series . . .[121]

and again, a little further on:

> [T]he MM-claim that a 'hockey stick' outcome in the PCs is an artifact of the MBH standardization procedure is incorrect . . .[121]

and yet again in their conclusions:

> The claim by [McIntyre and McKitrick] that a spurious 'hockey stick' climate reconstruction is introduced by data transformation is unfounded.[121]

Having set up their straw man, they then went on to discuss PC retention policies but also claimed that in their second paper they had shown that no matter how you standardised the data, you would still get a hockey stick provided you retained enough PCs. This repetition of Mann's Preisendorfer argument was also repeated in several places throughout the paper – the text, the summary and the abstract. The problem with this argument was that McIntyre was unable to respond to it because Ammann's other paper had not been published. McIntyre penned a letter to the GRL editor, James Saiers, complaining about what he called this 'pyramid scheme', and also pointing out that he and McKitrick had discussed the whole issue of PC retention in the *Energy and Environment* paper and not the GRL one anyway.[119] They could hardly be expected to defend their *Energy and Environment* paper in GRL.

Apart from these issues, Ammann and Wahl had presented several results as if they were revealing them for the first time, when in fact they were merely reiterating points made by McIntyre and McKitrick in their own papers. This failure to cite the Canadians also had the side effect of implying to the unwary reader that there had been an oversight in McIntyre's work. There were other bones of contention too. In fact, the appendix to McIntyre's letter listing all the issues with Ammann and Wahl's GRL comment ran to nine pages. This exhaustive treatment did seem to have the desired effect on Saiers though, and a couple of weeks later McIntyre received a letter informing him that Ammann and Wahl's comment had been rejected. At this point, one might have expected UCAR to withdraw or modify their press release, but this was not done, allowing Hockey Stick supporters to continue to claim that McIntyre's work had been refuted.

The paper
Meanwhile the second, longer paper ('the CC paper') had started its protracted path to publication at the journal *Climatic Change*.[122] This paper purported to be a replication of the Hockey Stick and confirmation of its scientific validity.

All McIntyre's previous attempts at creating an exact replication of the Hockey Stick had been hampered by lack of access to Mann's computer code. Fortunately, though, at the time they had issued their press release, Ammann had also published his own computer programs, which, given that he was claiming that he had exactly replicated MBH98, must almost certainly have been identical to Mann's. McIntyre therefore set to work to analyse Ammann's code and was able to make very fast progress. In fact, by the time the CC paper was submitted to *Climatic Change*, McIntyre had reconciled Ammann's work with his own.

Amazingly, Wahl and Ammann's emulation of the Hockey Stick turned out to be nearly *exactly* the same as his own. Therefore he could be quite sure that the CC paper suffered from exactly the same problem as the Hockey Stick itself: despite Mann's having told Marcel Crok that the Hockey Stick passed the R^2 verification test, the R^2 number was in fact so low as to suggest that Mann's creation had no meaning at all, although the RE was relatively high. Because McIntyre's replications of both Mann's and Ammann's papers had these low scores, far from proving the scientific integrity of the Hockey Stick, the CC paper actually confirmed one of McIntyre's main criticisms of it.

As the CC paper was critical of his work, McIntyre was invited to be one of its peer reviewers, and shortly after accepting the appointment he received a short letter from Ammann and Wahl. This letter was simply a formality, an invitation to McIntyre to contact them if he didn't receive a copy of the paper from the publisher.[123] However, this seemed like a good opportunity to break the ice a little, so McIntyre wrote a long, good-humoured reply, which brought Ammann up to date on the attempt to reconcile his own code with Ammann's, and outlining what he thought the main points of contention were going to be in the final reckoning. If McIntyre thought he was proffering an olive branch, Ammann had other ideas, declining to acknowledge the Canadian for a third time.

Some days later McIntyre wrote to *Climatic Change* to ask them to obtain a full set of verification statistics from Wahl and

Ammann – R^2, RE and a selection of others. He noted in his email that Ammann had emphasised the importance of reporting these figures and had indicated that they would be available.[123] Confirmation that the R^2 was close to zero would strike a serious blow at Ammann's CC paper, and it appeared that his two opponents understood this fact just as well as he did: their response was an outright refusal to release any of the numbers McIntyre had requested, and a suggestion that he might like to calculate the statistics himself using the code they had made available.[123] This was a clear flouting of *Climatic Change*'s rules, and moreover directly contravened the journal's stated position that peer reviewers were not expected to run code.* As a justification of their extraordinary action, Ammann and Wahl gave a lengthy exposition of the superiority of RE over other measures and also argued that, in their forthcoming GRL comment, they would rebut McIntyre's criticisms of Mann's RE benchmarking.[123] This was a remarkable statement because, as we've seen, the GRL comment had actually been rejected by the journal some days earlier, and besides, the availability of the code didn't absolve Ammann and Wahl of the duty to calculate and present their verification statistics.

At the end of June, with his review of the CC paper nearly complete, McIntyre took the opportunity to tactfully probe this point, by asking *Climatic Change* to obtain the publication date and content of this alleged rebuttal from Ammann.[123] Perhaps in order to maintain a modicum of decorum, the *Climatic Change* editor, Stephen Schneider, chose not to respond to this letter. Two weeks passed before McIntyre decided that he could wait no longer and sent off his review comments.[124] This was a lengthy letter, running to ten pages, in which McIntyre laid all of his cards on the table. The tone was one of frustration and exasperation that he was being forced to deal with a paper which was such a travesty.

---

*   See page 160.

Climatic Change should reject this article. First, the errors and mischaracterizations are so numerous and affect the central conclusions so severely that dealing with the required corrections will require a completely new article and rejection of the present article is mandated. Secondly, the authors have flouted a Climatic Change policy requiring authors to provide supporting data and calculations and have provided a highly implausible rationalization for their position. Finally and most importantly ——————————* in their Response Letter by citing a submission they knew had already been rejected, in support of a point it did not provide support for anyway.[124]

Twenty-four hours later, Schneider's assistant emailed through a letter from Ammann in which he admitted that the GRL comment had been rejected and with it the alleged refutation of McIntyre's RE benchmarking work. Ammann had gone on to say that the GRL editors had rejected the paper because he and Wahl had covered points made by other commenters, although he stressed that they would be submitting the comment elsewhere.

With the replication of the Hockey Stick in tatters, reasonable people might have expected some sort of pause in the political momentum. Seasoned observers of the climate debate, however, will be unsurprised to hear that, in practice, global warming promoters acted as if nothing had changed. The UCAR press release was not withdrawn and key global warming players such as the head of the IPCC's scientific assessment working group, Sir John Houghton, and Mann continued to cite the Wahl and Ammann papers as evidence that McIntyre and McKitrick had been refuted. In testimony before the US Senate in July 2005, Houghton cited Ammann's two papers (which McIntyre referred to as the Big Whopper and the Little Whopper) in order to counter a suggestion

---

* The redaction is in the online copy of the letter cited here, but presumably not in the original sent to Schneider.

that there was a problem with the Hockey Stick. He outlined Ammann's arguments in both the paper and the comment, stating that one was 'in review' and the other 'in press'.[125] But at the point at which he made this declaration, the CC paper was in publishing limbo and the GRL comment had already been rejected more than a month earlier.

The coup at GRL

Events soon took another surprising turn. In August 2005, McIntyre read an article in *Environmental Science and Technology*, a journal published by the American Chemical Society. The article was the latest in the long line of 'hit' pieces that McIntyre had had to endure since the publication of his new papers in *Energy and Environment* and GRL, and included all the usual attempts to connect him to oil companies, to question his credentials or to otherwise denigrate his work. What was interesting about the article was a short interview with the editor-in-chief of GRL, Jay Famiglietti. Famiglietti told the interviewer that he had decided to replace James Saiers as the editor-in-charge of the file for the McIntyre paper and its responses. In fact he was taking over the file personally. This was justified, he claimed, because of the high number of responses – four – that it had received, a turn of phrase that neatly avoided mention of the fact that two of those responses had been rejected already. This was a concern for McIntyre. Famiglietti had been quoted in the article as saying:

> If I had a student come to me and say, 'I found this one paper that proves that climate change is hogwash,' I'd say, 'Well, that's one paper out of how many? In science, you never look at [only] one paper.'[126]

leaving the question in readers' minds of whether it was McIntyre who had suggested something so scientifically ridiculous. It looked very much as though Famiglietti might not be entirely even-handed.

The worries over Famiglietti's intentions were quickly realised when, at the end of September, he wrote to McIntyre to inform him that David Ritson's comment, which Saiers had rejected out of hand, had not only been readmitted for review *but had been accepted*. This would appear to have breached the journal's own policies, which stated that a comment on another article had to be sent out for peer review accompanied by a response from the author of the original article. To add insult to injury, Famiglietti now belatedly invited McIntyre to prepare a response.

McIntyre was understandably annoyed and wrote a strongly worded letter to Famiglietti, pointing out that, in the light of his comments in *Environmental Science and Technology* and his multiple breaches of his own journal's policies, his impartiality could now be questioned. He asked that Famiglietti hand over the file for McIntyre's paper and its responses to another editor. Famiglietti replied almost immediately, asking to discuss the matter by telephone, and he and McIntyre arranged a conference call in which McKitrick would also take part. Famiglietti insisted, however, that he would only discuss the matter if the conversation was off the record. The only information that he would allow to be divulged was a statement that his comments in *Environmental Science and Technology* had not been directed at McIntyre and McKitrick. All details about his 'coup' and his treatment of McIntyre remain a secret to this day.

If McIntyre had any suspicions about the implications of the goings-on in the GRL editorial office, these must have been swept aside when, shortly afterwards on 29 September, Mann commented on his RealClimate blog that both of Ammann's papers were back in play. The CC paper and the GRL comment, he said, were 'pending final acceptance', although how he knew this is not clear. McIntyre checked the UCAR webpage for Ammann's GRL comment but there was no apparent change in its status. However, two days later, an observant Climate Audit reader noticed that the page had been updated and now showed that the GRL comment had been resubmitted on the 25th. As McIntyre acidly observed, Ammann's work had made remarkable progress through the peer review system between its resubmission on the 25 September and its position of 'pending final acceptance' just four days later. Both of the Wahl and Ammann papers were indeed back in play.

The IPCC submission deadline

As 2005 neared its end, two important events loomed large. The first was the year-end deadline for submission of papers for the IPCC's Fourth Assessment Report. Prompted by a Climate Audit reader, a possible reason for the goings-on at GRL gradually dawned on McIntyre and his supporters: did the IPCC *need* to have the Wahl and Ammann papers in the report so that they could continue to use the Hockey Stick to maintain the political pressure? Could this have been the reason for Famiglietti's coup at GRL? Had someone put pressure on the journal to ensure that Wahl and Ammann's papers remained on the record so that they could be used by the IPCC? The suspicion remains that this was the case. It appears that the price for resurrecting Ammann's GRL comment was to resurrect the 'goofy' Ritson comment too. Then the whole affair could be presented, however unconvincingly, as a 'clean start'. While this could be seen as 'conspiracy theorising' we will see in Chapter 11 just how important Ammann's work was to the IPCC project.

AGU 2005

The second important happening was the Fall Meeting of the American Geophysical Union, which would be attended by many of the big names in paleoclimate and at which both McIntyre and Ammann would be making presentations. McIntyre's plan was to use the question and answer session after Ammann's talk to once again press for his verification statistics. His question was to be: 'What is the value of the cross-validation R^2 for the fifteenth century MBH98 reconstruction?' Perhaps now, after requests to Mann, to Bradley, to the National Science Foundation, to *Nature*, to *Climatic Change*, and to Ammann himself, and even a demand from the US Congress (see Chapter 9), the figure might finally see the light of day.

Ammann's AGU presentation was pretty much as expected – there was a great deal of criticism of McIntyre and little new science to add to the record. When it came to the question and answer session McIntyre was finally able to confront Ammann with the fateful question. So what then was the R^2 figure for the fifteenth century?

Ammann still wasn't saying. When McIntyre put the question, Ammann prevaricated at great length, presenting an extended argument as to why the audience shouldn't have the long-awaited figure. This was essentially the same argument that he had given for his refusal to release the figures as part of the review process of the CC paper – that R^2 was an inferior measure, which didn't capture important features of the data. His evasions didn't go unnoticed by the audience though, which included many of the big names in climatology, von Storch and Zorita among them. However, Ammann extended his reply sufficiently to talk out the session and it was not possible for anyone to press him further.

As a student at Oxford, McIntyre had played some rugby and had developed an admiration for the rugby players' ability to enjoy a beer with the same people they'd been pummelling shortly before. He often tried to adopt the same approach in his face-to-face meetings with his climatological opponents. After the AGU session, therefore, he attempted to clear the air by inviting Ammann out to lunch, and was gratified, if rather surprised, when the younger man took him up on the offer.

McIntyre later explained to his Climate Audit readers what had happened. Under the circumstances, most of their time together seems to have been spent relatively amicably, the two men exchanging small talk and passing the time. Inevitably though, the discussion had turned to more serious matters such as the need to disclose the verification statistics:

> I urged him at lunch – in his own interests – to deal with
> the issue himself. In any enterprise, dealing with the bad

news is no fun, but you've got to do it and you're always better off dealing with it yourself, rather than having someone else hammering you with it. I pointed this out to him in the nicest possible way. I told him that, if he doesn't, it will be awfully easy for me to excoriate him for withholding these adverse statistics and that I would obviously do so. I asked him: why give me such an easy target? He was relatively young; I was trying to coach him.[127]

If Ammann understood that McIntyre was trying to help him, he certainly didn't seem to take the advice on board. He launched into another attempt to justify withholding the validation statistics, claiming that McIntyre had not published the equivalent figures for 'his reconstruction' – an argument which, apart from being fallacious, flew in the face of McIntyre's repeated statements that he and McKitrick were not offering up an alternative reconstruction but were merely demonstrating that Mann's was not robust. McIntyre reiterated this for Ammann's benefit, but the conversation seemed to be going nowhere, and eventually McIntyre threw in the towel and suggested that they return to the conference. As they were getting ready to leave, however, Ammann returned to the subject of their professional relationship, complaining vehemently about the way that McIntyre was dealing with him and Wahl. McIntyre was not impressed:

> As we were winding up, in fact, just as we were returning to AGU, Ammann screwed up his nerve to complain about getting roughed up and my tactics in doing so, which he didn't like very much. This is a guy who had used UCAR press facilities and distribution to issue a national press release on the very day that we're making a rare public appearance, announcing his submission of two articles supposedly debunking us and the horse we rode in on. This press release was then relied on by Houghton, Mann and

others in their evidence to the US Congress. Ammann had
given newspaper interviews and presented in Washington
and he's complaining about getting roughed up.[127]

Unimpressed as he was, McIntyre thought he saw a way to break
the impasse of each side firing critical papers across at the other
without any final resolution of their differences. He suggested to
Ammann that they write a joint paper outlining where they agreed
and where they differed and setting out possible approaches to
resolving those differences. Ammann, however, was non-
committal. It would, he said, interfere with his career advancement
to be so closely involved with McIntyre, although one wonders if
this linking of their names would have been quite as bad as the
alternative: namely, a possible reputation for withholding adverse
results. Regardless of this, a few days after returning to Canada,
McIntyre wrote to Ammann, formalising the proposal of a joint
paper but attaching an expiry date to the offer. Unfortunately for
everyone though, Ammann set out his stall very clearly by failing
to reply, and a McIntyre reminder on the expiry date likewise went
entirely unacknowledged. The Hockey Team was determined to
continue the dispute.

The Climatic Change paper is resurrected

While the AGU was meeting in San Francisco, things started to
move on the two Wahl and Ammann submissions. On 9 December,
GRL wrote to McIntyre, informing him that they had decided to
move forward with Ammann's comment and advising him to
prepare a reply. Then, just three days later, it was announced that
Ammann's CC paper had been 'provisionally accepted'. It wasn't
entirely clear *what* these provisions of acceptance were, but one
possibility may have been that Wahl and Ammann would be
required to include their verification statistics. Another was that
the editors at GRL must not reject the comment again, because, as
we've seen, it contained the statistical arguments to support the
assertions in the CC paper. All the time, the year-end deadline for

submissions to the IPCC was looming large and there was now precious little time remaining for Ammann's papers to meet it.

For Schneider, the *Climatic Change* editor, to move forward with the paper was remarkable, given that he had been presented with pretty clear evidence that Wahl and Ammann had misrepresented the status of their GRL comment. Scheider still had a problem though, which was that McIntyre's review comments on the paper were likely to be excoriating. In order to deal with this he sidestepped the issue by simple dint of not inviting the Canadian to review the second draft. It is perhaps worth remarking that Schneider has been in the forefront of efforts to bring global warming to public attention and is also a man who once said that 'Each of us has to decide what the right balance is between being effective, and being honest. I hope that means being both'.[128] Sceptics wondered if this was one of those occasions when Schneider's hopes had been dashed.

This then, was all that McIntyre knew of the CC paper until the New Year.

Formal reply to the GRL comment

Meanwhile there was the revised GRL comment to deal with. The new version was almost identical to the first, with the exception of a new section on verification statistics. In this, Wahl and Ammann claimed that there was a problem with the benchmarking exercise that McIntyre had used to rebut Peter Huybers' critique. They said that if you fixed this problem, which they said was due to the pseudoproxy PCs being statistically dissimilar to PCs generated by real data, it was possible to confirm that the correct 99% benchmark for RE was zero.

By the end of January 2006, McIntyre and McKitrick had prepared a new reply, which was duly submitted to GRL. It covered the same points that they had made in their letter to James Saiers the first time around* and it was worded very strongly – in the face

---

*   See page 204.

of Ammann's refusal to release his verification statistics and because
he had ignored the invitation to write a joint paper, McIntyre was
not inclined to play nicely any longer.

> [Ammann and Wahl] not only repeat results that we had
> previously published, but claim them as their own and then
> accuse us of having failed to report them. In their abstract
> and summary, [they] make claims that are unsupported in
> their text, then assert our results are 'unfounded', despite
> the fact that results from their own code yields validation
> statistics (unreported by [them]) that strikingly confirm
> claims in [our GRL paper] concerning spurious significance
> in [MBH98].[129]

Two pages of the six that comprised McIntyre's letter were a listing
of 'Misrepresentations and unsupported points' in Wahl and
Ammann's comment. If Famiglietti intended to publish, then the
Climatic Change readers were also going to hear exactly what
McIntyre's objections were. Meanwhile, McIntyre quietly filed a
complaint of academic misconduct against Ammann with his
employers UCAR on the grounds that Ammann had withheld the
adverse verification statistics in his submission to Climatic Change.
The pressure on Ammann was being steadily ramped up.

The CC paper and the verification statistics

It wasn't until March 2006 that there was any further progress on
the two Ammann papers. Then, without warning, the status of the
revised CC paper was changed to 'In press', meaning that the peer
review was complete and the paper was ready to go to print. In the
scientific publishing process this means that the game is over and
the paper is finalised. At this point it is common practice in
scientific circles for authors to make an online preprint available
and McIntyre was pleased to see that this was just what Ammann
had done.

TABLE 8.1: Verification statistics for Ammann's MBH98 emulation

Proxy network MBH periods	NH mean R^2 Calibration-period	NH mean R^2 Verification-period
1400–1449	0.414	0.018
1450–1499	0.483	0.010
1500–1599	0.487	0.006
1600–1699	0.643	0.004
1700–1729	0.688	0.00003
1730–1749	0.691	0.013
1750–1759	0.714	0.156
1760–1779	0.734	0.050
1780–1799	0.750	0.122
1800–1819	0.752	0.154
1820–1980	0.759	0.189

NH: Northern Hemisphere. Reproduced from Wahl and
Ammann's *Climatic Change* paper.[122]

The resubmitted version turned out to be almost identical to the old one, except that a new section on the statistical treatments had been added, presumably as a condition of acceptance.[122] Buried deep down at the back of the paper was a startling revelation. Wahl and Ammann had backed down and done the decent thing. They had presented a table of verification statistics and, as expected, these *completely* vindicated everything McIntyre had been saying over the previous two years. The figures are shown in Table 8.1. The important section is the right hand column where the verification R^2 is close to zero for most periods, including particularly the critical AD 1400 step. In the AD 1700 step it even proved necessary for Ammann to increase the number of decimal places used in order to prevent the R^2 from appearing as zero. These figures demonstrated finally and conclusively that the MBH98 reconstructions were not reliable.

Publication chronology of the CC paper

The CC paper's provisional acceptance date at *Climatic Change* was 12 December 2005, just a few days before the IPCC deadline, which stated that any papers cited had to be in press by the time of the lead author meeting on 13–15 December. However, after its provisional acceptance, the paper had been rewritten, with the new sections containing the revelations on verification statistics added. This new version was the one that was finally accepted by the journal and was dated 24 February 2006. This was before the IPCC's cut-off point for final journal acceptance, but the revelation of its failure of the verification statistics making it a very different paper, it really represented a new submission. In reality then, it had missed the IPCC's cut-off date for submission. It is also worth considering what peer review of the new sections of the paper took place between submission on 24 February and final acceptance four days later. It seems very likely that there was in fact no peer review at all. But all this activity around the cut-off date had another more remarkable side effect. As McIntyre put it:

> So under its own rules, is IPCC allowed to refer to Ammann and Wahl [2006]? Of course not. Will they? We all know the answer to that. When they refer to Ammann and Wahl [2006], will they also refer to its confirmation of our claims about MBH verification [R^2] statistics? Of course not. That information was not available to them in December. But wait a minute, if Ammann and Wahl was in press in December, wouldn't that information have been available to them? Silly me.[130]

In other words, the version of the paper that had gone forward to the IPCC didn't include the adverse verification statistics, but the version accepted by the journal did. The paleoclimatologists got their rebuttal of McIntyre and the journal got a fig leaf of respectability to hide behind.

The comment is rejected again

Now that the adverse R^2 statistic was out in the open, Ammann would have been struggling for a way to save the paper from ridicule. The only way he could do this was by arguing that the correct measure of significance was in fact the RE statistic. This was a pretty feeble case because, as we've seen, it was pretty clear from the statistical and even the climatological literature that a suite of verification statistics was preferred to any one measure. Ammann's problem was that in order to show that his reconstruction passed even the RE test, he needed to establish a low enough benchmark – as we have seen the correct benchmark appeared to be higher than the RE score of the Hockey Stick. So what was Ammann's RE benchmark in the new version of the CC paper? The paper stated:

> We consider the issue of appropriate thresholds for the RE statistic in Appendix 2, based on analysis and results reported [in the comment, in review with *Geophysical Research Letters*].

However, when the reader referred to Appendix 2 he would read essentially a restatement of this same position:

> In implementing this procedure, we found a technical problem that we reported in [the GRL comment].

Meanwhile, when Wahl and Ammann had discussed the actual level of benchmarking they had applied, they had once again referred to Appendix 2 and to the GRL comment:

> Numerically, we consider successful validation to have occurred if RE scores are positive, and failed validation to have occurred if RE scores are negative ([in the GRL comment]; Appendix 2).

So, the GRL comment was necessary to justify a benchmark level of zero for the RE statistic. Without it, the Hockey Stick was effectively broken. And unfortunately for Ammann, just a couple of weeks after the acceptance of the paper, there was to be another hiccup that would threaten his findings.

After all the shenanigans at GRL, with the replacement of the editor and the resubmission of letters, the journal now decided once again to reject Wahl and Ammann's comment. Without it, Ammann could not claim that his results were statistically significant and, since he had purported to have exactly replicated the Hockey Stick, this was potentially the end of Mann's creation too. Ostensibly the journal's decision was made on the grounds that the arguments were 'already out there', but it was more likely that there were so many holes in the statistical arguments as to make publishing them an embarrassment to the journal. McIntyre was extremely unimpressed:

> What a total waste of time. Famiglietti mouthed off to *Environmental Science and Technology* last August and replaced Saiers as editor in charge of our file. He then took the comments by Ritson and by [Ammann and Wahl] (already rejected by Saiers) out of the garbage can, told us that the Ritson comment was accepted, then he rejected the Ritson comment after he saw our reply. Likewise with Ammann and Wahl . . . [McIntyre appends a collective dismissal].[131]

9 The Hockey Stick in Washington

In politics, what begins in fear usually ends in folly.

Samuel Taylor Coleridge

Senator Barton takes an interest

When Michael Mann told Antonio Regalado of the *Wall Street Journal* that he would not release his code because this would be 'giving into the intimidation tactics' of his opponents, he can little have imagined how much trouble he was getting himself into. It cannot have crossed his mind that his comments would catch the attention of Representative Joe Barton, the Republican chairman of the House Committee on Energy & Commerce in the US Congress. Barton was a Texan with close connections to the oil business and a determined global warming sceptic, two characteristics which are enough to condemn him as irredeemably biased in the eyes of many environmentalists. He was brash, confident and outspoken, and quite unafraid to put noses out of joint as he fought to get his way. *The Hill News*, a newspaper dedicated to the goings-on in Congress, gives a flavour of the man:

> Barton has emerged rapidly as one of the toughest chairmen in the House, unafraid of rolling his shoulders and using his elbows when he thinks it is necessary to expand or protect his domain. He is helped in this by an apparent indifference to getting good press and by having seemingly absorbed a version of Machiavelli's dictum that it is more important for political leaders to be feared than to be liked.[132]

Soon after the *Wall Street Journal* article was published, Barton contacted McIntyre to enquire if he had spoken to the paper and if the article was true. When McIntyre confirmed the story, the congressman swung into action. In June 2005, Barton wrote letters to Mann, Bradley and Hughes, as well as to Rajendra Pachauri, the head of the IPCC, and Arden Bement, the head of the National Science Foundation (NSF), the body which funds much of the scientific research in the USA.[133] The letters were co-signed by Ed Whitfield, the chairman of the Subcommittee on Oversight and Investigations. This was the committee that had investigated earlier scandals like the downfall of Enron, and it had the power to demand evidence under oath. Barton was therefore upping the pressure in a considerable way and it is unlikely that he was unaware of what he was doing.

The letters

Barton's letters to Mann, Bradley and Hughes explained that the House Energy & Commerce Committee was concerned about the reports in the *Wall Street Journal* that the Hockey Stick could not be replicated. He also pointed to concerns over the independence of the IPCC reports (Mann having been the lead author of the chapter which assessed his own work), and also to the issue of sharing of data and code. There followed a long list of demands for information: everything from CVs to details of financial support and copies of grant agreements. Much of this was directly relevant to McIntyre's researches, for example:

> 6. Regarding study data and related information that is not publicly archived, what requests have you or your co-authors received for data relating to the climate change studies, what was your response, and why?
>
> 7. The authors McIntyre and McKitrick (*Energy & Environment*, Vol. 16, No. 1, 2005) report a number of errors and omissions in Mann et. al., 1998. Provide a detailed narrative explanation of these alleged errors and how these

may affect the underlying conclusions of the work, including, but not limited to answers to the following questions:

a) Did you run calculations without the bristlecone pine series referenced in the article and, if so, what was the result?

b) Did you or your co-authors calculate temperature reconstructions using the referenced 'archived Gaspé tree ring data,' and what were the results?

c) Did you calculate the R2 [sic] statistic for the temperature reconstruction, particularly for the 15th Century proxy record calculations and what were the results?

d) What validation statistics did you calculate for the reconstruction prior to 1820, and what were the results?

e) How did you choose particular proxies and proxy series?

The reaction

Within days, the great and the good of the climate science fraternity had been stirred into action. Barton was deluged with letters of protest; outrage and disgust were pronounced on all sides. The American Association for the Advancement of Science wrote to the congressman saying that his letters read as if he was seeking some way of discrediting Mann, Bradley and Hughes and encouraged him to rely instead on the 'multiple layers of peer review' through which the Hockey Stick had passed, a position that must have brought a smile to McIntyre's lips.

The BBC quoted paleoclimatologist Tom Crowley making the somewhat absurd speculation that biologists could be asked for the data and code that proved the theory of evolution.[134] In another interview, McIntyre was accused of sending threatening demands for data to Crowley, an allegation which was demonstrably untrue.*[135] Crowley also rather oddly accused Barton of being a mouthpiece for McIntyre.

---

* See page 280.

Meanwhile the *Washington Post* called it a witch-hunt and a few days later *Nature* thundered furiously in an editorial roundly condemning the letters. In fact, with such an avalanche of outrage from all sides, some observers can surely not have helped but think that maybe the scientific community was protesting just a little too much.

As the wave of fury grew, the situation was further inflamed when it started getting tangled up with Washington politics. The senior Democrat on Barton's Energy and Commerce Committee, Henry A. Waxman, wrote to Barton complaining about what he described as 'a transparent effort to bully and harass climate change experts'.[136] Meanwhile the head of the House Science Committee, a Republican called Sherwood L. Boehlert, demanded that Barton call off his global warming investigation, describing it as 'misguided and illegitimate'.[136] Boehlert, in contrast to Barton, was an advocate of restrictions on carbon dioxide emissions, so on the one hand the argument can be seen as a dispute between the two sides of the global warming debate, one trying to force the scientists to come clean about the reliability of the paleoclimate reconstructions, the other trying to keep things safely under lock and key. But on the other hand it can also be seen as a turf war: a dispute over which of these two powerful committees was going to 'own' the global warming issue. Either way, it was going to get complicated. Barton was not, however, going to back down in the face of a little criticism from his fellow congressmen, as the committee's spokesman made clear:[137]

> Chairman Barton appreciates heated lectures from Representatives Boehlert and Waxman, two men who share a passion for global warming. We regret that our little request for data has given them a chill. Seeking scientific truth is, indeed, too important to be imperiled by politics, and so we'll just continue to ask fair questions of honest people and see what they tell us. That's our job.[132]

One of the many letters Barton received at this time, and one of the less outraged ones, was from Ralph Ciccerone, the head of the National Academy of Sciences (NAS), the leading learned society in the USA. Ciccerone's position was, on the face of it, rather more helpful, pointing out that perhaps the House Energy and Commerce Committee was not best equipped to resolve a scientific issue, and offering the NAS's services to investigate the current state of paleoclimate on the Committee's behalf. Barton, however, was unimpressed. His spokesman explained:

> We can't evaluate the idea without having seen it, and maybe it's a darned fine one, but an offer that says, 'Please just go away and leave the science to us, ahem, very intelligent professionals,' is likely to get the reception it deserves. We get a lot of offers to butt out from folks who would rather avoid public scrutiny, and reputable scientists wouldn't feel comfortable in the company of most of them.[137]

Mann's reply: code

A few weeks later, Mann's reply, together with those of Bradley and Hughes, was delivered to Barton. Mann's response must have been something of a surprise, with the Hockey Stick's author insisting that he had already made public all his data and methods, a claim that would have been very surprising to McIntyre and McKitrick. Meanwhile, Mann was utterly unrepentant about his refusal to release the code:

> My computer program is a piece of private, intellectual property, as the National Science Foundation and its lawyers recognize. It is a bedrock principle of American law that the government may not take private property 'without [a] public use,' and 'without just compensation'.

> That notwithstanding, the program used to generate the
> original Mann et al. 1998 temperature reconstructions is
> posted at [my FTP site].[138]

Readers by now will recognize the standard Mann pattern of response – bluster followed by a partial tactical retreat. Whether Barton was taken-in by this is not clear.

Bradley and Hughes took a slightly different tack, with Bradley stating that he 'normally' archived his data and Hughes saying that he had complied with NSF policies. Neither he nor Bradley directly addressed the question of the availability of their code, and McIntyre meanwhile had evidence that showed plenty of gaps in their data archiving records.[139,140]

The citation of NSF policies by Mann and Hughes bears closer examination. We saw in Chapter 6 that McIntyre had been trying to get hold of the residual series for MBH98 and also Mann's computer code since right back in 2003. His correspondence with NSF had continued intermittently over the years, but his attempts to obtain the data of other climatologists, including Hughes, had largely been rejected by the science bureaucracy. However, this experience did at least mean that McIntyre had developed a good understanding of the guidelines that were used. It was fairly clear that NSF was indeed advising that scientists did not need to release their code to third parties. However, whether it was in their power to do this was not entirely clear. The universities (including Mann's own bases in Virginia and, before that, Massachusetts) included clauses in their employment contracts that reserved the title to intellectual property developed by their staff to the university itself, directly contradicting the claims of the NSF.

When McIntyre started to dig further into the question of NSF policy, he came across the following statement:

> Appropriate commercialization of the results of research
> will continue to receive encouragement by permitting
> [universities] to keep principal rights to intellectual

> property conceived under NSF sponsorship. The
> Foundation emphasizes, however, that retention of such
> rights does not reduce the responsibility of researchers and
> institutions to make research results and supporting
> materials openly accessible.[141]

This was remarkable when set against the way one of the senior
staffers at NSF had described the policy to McIntyre, namely that
computer codes belonged to individual scientists:

> On the question of computer source codes, investigators
> retain principal legal rights to intellectual property
> developed under NSF award. This policy provides for the
> development and dissemination of inventions, software
> and publications that can enhance their usefulness,
> accessibility and upkeep. Dissemination of such products
> is at the discretion of the investigator.[142]

And on the question of data, there was virtual unanimity among
the policies of the various funding programs that data should be
archived. In fact one of these, the Paleoenvironmental Arctic
Sciences Program, had among its steering committee members
none other than Hughes, while its policy on data had been co-
authored by Bradley. Both of these men seemed to have adopted
rather different approaches in their *own* work to the ones they
advocated for others.

Mann's reply: verification statistics

Mann's position on the availability of his code may have been
weak, but he had some other remarkable claims to make in his reply
to Barton. After the by now traditional potshots at the scientific
credentials of McIntyre and McKitrick and *Energy and
Environment*'s status in the firmament of scientific journals, Mann
answered Barton's specific questions on McIntyre's critiques. He
was keen to claim that his work was supported by that of Wahl and

Ammann. In fact, he was *extremely* keen, mentioning their alleged refutation no less than eleven times in his response to the committee. He didn't, of course, mention the fact that neither of the Ammann papers were published and that one of them had been rejected.

There was another surprise when Mann came to explain to the congressman whether he had calculated the R^2. As we have seen, the original MBH98 paper suggested that the figure had been calculated, and in fact, R^2 results had been presented for the AD 1820 step. The results for the other steps, however, were nowhere to be seen, although Mann had later told Marcel Crok that the Hockey Stick passed the R^2 tests. However, Mann's response to Barton was more nuanced, involving a paraphrasing of the question which gave it a slightly different meaning. While the committee had asked whether he had calculated the R^2, in his answer he only claimed that he had not relied upon it.

> The Committee inquires about the calculation of the R^2 statistic for temperature reconstruction, especially for the 15th Century proxy calculations. In order to answer this question it is important to clarify that I assume that what is meant by the 'R2' statistic is the squared Pearson dot-moment correlation, or $[R^2]$* (i.e., the square of the simple linear correlation coefficient between two time series) over the 1856–1901 'verification' interval for our reconstruction. My colleagues and I did not rely on this statistic in our assessments of 'skill' (i.e., the reliability of a statistical model, based on the ability of a statistical model to match data not used in constructing the model) because, in our view, and in the view of other reputable scientists in the field, it is not an adequate measure of 'skill'.[138]

---

*   Mann actually used the alternative notation r^2. See note on page 16.

This narrative would have left the committee none the wiser as to whether Mann had actually calculated the R^2 statistic for the earlier steps, although it seemed fairly likely that he had, in view of his remarks in the original paper and also to Crok. One might also wonder why the Hockey Stick authors had done this if they didn't consider it to be 'an adequate measure of skill'.

Mann's letter to Barton now appeared to be throwing into doubt whether the R^2 number had actually been calculated at all. However, as we have seen previously, Mann had been forced by the attention of Congress to release his code and while the Barton Committee was considering his response, McIntyre was busy working through pages of Fortran. It wasn't long until he found what he was looking for. There on pages 28–29 of his print out was the section of the program that demonstrated conclusively that Mann had calculated the R^2. Why would he have published the R^2 figure only for the AD 1820 step? Why withhold the others? The most likely explanation was that for the other steps, the R^2 was so low as to demonstrate that the temperature reconstruction was meaningless, and the IPCC's assertion that MBH98 had 'significant skill in independent cross-validation tests' was a fiction.

The NAS panel

While everyone was digesting the responses from Mann, Bradley and Hughes, the political complications continued to multiply. Having both been told to take a running jump by Barton, the NAS and Boehlert's Science Committee decided to hook up together: the NAS was going to perform the investigation that Ciccerone had proposed, but under the auspices of Boehlert's committee rather than Barton's. An announcement was issued, stating that the committee had asked the NAS to put together an expert panel to investigate the whole subject of paleoclimate. The panel was to be headed by Gerald North, an eminent atmospheric scientist from Texas A&M University. Boehlert had seized back the initiative on the global warming issue.

Boehlert had set out three specific areas for the committee to cover:

- What is the current scientific consensus on the temperature record of the last 1,000 or 2,000 years? What are the main areas of uncertainty and how significant are they?
- What is the current scientific consensus on the conclusions reached by Drs. Mann, Bradley and Hughes? What are the principal scientific criticisms of their work and how significant are they? Has the information needed to replicate their work been available? Have other scientists been able to replicate their work?
- How central is the debate over the paleoclimate temperature record to the overall consensus on global climate change? How central is the work of Drs. Mann, Bradley and Hughes to the consensus on the temperature record?

The first McIntyre and McKitrick knew of the committee's appointment was when a letter from North arrived asking if they could attend and give evidence. It certainly seemed welcoming enough, the NAS board describing the two men's participation as 'critical'. However, concerns soon mounted as the identities of the panel members started to leak out. One name that stood out was Doug Nychka, a colleague and collaborator of Ammann's at UCAR and a former co-author of Mann's. This hardly suggested someone who was likely to be neutral, and as he was the only statistician on the panel at that time, his appointment was a potentially serious issue. There was nobody on the panel who appeared to have expertise in critical areas like spurious regression. And in fact, Nychka wasn't the only UCAR employee on the panel. Bette Otto-Bliesner turned out to be a superior of Ammann's, who worked in the office next door to him at UCAR and who had also published alongside Bradley. Three other panel members, Karl Turekian, Robert Dickinson and North himself, were ex-UCAR men too.

Alarm bells were also set off by the published comments of another panelist, Kurt Cuffey, who had said that serious scientific debate on whether global warming was occurring was at an end.[143] Given that this was one of the issues the panel was supposed to be considering, it did rather suggest that his opinions were already set in stone.

There were clear rules laid down for the composition of NAS expert panels, which mandated that panellists should represent a mix of different views and also that they should have relevant expertise. It was rapidly becoming clear that this wasn't the case for the paleoclimate panel. However, the rules also allowed for interested parties to make formal objections to the appointments, and given the overwhelming preponderance of Hockey Team associates, it was certainly worthwhile giving this a try. McIntyre had little expectation that this would produce any changes, but if there was a whitewash, critics would be able to point to the composition of the panel, and the NAS couldn't plead ignorance. With only a month to go before the panel's hearings in Washington, there was little time to lose – in fact it appeared unlikely that the panel would have time to consider any proposed change before the scheduled start of the hearings in March, but it was worth the attempt and the letter was duly delivered to the NAS. In it, McIntyre and McKitrick protested at the appointments of Otto-Bliesner, Nychka and Cuffey, and also requested panellists with more relevant skills – someone who could understand issues of statistical significance in the peculiar circumstances of multiproxy reconstructions,* and someone with a background covering the areas of journals, software evaluation and statistical methods who could contribute to the panel's understanding of the replication issues. And, adding a little spice to the request, McIntyre also asked the NAS to include someone with expertise in the area of scientific misconduct.

---

*   Calculating statistical significance in multivariate models using highly autocorrelated time series is highly complex, with a strong risk of spurious significance.

When the final make-up of the panel was announced towards the end of February, it became clear that McIntyre's letter had been largely ignored. The only concession that had been made by the academy was to appoint a second statistician, Peter Bloomfield. But even this modest step was less favourable than it might at first have seemed. Bloomfield turned out to have assisted Keith Briffa on some of his papers, providing statistical guidance for the confidence interval calculations.* So now, *both* statisticians on the panel were to be associates of the Hockey Team. As McIntyre wryly asked his readers,

> Out of all the statisticians in the world, why would they pick one who consulted on confidence intervals for one of the Hockey Team studies?[144]

Why indeed?

Until that time, McIntyre had known only that he and Mann would be speaking before the panel, but at the same time as the panel announcement, the full list of speakers was also released to the public. The hearings were to spread over two days, each speaker having just 45 minutes to make their case. Day one would open with a pair of geophysicists, followed by some ice core experts (both closely associated with Mann), before they got on to the meatier matters, with presentations from the tree ring experts, Gabriele Hegerl and Rosanne D'Arrigo.** Hegerl and D'Arrigo were both Hockey Team members but they were second team rather than the first. The day would close with Hans von Storch, who, as we have seen, was a neutral, and then finally McIntyre and McKitrick. Day two had just two sessions – the first was Malcolm Hughes, while the honour of closing the event went to Michael Mann himself. The hearings were now less than ten days away.

---

*   This was a problem that was presumably somewhat challenging for Bloomfield given that the post-1960 figures had gone off at a tangent to the rest of the record. That is another story though.

**   Hegerl is, strictly speaking, a statistician.

Barton strikes back

Before the NAS panel could actually start its hearings, and just days after the announcement of the speakers, Barton made a determined effort to wrest back control of the global warming issue from Boehlert. He announced that he had asked an eminent statistician called Edward Wegman to form a second expert panel, which was to be tasked with examining the specific question of the statistics of MBH98.

Wegman had no connection to climate science and no stated position on global warming, so there was no easy way for anyone to criticize his appointment. His credentials as a statistician were unimpeachable. With two panels now due to report, McIntyre could at least get some reassurance that even if the NAS panel decided to whitewash the whole question of the Hockey Stick, something that appeared increasingly likely in the light of their flouting of their own rules on panel balance, Wegman might at least be expected to understand the statistical problems with MBH98. His views would therefore provide a valuable counterweight to anything the NAS might report.

The academy

At the start of March 2006, McIntyre and McKitrick travelled to Washington for the panel presentations. The panel was to meet at the NAS headquarters on the Mall, a classically inspired building set amid extensive wooded grounds, close by the Lincoln Memorial.

When the hearings opened on Thursday morning, Mann was nowhere to be seen. In fact he didn't appear at all until Friday, missing all of the presentations on the first day. It was almost as if the schedule for the hearings had been specifically designed to allow Mann to avoid McIntyre as much as possible.

Proxy studies

The presentations got underway in the Academy's lecture room, with talks on boreholes and corals as ways of estimating temperatures of the past. While some of the speakers were known for their strong support for the global warming hypothesis, it was remarkable just how cautious they were about what could be concluded from their own area of expertise. Geochemist Daniel Schrag said that it was very difficult to make an estimate of average temperature from instrumental data, let alone proxies, and that policymakers were demanding more than the scientific community could actually provide in practice. Richard Alley, an expert in glaciers, pointed out that there was no concerted effort to update paleoclimate data (see Chapter 13) and that what data there was had rarely been collected for the purposes of climate reconstructions. As he put it, the whole system was not set up to answer the type of question that was being asked of it. The scientific community, he said, 'had not really integrated [polar cores] in a coherent way' because this was 'not the highest priority of the scientific community',[145] a remarkable statement given the importance of the global warming question and all the billions of dollars the subject had received in research funding over the previous decade. Like Schrag, he too pointed to the squeeze from policymakers.

Data availability and the statement of task

While data availability had been one of the questions that Boehlert had asked the NAS to investigate, as the first day's presentations were drawing to a close, the subject had still not been aired. It wasn't until von Storch took his turn that the panel was forced to consider the way in which the climatology community appeared to be resisting independent verification of their work. Their attention was drawn to the matter in a way that was very hard for them to ignore, when von Storch said that data should be made available to everyone, including 'adversaries'. To reinforce his point he quoted a notorious statement by the Hockey Team player Phil Jones,* who had rejected a request for data by an Australian researcher by saying:

---

*   See page 61.

> We have 25 or so years invested in the work. Why should
> I make the data available to you, when your aim is to try
> and find something wrong with it?*

Von Storch had some more surprises for the committee too. As he drew his presentation to a close, he showed a last summary slide, setting out his answers to the questions with which Boehlert had requisitioned the NAS panel and report. This was a useful way of tying the various strands of his talk together, leaving the panel with straightforward answers to the questions they had been asked to investigate.

To everyone's astonishment the arrival of the slide, entitled 'Rep Boehlert's Questions', appeared to create confusion among the panel members, who broke into an animated discussion. It eventually emerged that the panel members had not actually been told anything about the Boehlert questions. Between receiving the Boehlert questions and briefing the panel, Ciccerone had apparently rewritten the statement of task, redirecting the panel to somewhat less controversial ground.

The panel members were now forced into the uncomfortable position of having to decide, in full view of the witnesses and the other onlookers, whether the Boehlert questions were actually within the scope of their task or not. The rewording was not insignificant. For example, where Boehlert had asked about MBH98 – what the criticisms of the paper were, whether the information required to replicate it had been available and whether others had actually managed to replicate it – the NAS statement of task avoided mentioning MBH98 at all. Likewise the whole subject of replication was surreptitiously dropped.

Were the changes made deliberately or by accident? We can never be certain, but our views on this question might be coloured

---

*   Jones' comments were made in an email to the Australian researcher, Warwick Hughes. Hans von Storch apparently confirmed with Jones that this was a true representation of his position before quoting him to the NAS panel.

by the observation that the bureaucrats at NAS also forgot to attach Boehlert's original requisition to the statement of task sent out to the panellists. It looked very much as if Boehlert had been outmanoeuvred by the scientific bureaucracy. Why would they do this?

McIntyre was fascinated by the political manoeuvring he was seeing unfold before him. Now that their omission was out in the open, the NAS panel was in a quandary. If they failed to answer the Boehlert questions, the House Science Committee would look very foolish – people would assume that they had simply been outwitted by the wily mandarins in the NAS. But what was worse for the Science Committee, Barton's Energy and Commerce Committee could now start holding their own hearings. They could merely point out that although the NAS had asked to address the issues raised in Barton's original letters, having been given the opportunity to do so, the academy had failed to answer either these questions or indeed those in the Boehlert requisition.

Quite what the panel would decide to do was unclear. However, McIntyre noted that at Mann's presentation the following day, with the revelations over the Boehlert questions fresh in their minds, none of the panel members saw fit to question Mann on the Boehlert question most pertinent to his work – 'has the information needed to replicate your work been available?' It didn't bode well.

Cherrypicking

When Rosanne D'Arrigo stepped up to the microphone to give her presentation on tree ring studies, the panel may well have been unprepared for the bombshells she was about to drop. The earlier presentations by Schrag and Alley had already surprised some observers with their lack any of the alarmist language that is so common in climatology, but these two were to be nothing compared to D'Arrigo.

D'Arrigo worked at the Lamont-Doherty Earth Observatory, part of New York's Columbia University. She had a long and distinguished publication record, with many of the big names in paleoclimatology having been her co-authors at one time or another, Mann, Cook, and Jacoby among them. There was no inkling that she might be about to make fools of the whole of the paleoclimate community.

Her first bombshell was a slide in which she discussed the issue of 'cherrypicking' – a term used to describe scientists examining the data records before processing and removing those which might give the 'wrong' answer. In other words, a researcher bent on producing a hockey stick shaped temperature reconstruction could simply introduce only hockey stick shaped series (like the bristlecones and Gaspé) into the algorithm. Of course, nobody would fall for such an obvious travesty of the scientific method, at least if it was reported in those terms. Still, to some extent, Mann's short-centred PC algorithm could be seen as simply an automated way of achieving exactly the same effect. There was no longer a need to examine every data series individually; now it was enough to let the short centring process extract hockey sticks from the data.

D'Arrigo was startlingly straightforward on the subject. Cherrypicking, she said, was *necessary* if you wanted to make cherry pie.[146] In other words, she appeared to be suggesting that you needed to peek at the data to get the result you wanted. The panel must have been stunned by this admission, but it appears that nobody took her up on it.

In fact, D'Arrigo was not alone in her apparent belief that it is scientifically acceptable to cherrypick data. She and her close collaborator, Gordon Jacoby, had published a widely cited paper in which they selected ten sites from a total of 36 studied, justifying the omission of the other 26 on the grounds that they had selected only the 'most temperature-influenced'.* What made it worse was

---

* If you can't see why this is so egregious, consider the trials of a new drug in which only the results of the 'ten best-responding patients' are reported – it would be illegal in most places in the world. However, in climatological circles, this kind of behaviour appears to be readily accepted.

that Jacoby and D'Arrigo had refused to archive the data from the 26 eliminated series, arguing that because they didn't have a temperature signal, they were better left out of the archive. When McIntyre had written to the journal concerned, asking that they obtain the missing data on his behalf, Jacoby promptly refused the request.

> If we get a good climatic story from a chronology, we write a paper using it. That is our funded mission. It does not make sense to expend efforts on marginal or poor data and it is a waste of funding agency and taxpayer dollars. The rejected data are set aside and not archived.[83]

These extraordinary admissions, which were not available to the panel, show that the practice of cherrypicking is not uncommon among paleoclimatologists. An important issue for the panel to report upon, or so you might think.

Divergence

D'Arrigo also spent part of her talk discussing a new paper that she had recently published. Her presentation included a graph of her temperature reconstruction and this attracted the attentions of Kurt Cuffey. Although McIntyre had complained about his presence on the panel, Cuffey had approached him before the hearings to explain that he was quite capable of separating his overall views of global warming from his duties as a panel member. True to his word, he turned out to be one of the most inquisitive panellists. What had attracted Cuffey's attention was the fact that the figures D'Arrigo had forecast for the late twentieth century and the actual temperatures that had been observed were wildly different. This was of course the so-called 'divergence problem', which we touched on in Chapter 3.* Let us recap.

You will remember that estimates of temperature reconstructions can be made both from tree ring widths and from

---

*   See page 63.

the wood density. The divergence problem referred to the simple fact that while the instrumental records all showed a sharp late-twentieth century warming, tree rings mostly resolutely refused to respond. Despite the higher temperatures, neither ring widths or densities seemed to have been affected and in fact, if anything, there appeared to have been a widespread *decline* in ring widths and density over recent decades. Why didn't the warming show up in the tree ring data? This was an enormous issue for temperature reconstructions and could conceivably undermine the whole approach. If tree rings didn't pick up the warming now, how could anyone be sure that they had picked up earlier warmings like the disputed Medieval Warm Period?

The issue had been recognized for some years. In fact, right at the very start of McIntyre's researches into MBH98, he had come across the extraordinary pair of papers by Briffa that demonstrated how the paleoclimate community had dealt with the problem. The first of these, Briffa 2000, was a study of tree rings across the world and presented a picture of how their growth patterns had changed across the centuries.[147] His results clearly showed a marked decline in ring width density in the twentieth century.

In the second paper, Briffa created a Northern Hemisphere temperature reconstruction from similar data, which again suggested a recent decline in temperatures, on the basis of declining tree ring densities.[148] However, this time, Briffa had gone a little further. He created a 'spaghetti chart', a graph of several temperature reconstructions overlaid on top of each other in an attempt to show how well they matched up against each other (or not). Here though, he stepped out of line and did something that was highly dubious in scientific terms: he truncated the chart for his own series at 1960 so that the 'inconvenient divergence' disappeared. This is not to suggest that he did this in a secretive way. In fact he explained his reasoning as follows:

> [I]n the absence of a substantiated explanation for the decline, we make the assumption that it is likely to be a response to some kind of recent anthropogenic forcing. On

the basis of this assumption, the pre-twentieth century part of the reconstructions can be considered to be free from similar events and thus accurately represent past temperature variability.

This is an explanation that would appear wholly inadequate in most other areas of science. The hard fact is that tree rings and temperature records are diverging in the modern era, the one period when both can be directly observed. The only reasonable conclusion that can be taken away from this observation is that these tree rings are not capable of detecting warming trends. Instead, Briffa had simply assumed that the divergence didn't happen in earlier periods and that the lack of a trend in tree rings in the past meant that there were no warm periods either. What is more, despite the fact that this hypothesis cannot even be tested, Briffa's thinking is widely accepted among paleoclimatologists.

The truncation of the divergence in Briffa's paper wasn't the end of the story either. In its Third Assessment Report in 2001, the IPCC presented another spaghetti chart, which included Briffa's reconstruction. Like Briffa, they chose to truncate his series at 1960 to eliminate the divergence, but shockingly they did so *without discussing what had been done*. Whether by chance or design, the truncation point of 1960 had one huge advantage for the IPCC: at that point, the lines on the spaghetti chart were all bunched together making it hard to see that one of the lines which had gone into the bunch had failed to emerge at the other side. The truncation ended up neatly disguised.

The citation given by the IPCC report was, interestingly, to Briffa et al 2000, the original paper, which hadn't been truncated. We can only hope that this mis-citation was not deliberate. However, if the full data series had been included the chart would certainly have told an entirely different story, and the whole case that the current warming was unprecedented would have been undermined. It is perhaps worth remembering that the author of this section of the report was Mann.

To return to the NAS panel, D'Arrigo passed up the opportunity to explain the divergence problem in any great detail, noting only that it had been discussed by Briffa and arguing that it only applied to a few sites. In the light of this omission, McIntyre and McKitrick made some rapid changes to their presentation adding some new slides to explain the inadequacy of Briffa's explanations and demonstrating that the divergence problem was extremely widespread. In fact divergence was probably the norm, rather than the exception, as Briffa had noted in his original studies when he had described it as a 'widespread phenomenon'.

Hughes also addressed the subject of the divergence problem and put forward two possible explanations for it, neither of which can have seemed terribly persuasive. One idea was that increased snow pack had caused a delay in growth each year, while the second was a speculation by Briffa that trees were being damaged by atmospheric ozone. One of the panellists enquired if a similar divergence might have happened in the past, to which Hughes made the surprising response that this was a third explanation that he had hoped to avoid discussing.[149] This appeared then to be an admission that there was a real possibility that temperature signals could not be reliably extracted from tree ring records at all.

Bristlecones

The question of the reliability of the bristlecones was a clean sweep for McIntyre and McKitrick. Their presentation laid out the full gory detail of the problem, avoiding none of the controversial angles to the issue. Among these was the existence of the CENSORED directory, which, you may remember, showed that Mann was fully aware that a reconstruction prepared without the bristlecones had a prominent Medieval Warm Period. The panel cannot have failed to have noticed the contrast between this evidence and MBH98's claim that their findings were 'relatively robust to the inclusion of dendroclimatic indicators', in other words that you could remove any tree ring data you liked and still get no Medieval Warm Period. It is perhaps not surprising therefore that none of the speakers made any serious attempt to defend the use of bristlecones. Only Mann broached the subject, and then only in a rather oblique fashion, referring to the south-western USA as a 'sweet spot' for creating Northern Hemisphere temperature reconstructions. This slightly bizarre explanation was essentially a reference to his claims that it remained valid to include a PC4 representing the bristlecones, and have these drive the shape of the final reconstruction, even though the pattern only represented a pair of tree species in a pair of small mountain ranges in the western USA. Here then was an issue where the panel should have been able to draw clear conclusions.

Verification statistics and confidence intervals

McIntyre was just as pointed in his remarks about Mann's withholding of the adverse verification statistics. He explained to the panel how Mann had reported in MBH98 that he had calculated the R^2 for the Hockey Stick, but had withheld the fact that the results had indicated that his reconstruction was unreliable. McIntyre went on to demonstrate how the IPCC had later misrepresented the Hockey Stick as having significant 'skill'. Having dramatically failed the verification R^2 test, the confidence intervals for the Hockey Stick were, in the words of Hegerl, 'from floor to ceiling'.[150] In other words, you could have no confidence in the result at all.

This was a very damning set of accusations and one which promised some fireworks when Mann came to speak the following day. In the event though, absolutely nothing happened. John Christy, who was seen as the lone sceptic on the panel, asked Mann about his R^2 score. Mann tried to evade the question by denouncing its usage in general, but Christy pressed him further, asking whether he had in fact calculated the figure. Mann's reply was sharp and to McIntyre, at least, breathtaking:

> We didn't calculate it. That would be silly and incorrect reasoning.[151]

This was an extraordinary statement. Mann's paper clearly stated that 'For comparison, correlation (r) and squared-correlation (R^2) statistics are also determined'. He had presented R^2 information in the paper. The commands to calculate the R^2 were in the code he had submitted to Congress. He had told Crok that the Hockey Stick passed the R^2 test, something he could only have determined if he had calculated its value in the first place. Not only did Mann's statement fly in the face of everything he had said previously, but it also contradicted the evidence McIntyre had given the day before. It was also laughably wrong from a statistical perspective. There were two qualified statisticians on the panel: Nychka and Bloomfield. Here then was the moment for them to step up to the mark and prove their independence and worth. . .

Nychka and Bloomfield, however, said absolutely nothing and indicated that they had no interest in questioning Mann on the issue. The conversation moved on.

Later on during Mann's time before the panel, the questions returned to the subject of the verification statistics , and he again made some remarkable statements that should have been followed up by the two statisticians. As before though, Nychka and Bloomfield failed to follow up in even the most rudimentary fashion. For example, Mann said that he had calculated another verification statistic (the CE) but neither Nychka or Bloomfield thought to ask whether the Hockey Stick had actually passed the test (it hadn't). Mann also said that the R^2 was 'not good' and 'not sensible' and even that statisticians didn't use it – an idea that was outlandish to say the least – and each time he was left unchallenged. Even a statement by Mann that he was 'not a statistician' seems to have left them unmoved.

Mann, of course, was vigorous in his own defence. He asserted that McIntyre's work was without 'statistical or climatological merit', citing Wahl and Ammann in his support. At this point it was incumbent upon Nychka to state an interest – he was credited in Ammann and Wahl's paper for giving advice on statistical matters – but again, he chose to remain silent. Mann went on to repeat his allegation that McIntyre's 'reconstruction' failed verification tests. This again directly conflicted with evidence given by McIntyre and McKitrick the previous day. McIntyre had stated plainly that he and McKitrick had *not* presented their own reconstruction, but had only demonstrated that MBH98 was not robust. Once again, however, the panel failed to follow through and question a speaker on the apparent contradictions in the evidence being presented.

Credibility of MBH

With the statistical arguments largely won by McIntyre and there being no plausible defence of the bristlecones, the best possible line for Mann's defenders to take was to argue the Hockey Stick's consistency with other studies, among them the new papers from Hegerl and D'Arrigo. The panel was treated to presentations from Hegerl and D'Arrigo showing these new reconstructions. The implicit argument was that, whatever the outcome of the debate over the data and methods used in MBH98, its hockey stick shape was still broadly in line with other studies and that it was therefore probably correct.

McIntyre was not unaware of this line of reasoning and over the previous three years he had attempted to replicate some of these other studies too. This had been a long and difficult process. At every step of the way his attempts to obtain data and code had been blocked by the climate scientists, and, as with MBH98, journals and funding agencies had refused to help. However, he had developed a good understanding of how most of these secondary studies had achieved their results, and what he told the panel should have given them considerable cause for alarm.

The subject of these confirmations of the Hockey Stick will be examined in detail in Chapter 10, but in essence it was clear was that the 'independent' confirmations were actually nothing of the sort, their alleged independence being nonexistent. Millennial temperature reconstructions were all produced by a small group of people closely linked to Mann. There were few people in the field who had not been co-authors either with Mann, Bradley or Hughes, so it was false to argue that these researchers were independent in the normal sense of the word. The data used in these studies was likewise problematic. Nearly all of the papers were based on the same small set of flawed proxies, most of which were also used in MBH98. It was therefore no surprise that these other papers reached similar conclusions, despite the fact that the exact methodology varied from study to study. Nor was it clear that they were any more statistically robust than MBH98. As McIntyre pointed out in his presentation, nearly all the reconstructions relied on a procedure called variance rescaling. In order for this method to yield correct results, it is necessary first to show that the reconstruction passes a statistical test called the Durbin–Watson statistic. McIntyre pointed out that nearly all of the other reconstructions had not performed the Durbin–Watson test, and that if they had they would have failed it, indicating that their uncertainties were much higher than reported. He also pointed out that nearly all of them spectacularly failed the verification R^2 as well.

Here then was an interesting dilemma for the panel. On the one hand they had Hegerl and D'Arrigo telling them that Mann's work was supported by other studies, while on the other hand they had McIntyre showing them that the other studies were potentially just as flawed. How would the panel report this difference of opinion? Would they jump one way or the other, or would they hedge their bets, reporting that opinion was divided?

Follow-up

In view of some of the surprising replies given by Mann and the failure of the panellists to probe his answers, McIntyre and McKitrick decided to make some follow-up comments in a written submission that they sent to the panel a month after the hearings.[152] Their language was stark and to the point. For example, they addressed Mann's claim that he had not calculated the R^2 statistic:

One of the panellists asked Mann for the value of his verification [R^2] statistic for the 15th century step. Mann said that they did not calculate this statistic. This was untrue, based on the text of MBH98, as reported in McIntyre and McKitrick [NAS *Panel* 2006].

There were plenty of other issues too. The latest incarnation of Ammann's CC paper had appeared shortly after the hearings, and McIntyre took the opportunity to bring its catastrophic failure of the R^2 test to the panel's attention. They also pointed out the recent rejection of Ammann's GRL comment and the effect this had on the CC paper's claims of statistical significance.

One of their most interesting points, however, was the way they addressed the cherrypicking issue. Having expressed their concern over how the panel had been presented with a series of proxies and reconstructions with twentieth century upticks, they argued that it was quite possible to come to different conclusions simply by selecting a different set of proxies. By selecting only proxies with warm medieval sections and calculating an average using a standard multiproxy approach, they showed that they could produce a 'reconstruction' that showed both the twentieth century uptick and a medieval warm period. This 'applepicking' reconstruction was tongue-in-cheek, of course, but demonstrated powerfully that it was possible to get the answer you wanted, simply by picking the right proxies. Whether this would have any effect on the outcome was another question.

The report

Back home in Canada, rumours started to reach McIntyre and McKitrick that the NAS report was going to be 'two-handed': the panel would accept some of their arguments but not others. This was consistent with the impression McIntyre had gained during the hearings, with all the most searching questions being studiously avoided by the panel.

The report was finally published in June, with North, Cuffey and Bloomfield delivering a press conference in Washington to the assembled media.[153] The overall thrust of the report was that temperature reconstructions for the last 400 years were reliable, but there was 'less confidence' in earlier periods. How much less confidence was not made clear. The good news was that the panel had accepted McIntyre and McKitrick's statistical criticisms of MBH98. They also accepted the arguments against using bristlecone pines, going as far as to say that these should be avoided in temperature reconstructions. But their support appeared very grudging: their acceptance of McIntyre and McKitrick's critique was outlined in the main text but was entirely absent from the summary at the start of the report. Here instead they gave prominence to an argument that, despite all its flaws, the Hockey Stick was 'plausible' because of its similarity to other temperature reconstructions.

The panel appeared to get completely muddled up in their attempts to support this position. They referred to a study by Osborn and Briffa, which looked at a number of proxy series and showed that more of them had warm deviations in the twentieth century than had cool ones.[154]* The argument in the paper was that by looking at the proxies in this simple way there was no need to engage in the statistical brouhaha that had plagued MBH98. The problem with the panel's citation was that, of the 14 series cited in Osborn and Briffa, one was Mann's NOAMER PC1 (coyly referred to as 'W. USA, regional') and another was a foxtail series (foxtails, you will remember, are very similar to bristlecones). In fact, there were problems with most of the series used in Osborn and Briffa, and McIntyre had described the article as being little more than a compendium of every dubious proxy series in the archive with a few others added in to make some 'noise'.

The report also included a spaghetti chart, showing how MBH98 compared to other major multiproxy studies. These all had broadly similar hockey stick shapes but incredibly, all except one

---

*   The paper is examined in more detail on page 298.

included bristlecone pines in their proxy rosters. It is hard, if not impossible, to reconcile the panel's acceptance of a group of reconstructions based on bristlecones with their concurrent statement that bristlecones should not be used in temperature reconstructions. They were certainly aware of the issue, since McIntyre had raised the subject in his written submissions to the panel.[155]

Elsewhere, the panel reported that they rejected Mann's idea that a single validation statistic was acceptable and said that a reconstruction with a low CE score was 'unreliable', while failing to point out that MBH98 fell into this category. They also recommended the use of a Durbin–Watson statistic as a test of validity. This statement directly contradicted their position on the 'independent' reconstructions, because all of those other studies, which the panel alleged gave broadly the same answer as the Hockey Stick, failed the Durbin–Watson statistic (as well as the R^2). The panel cannot have been unaware of this fact because, as we have seen, McIntyre had already pointed it out to them in his presentation.

Although McIntyre said he accepted the integrity of the panel, and indeed wrote to thank Ralph Ciccerone for giving him and McKitrick a fair hearing, the internal contradictions in the panel's report were very frustrating, and led McIntyre to describe the report as 'schizophrenic'. Just as annoying was the failure of the panel to address many of the controversies at all: the difficult questions that had been asked by Barton appeared to have been been quietly shelved. Had Mann withheld or misrepresented adverse results? The panel wasn't saying, although in the press conference they all agreed that they had 'seen nothing that spoke . . . of any manipulation', a surprising position when McIntyre had pointed out to them that Mann had calculated the R^2 but had not reported it. Had Mann withheld his data and code? They would only opine that researchers *should* make these materials available, a position that would pass muster as a statement of the patently obvious, but hardly addressed Boehlert's question of whether it had been available in practice. At the press conference, Gerald North explained that they had found the whole subject too large to deal with and announced that a new panel would be formed specifically to look at the issue of scientists making data and materials available, an approach that looked remarkably like ducking the question and hoping that everyone lost interest in the meantime. The new panel was supposed to report in mid-2007, but was strangely delayed, finally appearing only in 2009.*

At the press conference for the launch of the report, McIntyre sensed a certain amount of unease among the assembled journalists. The Boehlert questions had clearly not been answered, the flaws in the Hockey Stick seemed to have been acknowledged and there was a growing recognition that Mann's iconic paper had been oversold both to the press and the public. Some of the questions were therefore quite searching and there was plenty for the representatives of the panel to deal with. Some of their replies were very controversial. For example, Peter Bloomfield, the statistician who had so signally failed to address any of the contradictions in Mann's statistical arguments, explained to the assembled media the panel's findings on Mann's use of PC analysis. The usage in MBH98, he said, was 'unconventional' and 'problematic' and 'introduced certain distortions', this presumably being a coded way of saying 'wrong'. Like so many before him, he managed to discuss the whole subject without once mentioning McIntyre and McKitrick. As he continued, Bloomfield made a curious claim: if you averaged the proxies in MBH98, he said, you got the same answer – a hockey stick. This directly contradicted a slide in McIntyre and McKitrick's presentation, which demonstrated just the opposite: the average of the proxies looked nothing like the Hockey Stick at all, and if

---

*   When the report finally appeared, it turned out that the panel had largely ducked the question of whether Mann's data had been available. While mentioning briefly that Mann had 'resisted' requests for his data, the panel preferred to discuss what it saw as Barton's 'intimidation' of the Hockey Stick authors. Discussion of the Hockey Stick was limited to just two pages of the report.[156]

anything had a downtick in the twentieth century (see Figure 9.1). However, what was particularly odd about Bloomfield's claim is that the panel had made it quite plain that they had not performed any verification work of their own. So how Bloomfield had arrived at the conclusion he did is unclear, but it is a claim that has been repeated since.

Schizophrenic as it was, the report had something for everyone. Sceptics could point to its acceptance that Mannian short centring was wrong while greens could revel in its conclusion that the secondary studies gave the same answer. Most observers thought that the Hockey Team had done best out of it.* They could potentially have ended up with their whole field of study in tatters and their most prominent scientific result held up to ridicule. As it was, the Hockey Stick was still in play, even if the panel had been forced to rely on the somewhat dubious argument that Mann had used incorrect methods and inappropriate data but had somehow still managed to reach the right answer.

The press reported the panel's findings very much according to their prejudices. *Nature* published an editorial, its headline declaring 'Panel affirms hockey-stick graph' while the BBC said that the panel had given its 'backing for [the] hockey stick graph'.[158,159] On the other hand the *Washington Times* said that 'the NAS report has re-established the [Little Ice Age and the Medieval Warm Period], and broken the Hockey Stick'.[160] Perhaps only the *Wall Street Journal* really understood what the panel had done, reporting to its readers 'Panel study fails to settle debate'.[161]

The Wegman report

The dust was still settling on the NAS report when the excitement was ramped up again with the release of the Wegman report. On the day of its publication, 14 July 2006, Barton also issued a press release to the effect that the Oversight and Investigations

---

* Climate policy academic Roger Pielke Jnr described it as 'a near-complete vindication for the work of Mann et al'.[157]

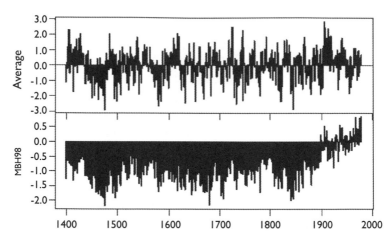

FIGURE 9.1: The average of Mann's proxies compared to the Hockey Stick

Top: Simple average of Mann's proxies. *Bottom*: The Hockey Stick.
Reproduced from McIntyre and McKitrick's presentation to the NAS panel.

Subcommittee, sometimes referred to as the Whitfield Subcommittee, would shortly be holding hearings on the subject of Barton's original questions. Clearly, the failure of the NAS to address the Boehlert questions had left Barton with no choice but to investigate the Hockey Stick himself.

Rumours circulated around the climate blogs as to who had been invited to the hearings. It was said that some climatologists were refusing to attend, which, if true, was a dangerous course to take, since the Whitfield Subcommittee could enforce attendance through a subpoena. A few days later it was announced that the witnesses were to be the heads of the two panels, Gerry North and Ed Wegman, together with four researchers: Thomas Karl and Tom Crowley, who could be expected to speak for the Team, with McIntyre opposing and von Storch occupying his customary position of honest broker. Mann was apparently unavailable, but the Committee had invited him to attend the following week, thus obliging all the other witnesses to come to Washington again. The first hearings were just a week away.

Before the hearings could be held, the newspapers got hold of the details of the Wegman report: an article appeared in the *Wall Street Journal* which showed that Wegman was going to come out almost entirely on McIntyre and McKitrick's side of the argument.

> The report commissioned by the House Energy Committee, due to be released today, backs up and reinforces that conclusion. The three researchers – Edward J. Wegman of George Mason University, David W. Scott of Rice University and Yasmin H. Said of Johns Hopkins University – are not climatologists; they're statisticians. Their task was to look at Mr. Mann's methods from a statistical perspective and assess their validity. Their conclusion is that Mr. Mann's papers are plagued by basic statistical errors that call his conclusions into doubt. Further, Professor Wegman's report upholds the finding of Messrs. McIntyre and McKitrick that Mr. Mann's methodology is biased toward producing 'hockey stick' shaped graphs.[162]

McIntyre had been supplied with a copy of the Wegman report ahead of the hearings, and it was just as supportive as the *Wall Street Journal* was suggesting.[15] Not only did the authors share the NAS panel's view that Mann's short-centred PC calculation was biased, but they had also said so in language that eschewed the bureaucratic double-speak adopted by North's panel.

> In general, we found MBH98 and MBH99 to be somewhat obscure and incomplete and the criticisms of [McIntyre and McKitrick] to be valid and compelling.[15]

Wegman had been able to replicate McIntyre's work completely and had accepted all of his arguments. Wegman and his colleagues had also gone beyond a straightforward analysis of the MBH98 data and methods and had tried to analyse the reasons why a paper as flawed as Mann's had managed to slip through the peer review process and then had managed to reach a position of such huge importance for the policy-making community. Why was it that it had fallen to a retired minerals consultant and an economist to expose its errors? Why had the paleoclimate community defended it so vociferously?

In order to get to the bottom of these issues, Wegman and his coauthors had performed an analysis of the links between paleoclimate researchers. This showed that most of them were indeed joined to each other by ties of co-authorship and, moreover, that Mann had links to almost everyone else – he was like a spider at the centre of a web of collaboration, in a position to influence everyone around him. Wegman's conclusion from this 'social network' study was that the paleoclimate community was too insular, too self-contained, too close-knit. Their insularity and the fact that Mann was at the very centre of the community meant, said Wegman, that no effective independent review of Mann's work was likely (see Figure 9.2) and where mistakes were made, it was difficult for climatologists to correct their work:

> [O]ur perception is that this group has a self-reinforcing feedback mechanism and, moreover, the work has been sufficiently politicized that they can hardly reassess their public positions without losing credibility.[15]

But his criticisms went further: paleoclimatologists, he said, believed too passionately in the anthropogenic global warming hypothesis and they had failed to enlist the help of statisticians in their work:

> It is important to note the isolation of the paleoclimate community; even though they rely heavily on statistical methods they do not seem to be interacting with the statistical community.

Here at last, after three years of being accused of incompetence and dishonesty, was complete vindication for McIntyre. It was, he said, immensely gratifying. What would the politicians now make of Wegman's findings when they were discussed at the hearings, only a matter of days away?

Barton Committee hearings

The hearings were held in in Room 2123 of Rayburn House, the House of Representatives' huge office building on Washington's Capitol Hill, overlooking Independence Avenue. Barton's Energy and Commerce Committee had delegated responsibility to its Subcommittee on Investigations and Oversight, whose chairman, Ed Whitfield, would preside over the proceedings.*

The hearings opened with statements from the politicians. From the start it was clear that they were more interested in fighting over political territory than in understanding the lessons that should be learned from the Hockey Stick affair. Republicans were keen to protect the findings of the NAS and Wegman panels from attack, while Democrats tried to limit the impact of the reports on the wider global warming argument. Each and every member of the subcommittee needed to have their say, most opting simply to state their position on global warming. The witnesses all had to sit patiently while the members of the committee held forth.

For the attentive, these opening statements did contain hints of some of the lines of questioning the Democrat members were going to take. Bart Stupak, the representative from Michigan and the senior Democrat on the committee, called the reports 'irrelevant' and said that it was difficult to assess Wegman's work because it wasn't peer reviewed. Stupak's colleague Jay Inslee referred to the hearings as a snipe hunt. There were attempts to defend Mann by arguing that MBH98 was the first study of its kind

---

*   The story of the hearings and the quotations below are based largely on the official transcript.[163]

FIGURE 9.2: Mann at the the centre of a paleoclimate web

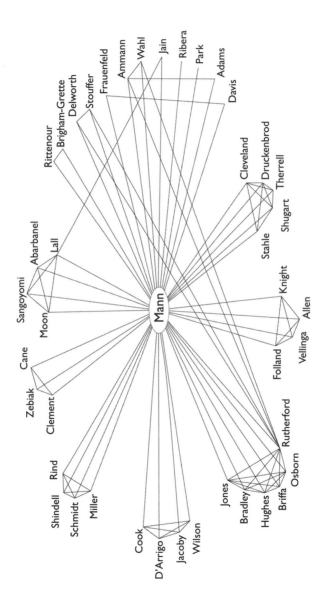

Lines indicate links of co-authorship. While each clique is largely self-contained, Mann has worked with all of the authors shown. Adapted from The Wegman Report

– although it wasn't* – and that Mann should therefore be given a certain amount of leeway. There was much talk of melting glaciers and record-breaking temperatures. It was therefore a relief when Whitfield finally wound up the political venting and called the first two witnesses.

The committee was to hear opening statements from Wegman and North, each outlining the contents of their reports, before cross-examining them on their findings. The Democrat members had very few cards to play. The NAS panel seemed to have backed the criticisms of Mann's work, and now Wegman was saying the same thing, but louder and more clearly. Their situation was made worse when North made his agreement with Wegman on the statistical findings absolutely clear, although he maintained his line that, despite Mann's mathematics being wrong, there was still a possibility that his conclusions were correct.

> CHAIRMAN BARTON: Dr. North, do you dispute the conclusions or the methodology of Dr. Wegman's report?
>
> DR NORTH: No, we don't. We don't disagree with their criticism. In fact, pretty much the same thing is said in our report. But again, just because the claims are made, doesn't mean they are false.

Barton tried to see off some of the potential lines of attack before they happened. He told the committee that he had 'heard' that Wegman had voted for the Democratic candidate, Al Gore, in the last presidential election, and Wegman indicated that this was indeed the case, neatly preventing any suggestion of political bias. Barton also asked Wegman about his previous contacts with the committee and with McIntyre, again preventing the Democrats from insinuating that Wegman might suffer from a lack of independence. The questioning of Wegman was at times somewhat

---

* MBH98 referred to Bradley and Jones 1993 and several other multiproxy studies.

aggressive but all very peripheral to Wegman's core argument that Mann's work had not correctly implemented a PC calculation. There was close questioning of his social network study and much discussion of the IPCC's 1990 chart showing the Medieval Warm Period,* which Wegman had reproduced in his report, although it is not clear what the Democrat team hoped to achieve by doing this. Wegman was also asked some entirely irrelevant questions about his views on the larger question of global warming. Representative Jan Schakowsky of Illinois, for example, asked him if he wasn't concerned that his results would be used by sceptics to discredit the global warming hypothesis. Von Storch, in evidence given later, declared himself 'shocked' by Schakowsky's implication that Wegman should have written something other than the truth if that was useful for the policy process.

North, meanwhile, had a much easier time of it. Having conceded that Mann's statistics were wrong, he was still maintaining his line that the independent confirmations suggested that Mann's findings were correct regardless. During his time at the microphone, North was questioned on the subject by the Democrat Henry Waxman, and these exchanges should have thrown some light on how the panel dealt with McIntyre's evidence that the other studies were just as contaminated by use of bristlecones as was Mann's.

> DR NORTH: But as I have said, it is only one of several lines of evidence that are used in drawing those conclusions.

> MR WAXMAN: And so therefore you have further studies that seem to come to similar conclusions?

> DR NORTH: There are other studies, and they were shown on the graphic that I showed you.

> MR WAXMAN: And they weren't based on the Mann studies, were they?

---

*   See page 25.

DR NORTH: They were not based on the Mann studies. Now, there are cases where they use the same data so there is some correlation and that is what I think Dr. Wegman referred to and that is correct. See, there is only a limited amount of data, so. . .

North's final statement shows that he knew at least something about the proxies used in the spaghetti graph studies – clearly McIntyre's evidence *had* been considered. Some further light was shone on this question some weeks later, when North took part in an online colloquy about the Hockey Stick and McIntyre took the opportunity to question him further.

MCINTYRE: The [NAS] Panel stated that strip-bark tree forms, such as found in bristlecones and foxtails, should be avoided in temperature reconstructions and that these proxies were used by Mann et al. Did the Panel carry out any due diligence to determine whether these proxies were used in any of the other studies illustrated in the NRC spaghetti graph?

DR NORTH: There was much discussion of this matter during our deliberations. We did not dissect each and every study in the report to see which trees were used. The tree ring people are well aware of the problem you bring up. I feel certain that the most recent studies by Cook, D'Arrigo and others do take this into account. The strip-bark forms in the bristlecones do seem to be influenced by the recent rise in CO_2 and are therefore not suitable for use in the reconstructions over the last 150 years. One reason we place much more reliance on our conclusions about the last 400 years is that we have several other proxies besides tree rings in this period.[164]

If the NAS panel didn't look at 'each and every' study in terms of the proxies used, then the implication is that they looked at least at *some* of them. This is the only explanation that would also be consistent with the evidence North gave to the Whitfield Committee – that there was some commonality of data between the studies. The inescapable conclusion is that the panel must have been aware that bristlecones were used in at least some of the 'independent' confirmations. It remains a mystery why the NAS panel didn't at least raise this as a question mark over the integrity of the purported confirmation – something that would require further assessment so that the impact of the flawed proxies could be properly assessed. Either way, McIntyre was not impressed, commenting to Climate Audit readers:

> I've said over and over how frustrated I am that the due diligence of the NAS panel was so negligible and slight and that they relied on mere literature review for so much of their study. It's ludicrous for them to say that bristlecones should be 'avoided' in temperature reconstructions and then to 'bring in other evidence' – a 'half-dozen other reconstructions' that use bristlecones – without testing for the impact of bristlecones on these reconstructions. I'll do the testing of the impact of bristlecones on the other reconstructions, but the NAS panel should have done it themselves.[165]

With the interrogation of the heads of the panels complete, Whitfield's committee moved on to the scientists, but in fact there was very little of substance discussed that hadn't already been said at the NAS panel. McIntyre reiterated his criticisms of Mann, and set out the problems with the peer review process and the failure of scientists to archive data and code, but it was clear by this time that the panel was rapidly losing interest in the subject. Even when Mann himself took the microphone a week later, the contradictions between his claims and those that McIntyre had given were brushed aside by the assembled politicians. The committee had moved on: the Republican side had its condemnation of Mann's paper and had concluded that they had won this particular battle. The Democrats, meanwhile, had got North and Wegman to state that nothing in their reports affected the overall case for global warming. The tide of the wider war was still flowing their way.

An interlude: Dr Thompson's Thermometer

During the interrogation of Wegman and North there was an interesting exchange which, while not directly relevant to the story, does tell us something about the importance of the Hockey Stick to politicians. During the questions to North and Wegman, one of the Republican congressmen, Cliff Stearns of Florida, started to discuss the importance of the Hockey Stick and how it seemed to crop up everywhere after the IPCC report of 2001. He pointed particularly to its use in Al Gore's movie, *An Inconvenient Truth*, and the book of the same title. There is indeed a mention of the Hockey Stick in *An Inconvenient Truth*, when Gore makes the case that the Hockey Stick is supported by ice core records:

> [S]o-called global warming skeptics often say that global warming is really an illusion reflecting nature's cyclical fluctuations. To support their view, they frequently refer to the Medieval Warm Period. But as Dr Thompson's Thermometer shows, the vaunted Medieval Warm Period (the third little red blip from the left below) was tiny in comparison to the enormous increases in temperature in the last half-century – the red peaks at the far right of the graph. These global-warming skeptics – a group diminishing almost as rapidly as the mountain glaciers – launched a fierce attack against another measurement of the 1000 year correlation between CO_2 and temperature known as the 'Hockey Stick', a graphic image representing the research of climate scientist Michael Mann and his

colleagues. But in fact scientists have confirmed the same basic conclusions in multiple ways with Thompson's ice core record as one of the most definitive.[166]

'Dr Thompson' was a reference to Lonnie Thompson, a distinguished paleoclimatologist who recreated temperatures from ice core records; his 'thermometer' was simply a reference to these temperature reconstructions. So according to Gore, Thompson's ice core reconstruction confirmed Mann's work – another independent confirmation to add to those shown by North in the NAS report.

When Stearns raised the subject of Gore's citation of the Hockey Stick, he was interrupted by the Democrat Schakowsky, who asked him to yield the floor so that she could make a point. Stearns was reluctant to do so, but Schakowsky was extremely insistent, and she was joined by her colleague from the Democratic side, Bart Stupak. Even with two people asking him to give way, Stearns still refused, insisting that Wegman should comment on the inclusion of the Hockey Stick in the movie, apparently trying to score a political point by linking the flaws in Mann's paper to the work of Al Gore. Wegman started to reply, saying that there was some ambiguity in Gore's citations and for a moment confusion reigned. However, after a few moments, calm was restored when Wegman concluded that Gore had in fact referred to the ice core studies. Stearns then finally gave way to Schakowsky, who had by then tried no less than five times to get the microphone.

> Ms SCHAKOWSKY: Thank you. I just want to read to you from that same – it says 'But as Dr. Thompson's Thermometer shows,' and so it is not based on Dr Mann. This is a different source which our staff had confirmed with Al Gore. I just want to make . . .

> MR STEARNS: I respect that.

Ms SCHAKOWSKY: . . . that point. I know, but your question wanted to reinforce the notion that this was based on this false or inaccurate Dr Mann study . . .

MR STEARNS: Well, I think . . .

Ms SCHAKOWSKY: . . . and it is not.

MR STEARNS: Okay.

And that was how the subject was left – with all parties concluding that the hockey stick shaped chart in Gore's book and movie were based on ice core records, thus demonstrating support for the conclusions of the Hockey Stick. They knew that it was an ice core study because Schakowsky had checked with Gore's office.

More than a year later, McIntyre started to ponder the subject of Gore's hockey stick. As he studied the graphic used, a number of things came to his attention. For a start the resolution* appeared to get higher in the nineteenth and twentieth centuries. This was peculiar because ice core studies should give the same resolution for all years. How had Thompson managed to get greater detail in the modern era? Moreover, in the twentieth century, the chart appeared to show positive and negative values simultaneously. Even the style of the chart didn't seem quite like anything else that Thompson had produced before. McIntyre was intimately familiar with everything Thompson had published and there was nothing in his papers that looked quite like the chart in Gore's movie. Where, he wondered, could it have come from? Bemused, he did what he often did when completely stuck for an answer and asked his Climate Audit readers if they had any ideas.

It was perhaps a surprise that the first comment was by one of his chief internet opponents, an Australian computer scientist and

---

*   The resolution is simply the smallest time period that can be distinguished in a proxy or a reconstruction. Tree rings give annual resolution, while other proxies might only show much longer periods.

scourge of the sceptic community called Tim Lambert. It was even more surprising that Lambert knew *exactly* where the graph had come from. Unfortunately he wasn't letting on, at least not immediately. As the Climate Audit readers studied the chart, more strange things were noticed. The temperature axis was upside down, with negative values at the top and positive ones at the bottom, implying that the world was actually cooling. The *x*-axis didn't meet the *y*-axis at zero. Finally, Lambert asked if McIntyre was only kidding that he didn't know the source study for the graph, and this gave some of the readers enough of a clue to work the answer out. What Gore called 'Dr Thompson's Thermometer' was in fact Mann's Hockey Stick itself, recoloured, placed on different axes and given a new name. It was no wonder that he could claim that his graph looked remarkably like the Hockey Stick: it *was* the Hockey Stick. Its appearance in place of Thompson's ice core graph turned out to be a copying error, which neither Gore, nor his staff, nor Thompson, their scientific adviser, had noticed. The splice of reconstructed and instrumental data was not acknowledged in *An Inconvenient Truth*, but if you referred to Lonnie Thompson's original papers, it was possible to see the version of the Hockey Stick Gore had used, with the splice clearly shown.

Gore was famously awarded a Nobel Prize for his work on global warming. It was, as McIntyre said, too funny. Thompson was later asked in a public meeting what he had done to correct this mistake. He replied that he had no responsibility to make the error known to the public.[167]

Ritson tries again

As the dust settled on the NAS report and the brouhaha over the Barton hearings faded away, the press and the participants, and some of the observers, started to set down where they thought the arguments had got to. McIntyre and McKitrick had won at least a partial victory with both the NAS and Wegman panels agreeing that Mannian short centring was biased. There was still the NAS's mystifying use of bristlecone-infected studies in support of the Hockey Stick, but there was still much cause for satisfaction.

Meanwhile the Hockey Team was going to make one last attack on the findings of the two panels. At the end of August, a posting appeared at RealClimate called 'Followup to the "Hockey Stick" hearings'.[168] The posting pointed to some of Mann's answers to follow-up questions from the House Committee, one of which proved rather surprising. As the RealClimate posting put it:

> Among the more interesting of these documents are a letter and a series of email requests from emeritus Stanford Physics Professor David Ritson who has identified significant apparent problems with the calculations contained in the Wegman report, but curiously has been unable to obtain any clarification from Dr. Wegman or his co-authors in response to his inquiries. We hope that Dr. Wegman and his co-authors will soon display a willingness to practice the principle of 'openness' that they so recommend in their report . . .[168]

We have, of course, met David Ritson already, as the author of the 'goofy' comment on McIntyre's GRL paper, which was rejected twice by the journal.* Ritson, it seemed, had been trying to obtain the data Wegman had used to perform the calculations in his report, and had emailed three times without getting a response, a result which he believed amounted to a blanket refusal. Noting that some of his own data requests were still outstanding after three *years*, McIntyre suggested on his blog that Ritson might be somewhat premature to jump to this kind of conclusion after three *weeks*.

The significant problem that Ritson claimed to have found was an error in the way Wegman had modelled the statistical properties ('the autocorrelation structure') of proxy series in his

---

*   See page 200.

simulations. The errors were, said Mann, the same as those that McIntyre had made in his own Hockey Stick studies and were 'so basic that they would almost certainly have been detected in a standard peer review'.[169] It was surprising for a self-confessed non-statistician to accuse one of the world's leading exponents of that subject of making a mistake in his statistical workings, particularly in those terms. What was even more amazing though was that to carry this claim off, Mann was going to have to show not only that McIntyre and Wegman were wrong, but also von Storch, Huybers and the NAS panel, all of whom had concluded that Mannian short centring was biased. It was also extremely odd that Ritson's original comments on McIntyre's GRL paper contained no mention of these allegedly 'basic errors'. A few weeks later, the Hockey Team's clutching at straws became downright embarrassing when McIntyre pointed out that the way he and Wegman had determined the statistical structure of the simulated data was *identical* to that used by Mann. It was, he pointed out, an extraordinarily weak point for the Hockey Team to make a stand on, and sure enough, Ritson promptly made a diplomatic retreat and dropped the subject.

Taking wing

Before we move on to the next chapter there is one final aspect to the story of the NAS panel that bears repeating. In the aftermath of the report Gerry North was in much demand, and one of the lectures he gave on the panel's work gave some interesting insights into the nature of their review. These will be important when we reach this book's final chapter, when we look at the implications of the Hockey Stick affair for the global warming debate and for science in general.

In a talk he gave at his own Texas A&M University, North explained to his audience the way the panel had worked.

> We didn't do any research in this project, we just took a
> look at the papers that were existing and we tried to draw
> some kinds of conclusions from them. So here we had

> twelve people around the table, all with very different
> backgrounds from one another and we just kind of winged
> it to see . . . so that's what you do in that kind of expert
> panel . . .[170]

North said these words, not with any sense of dissatisfaction or of concern. His tone was matter-of-fact; this was just the way things were in expert panels. It was just one more dismaying revelation from the Hockey Stick affair – faced with the most important scientific questions for decades, asked to study and report on a subject of incalculable economic, political and social importance, a group of distinguished scientists got round a table, talked about some papers and just 'kind of winged it'.

10 Zone Defence

Great spirits have always found violent opposition from mediocrities. The latter cannot understand it when a man does not thoughtlessly submit to hereditary prejudices but honestly and courageously uses his intelligence.

Albert Einstein

The NAS panel had concluded that short centring was biased and that the bristlecones were flawed, but they also said that the similarity of the MBH98 reconstruction to subsequent papers showed that Mann had managed to get the right answer anyway. This was not a new idea. From the time of the Third Assessment Report (TAR) back in 2001, the IPCC had attempted to bolster the position of the Hockey Stick with spaghetti charts – graphics showing all of the temperature reconstructions together. In TAR the spaghetti chart had shown MBH99 alongside a couple of other studies – Briffa's infamous truncated reconstruction and another by Phil Jones. By the time of the Fourth Assessment report, other reconstructions had appeared as well – Moberg et al, Jones and Mann, Esper et al, Crowley and Lowery, and others too.

McIntyre had filled many hours with the detailed analysis of each of these allegedly independent verifications of Mann's work and, before we go on to look at the IPCC's Fourth Assessment Report itself, we need to understand what he had found so that we can consider just how independent they really are. We have already seen that many of these studies include bristlecone proxy data, strongly suggesting that they are just as unreliable as MBH98. We have also seen that many of them fail key statistical tests like the R^2, CE and the Durbin–Watson statistic. However, over the following years, McIntyre's researches had unearthed many more surprising details that cast further doubt on the wisdom of using these 'independent' studies to support important government policy decisions. First though, we need to ask ourselves just what we mean by a 'confirmation'. What do we look for when another paper is said to confirm a study like MBH98? What factors make it a good confirmation and what would an inadequate one look like?

A temperature reconstruction can be questioned firstly on the basis of its data: were the proxies appropriate, were they distorted by factors other than temperature, was the measurement data processed correctly and so on? The other area that needs to be considered is the methodology: was it appropriate to answering the question asked? Was it correctly applied?

As far as the data is concerned, there are a host of issues that could reduce our confidence in the validity of the temperature reconstructions that appeared in the years after MBH98. We have seen throughout this book that there are enormous question marks over the validity of tree rings as a proxy for temperature, so we would presumably consider a confirmation of Mann's work that relied on tree rings less convincing than one that was based on other proxies, such as ice cores or speleothems.* We would presumably be still less impressed by a study that was not only based on tree rings but also used inappropriate trees such as bristlecones. We would likewise not wish to rely on studies whose data was not publicly archived, was outdated or whose inclusion was not clearly justified – in other words there should be no question of cherrypicking.

When it came to the methodology, we would expect that a valid confirmation would avoid the incorrect short-centring methodology that Mann used in his PC calculations, and it should also avoid using ad-hoc statistical methods that had not been thoroughly tested by the statistical community. Beyond that, we would have to consider each methodological choice on its merits.

---

* See page 43.

However, a study that reached the same conclusions as Mann while passing all of these tests would represent strong support for the NAS panel's case that Mann had reached the correct answer regardless of his incorrect method and inappropriate proxies. There is, however, another criterion that we should consider – one that might colour our view of the reliability of the paper as a confirmation. Mann and the NAS had been at pains to point out that the confirmations of MBH98 were 'independent', and we would therefore need to consider just how independent they truly were. What was the relationship between each of the authors and Mann, Bradley and Hughes? While it is important to say that co-authorship wouldn't make a study incorrect – this would be the same logical fallacy that assumes that any study funded by oil companies is incorrect – we would surely find a confirmation by a close colleague of Mann's, such as Ammann or Crowley, less convincing than one by an opponent.

The purpose of this chapter, therefore, is to survey some of the studies that appeared in the IPCC's Third Assessment Report of 2001 and in the years thereafter, so that we can assess their reliability. While this might seem a rather dry subject after the excitement of the earlier parts of Mann's story, don't skip on, because there are tales here that are at least as amazing as those that have been told earlier in this book.

Jones et al 1998

Jones et al 1998 was and remains one of the most important multiproxy studies, published by a British team at around the same time as the original Hockey Stick paper.[171] McIntyre had been looking at it for almost as long as the Hockey Stick itself. In some ways he saw Jones and his co-authors as being in competition with Mann to be the first team to 'get rid of the Medieval Warm Period', but the British team's attempt to achieve this feat had been far less ambitious than Mann's. Their final reconstruction was much less dramatic, with the Medieval Warm Period still clearly in evidence, although it peaked at levels lower than those reached in the twentieth century portion of the reconstruction. Jones et al could therefore be used to support the idea that the Medieval Warm Period was a weak and maybe localised phenomenon, but it didn't really help Mann's case, which was that the Medieval Warm Period never happened at all.

We have come across Phil Jones a couple of times already during the course of this story, and his co-authors on the 1998 paper included another familiar name, that of Keith Briffa. Both men had worked with other members of Mann's team and are seen as core members of the Hockey Team, so without even looking at the detail of the paper, it is clear that Jones et al 1998 was not strictly an *independent* confirmation. However, this is a minor point; the details of the paper are considerably more damning.

McIntyre's first step in trying to replicate a paper was to collate the data. While data might be cited correctly and accurately in the papers, it was always possible that what had actually been used was different in some way to the official versions, whether due to an error in the archive or one made by the authors. Because of this possibility, McIntyre's approach was to try to obtain the data, as used, direct from the authors. This could then be checked by him against the archived data in exactly the way he had done for MBH98.

However, having approached Jones for the figures, he was disappointed to discover that the Englishman was even more reluctant than Mann to supply the full details of his research, only agreeing to supply two 'grey' data series. (Grey series are versions of data that are different to the official versions in some way and are passed between authors without ever being archived. This makes a study that uses grey series virtually impossible to replicate and indeed the use of grey data is either forbidden or at least frowned upon in many other disciplines.) The remaining series, Jones said, McIntyre should take from the archives.

While this failure to supply all of the data didn't prevent progress, it *was* unhelpful and raised the possibility that McIntyre's time would be wasted in identifying differences between the archived versions and the versions used in the paper. However, undeterred, McIntyre set about his task.

The proxy roster in Jones et al consisted of just 17 series, a strikingly smaller number than the more than 400 used in MBH98. Much of Mann's claim to scientific rigour was based on the huge size of the proxy roster, and it is fair to say that the much smaller number used by Jones was more the norm for paleoclimate reconstructions. It is of course questionable just how reliable any temperature reconstruction can be when it is based on such a small quantity of data, especially as we know how tree rings vary from site to site and even within sites. One of the issues that was constantly raised about paleoclimate reconstructions was their lack of confidence intervals, which would allow readers to assess this issue of reliability in a scientific way. Jones et al was no exception.

In terms of the actual series behind the paper, no less than 13 of the 17 series used by Jones were also used in MBH98 and MBH99, which again throws doubt on the independence of the study, but this time in relation to its data rather than its authorship. On the positive side, however, none of these series had been extracted from bristlecone pines. That was the good news. Of the 17 series though, how many had a hockey stick shape? Was this shape coming from the majority of the data, or was it coming from just a few of the series, as was the case in Mann's papers? With so few series to examine, it was simple to check the graphs, and it was quickly apparent that there were very few hockey stick shaped series at all: in fact the relatively low temperatures in Jones' medieval sections could be ascribed to just a single series: the one called Polar Urals.

Briffa and Polar Urals

Polar Urals was one of only three series that covered the medieval period in Jones et al. Because of the length of the record it was a very popular series among paleoclimatologists, appearing in most of the reconstructions published up to that time. It had first come to prominence in a series of papers that Briffa had published earlier in the 1990s, in which he made the startling claim that 1032 was the *coldest* year of the millennium, at least in the area of Polar Urals. If true, this would have completely overturned climate history, implying that the Medieval Warm Period was non-existent or at best a local phenomenon.

In the published scientific literature, however, and unacknowledged in the Briffa papers, there was actually strong evidence that this claim was mistaken. A number of authors had noted telltale changes in certain environmental indicators, such as the regeneration of larch and a move of treelines to higher altitudes, which were strongly suggestive of Polar Urals experiencing a *warmer* climate in the eleventh century.[172,173]

With this contradictory evidence in mind, it was important to assess the impact of Polar Urals on the Jones reconstruction. McIntyre was able to show that by removing this one series from the proxy roster it was possible to make the medieval period appear warmer than the twentieth century. This was an important result. If these other indicators were suggesting that the Polar Urals region was actually *warm* in the eleventh century, did that mean that Briffa's Polar Urals proxy series was not actually representative of the region? It was going to be necessary to dig further into Briffa's work.

Like so many other studies of its kind, the early years of the Polar Urals chronology were distinguished by sparsity of data. In essence there just weren't enough trees on the site that were of suitable antiquity and in an adequate state of preservation to allow reliable cores to be taken. In fact, it turned out that the claim about the relative coldness in the year 1032 was based on cores taken from just *four* trees. This lack of data was troubling enough, but when McIntyre started to look at the measurement data – the core samples taken from these four trees – he was shocked by what he found.

The problems McIntyre discovered related to the quality control procedures used when processing the cores. Dendrochronologists usually have carefully defined criteria for assessing the reliability of their measurements. Briffa was no exception and in earlier studies he had calculated a measure which he called the 'subsample signal strength'. This figure was simply an assessment of how many cores had been extracted and also how well the graphs of the growth in individual cores matched up against each other. What was peculiar was that in the Polar Urals study, Briffa had been silent on what quality control measures he had used. Intrigued, McIntyre calculated the numbers himself and he was able to show that, in periods prior to 1100 at least, Polar Urals failed the subsample signal strength test – a finding that was hardly surprising in view of the small number of samples.

In fact, the quality control procedures adopted for the Polar Urals series seem to have been beset with difficulties. The tree cores used were in a very poor state, with some of them having gaps as long as 59 years and others having as many as seven gaps. In the 93 cores that made up the chronology, there were 41 gaps, suggesting a real problem with the quality of the data. Jones, who had worked with Briffa on the Polar Urals study, explained to McIntyre that the reason for the problem was that it had proved necessary to cut the cores into pieces to get them into the x-ray machine which measured the wood density. This had apparently caused some of the rings to be unmeasurable. This seemed implausible to McIntyre, who had noticed that there were some sections of the core which only consisted of three rings – far too small for Jones' explanation to hold. He therefore decided to get a second opinion from Douglas Larson, an experienced dendrochronologist and a colleague of McKitrick at the University of Guelph. Larson had taken an early interest in the work McKitrick was doing with McIntyre, and was therefore happy to help. He didn't mince his words:

> When one breaks a core, it fractures easily along a spring wood boundary because that wood is weaker than summer wood, with small cells. No wood actually falls away when a core is broken unless you use your teeth to break it. Or a hammer. If [Briffa and his team] have more than one

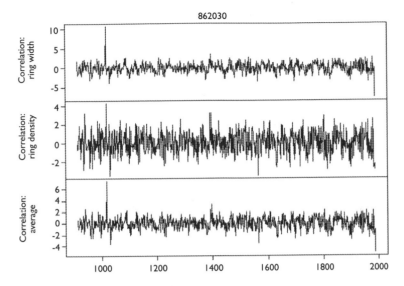

FIGURE 10.1: A well-dated tree

Top: Correlation based on ring widths. *Middle*: Correlation based on density.
Bottom: Average correlation.

missing ring at each end of a break, the series should not be
used at all. If there are 'lots' of breaks to allow for the
reorientation of the series in the radiograph, then that
means that they were sloppy when they took the core and
they were nowhere near the pith, so the core is a tangent
instead of a radius.[174]

Not only was there a problem with the quality of the data, but the
cross-dating appeared to be highly dubious too. Cross-dating is the
way dendrochronologists work out which year to assign to each ring
of the tree. The principle is relatively simple. Imagine a graph of
ring widths for a well-dated tree; say one that's still alive. You can
assign rings to years simply by counting back from the outside of the
tree towards the inside. Now say you also have a graph of ring
widths from a dead tree which you want to date. You can't count

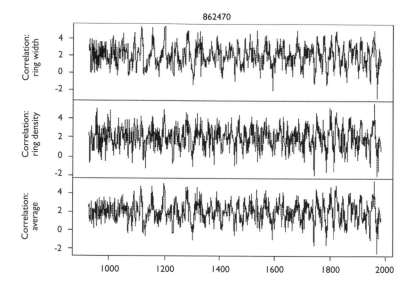

FIGURE 10.2: Tree with uncertain dating

back rings as you did for the live tree, because you don't know when it died. However, in order to get an accurate dating, all you have to do is to slide the graph of the dead tree rings against those from the live one, one ring at a time, and measure the correlation after each shift of one year. The idea is that when the rings match up – peak matching peak and trough matching trough – there will be a sudden spike in the measured correlation. When you shift one step further the spike will just as quickly disappear again. In this way it's possible to get very precise datings.

Figure 10.1 shows McIntyre's simulation of this effect. The three charts show three different ways of calculating the correlation between a tree with an unknown start year (No. 862030) and one where the start year is known, based on this approach of sliding the graphs along each other until the peaks and troughs match up. The spike at the year 1015 indicates that the correlation is best when this start date is assigned to the unknown tree. The fact that the spike in the correlation appears at the same date in all three charts gives the researcher confidence that year assigned is correct.

The problem with the cross-dating on the Polar Urals trees, or at least those four trees which supported Briffa's claim about how cold it was in 1032, was that the spikes, such as they were, did not provide the necessary certainty in the dating. One of the trees did have a clear correlation spike – in fact this was the tree we saw in Figure 10.1 – but the other three gave no clear indication of how the dates should be assigned. One of these, Tree 862470, is shown in Figure 10.2.

This, then, presented McIntyre with a mystery to solve. How had Briffa managed to assign a date to these trees without a correlation spike? There was no way of knowing for sure, but it did look very much as if the lowest density ring had simply been assigned to the required date of 1032 in order to back up the claim that this was the coldest year of the millennium.

Fortunately, an opportunity presented itself to show that these four trees were indeed misdated. In 2005, McIntyre discovered that an update to the Polar Urals chronology had been collected in 1999. By a stroke of good fortune he was able to obtain the details, and was gratified to see that the new figures showed an entirely different story to the old ones. Temperatures in the eleventh century now appeared to be *higher* than those in the modern era. This, together with the revelation of the poor cross-dating, appeared to be conclusive evidence that the cold eleventh century was an artefact of poor data quality rather than a genuine climatic effect. It was now likely that the Polar Urals series was entirely unreliable and this meant that doubt was cast on *all* of the multiproxy reconstructions of which it had formed a component.

This must have represented something of a problem for the Hockey Team, but in the end the solution was simple enough: the issue was bypassed by the simple expedient of not publishing the update. There was another potential problem though: any new studies by the same authors would now have to avoid using the update, making hockey sticks much harder to manufacture: there

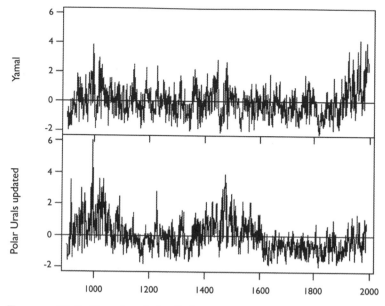

FIGURE 10.3: Yamal and the Polar Urals update

just weren't that many hockey stick shaped series to choose from. Fortunately for the Hockey Team, there turned out to be one that was suitable.

The Yamal substitution

With Polar Urals now unusable, there was a pressing need for a hockey stick shaped replacement. The solution came in the shape of a series from the nearby location of Yamal, which replaced Polar Urals as the representative of this region in Briffa's next paper[147] and indeed in pretty much every paleoclimate reconstruction thereafter.

Presumably not wanting to draw attention to the substitution, Briffa didn't discuss the use of Yamal rather than Polar Urals in the text of the paper, but the change was significant. The two series – Yamal and the Polar Urals update – are reproduced in Figure 10.3 and considering they are from sites just 100 km or so apart, the difference in their shapes is remarkable. Even a cursory examination of the two charts makes the reasons for the Yamal substitution clear. Yamal is pure hockey stick, with only a hint of a Medieval Warm Period, while the Polar Urals update is just the opposite, with strong growth in the eleventh century and no twentieth century growth spurt. Polar Urals rarely saw the light of day again in a paleoclimate study.

Briffa's Tornetrask series

Briffa was also responsible for one of the other series used in Jones et al 1998. This was a tree ring study based on samples taken at Tornetrask in northern Sweden. Like Polar Urals, the Tornetrask series was much used in paleoclimate temperature reconstructions, appearing in nearly every multiproxy temperature reconstruction at that time.*

Although Briffa had not archived any numbers, McIntyre was eventually able to get hold of the underlying data since a later researcher *had* done so. McIntyre found that he was able to emulate the chronology quite closely by processing the raw ring density measurement data through the various steps that are used to standardise tree rings (see Chapter 2).

Briffa had published his Tornetrask findings in a series of papers in the 1990s.[175–177] However, these papers turned out to have a serious problem: the tree ring density was falling in the twentieth century, suggestive of lower temperatures. The ring widths meanwhile were getting wider, suggesting higher temperatures. This would obviously have been a bit of a headache for Briffa and his approach to dealing with the issue was remarkable. Noting that prior to the twentieth century divergence the ring widths and ring densities had tracked each other fairly well, he simply asserted that

---

*   For example, Bradley and Jones 1993, Hughes and Diaz 1994, Jones et al 1998, MBH98, MBH99, Crowley and Lowery 2000, Briffa et al 2000, Bradley, Hughes and Diaz 2003, Mann and Jones 2003, Jones and Mann 2004.

in the absence of any better explanation it was reasonable to conclude that the divergence was not climate-related. Then, and without providing any further justification, he simply adjusted the ring density figures to bring them into line with the ring widths. Essentially, he bent the diverging line back up to where it 'should' have been. When this adjustment was carried through to the temperature reconstructions, it had the effect of lowering the temperatures in all periods prior to 1750, introducing an artificial warming trend into every one of the multiproxy studies in which it was later used. Briffa's only defence of his actions was to point out that it slightly improved his verification statistics.*

Like Polar Urals, the story of Tornetrask also features the impact of an update to the original study. While McIntyre had been aware of this update for several years, it wasn't until the middle of 2007 that he was finally able to get hold of the data. Prior to that, his every attempt to see the figures had been obstructed by Briffa. Eventually the details of the update emerged in the PhD thesis of Håkan Grudd, a scientist working at the University of Stockholm.[178]

Grudd had updated the proxy record for Tornetrask up to the year 2004 (it previously stopped in 1980) and he had found that with the updated data, the divergence of ring widths and densities disappeared. That was the good news. The bad news was that the chronology now showed, in Grudd's words, 'a long medieval warm period, centred on AD 1000' and also several other warm periods with temperatures in excess of those in the twentieth century. In fact Grudd's updated version looked almost identical to Briffa's version before the adjustments were made. So it was pretty clear that Briffa's original ad-hoc solution was not only unjustified, but unnecessary too.

---

*   Justifying adjustments by reference to verification statistics is considered incorrect methodology by statisticians because it essentially makes the verification period an extension of the calibration period. This, though, is a story that is beyond the scope of this book.

So much for Tornetrask. So much for Jones et al 1998. We will now move on and look at another important millennial temperature reconstruction in the shape of the work of Tom Crowley.

Crowley

Tom Crowley and his wife Gabriele Hegerl (who we met at the NAS panel hearings) were based at Duke University in North Carolina. Crowley was one of the very senior figures in paleoclimatology, with a list of publications that would pass muster among the most prolific scientists in the world. He was also an important member of the Hockey Team, having published papers with Mann, Bradley, Hughes, Ammann, Osborn and Briffa. His contribution to the spaghetti charts was encapsulated in two papers published in 2000. The first was Crowley and Lowery;[179] the second, published six months later, was known simply as Crowley 2000.[180]

The reconstructions differed somewhat between the two papers. Crowley and Lowery showed a long decline in temperatures from 1000 to the mid-nineteenth century followed by a twentieth century uptick, which had declined to sub-medieval levels by the end of the record. By the time of the second paper, however, the reconstruction was rather different, showing a long gentle decline from the year 1000 to the end of the nineteenth century and then a twentieth century uptick, which was remarkably similar to MBH98, and which was still heading upwards at the end of the record in the late 1980s.

As ever, McIntyre's first step was to try to get the data. From the first email, right back in December 2003, this quickly turned into a long and rather unpleasant saga: months of requests for data stonewalled and evaded until everyone was roundly sick of it. The length of the correspondence clearly annoyed everyone involved, with Crowley, as we saw in the last chapter, accusing McIntyre of using 'threatening language' soon after the publication of MM05(GRL).* At the same time, Crowley wrote an article in the

---

*   See page 222.

journal *Eos* in which he described what he saw as McIntyre's unacceptable way of dealing with him:

> I can attest that his initial message was of a somewhat peremptory character, requesting all my files, programs, and documentation, and that a quick followup by him had a more threatening tone, implying that the director of the US National Science Foundation (NSF) would be contacted if I did not comply.[181]

McIntyre responded to these accusations by posting up all of his correspondence with Crowley. The initial message with the 'somewhat peremptory character' was as follows:

> Dear Dr. Crowley,
> I am interested in examining the actual proxy data used in Crowley–Lowery 2000, which was referenced by IPCC. I have been unable to locate the data, as used, at the World Data Center for Paleoclimatology. Can you direct me to an FTP location where you have archived this data or otherwise make the data available. Thank you for your attention.
> Yours truly, Stephen McIntyre[182]

There was clearly no mention of code or documentation. There is also no mention of project funding in any of the correspondence, and in particular no mention of the NSF. In fact, it was Crowley himself who first raised the subject of funding. In his first reply to McIntyre, some six months after the initial data request and two months after McIntyre had made a formal complaint to *Ambio*, the journal in which Crowley and Lowery 2000 was published, Crowley explained that the project was not federally funded, the implication presumably being that he did not have to release his data. McIntyre responded to Crowley as follows:

> Both Lonnie Thompson and Phil Jones – the stated sources
> for your [article's] data – have been supported by US
> federal funding and obligations might well ensue from
> obtaining the data from these authors.[182]

It's hard to see this as anything other than a reasonable response. It is also striking that this email, which Crowley had implied followed straight on behind the initial request for data – 'a quick followup' in Crowley's own words – was actually sent more than seven months after the first one.

Even with a formal complaint to the journal in place, there were continuing delays, with Crowley first demanding McIntyre's own data and code (it was already public), then saying that he was in Europe and unable to post the data, then that he had been sick. It was October 2004 before some data was finally dispatched to McIntyre. Even then it was not actually what was requested. Instead of the original data series, Crowley sent a smoothed and transformed version. The original data, he explained, had been mislaid when he moved his place of work from Texas A&M to Duke. It was, as one Climate Audit reader memorably put it, 'the scientific equivalent of "the dog ate my homework"'.

Even without the original data sources, a certain amount of analysis of Crowley's results could still be performed. The proxies – just 15 of them – included, as expected, many of the usual suspects: bristlecones, Polar Urals and Tornetrask. There was also a Chinese ice core series called Dunde, prepared by Thompson (of 'Dr Thompson's Thermometer' fame – see page 259). A little further digging quickly revealed that these four components accounted for all of the hockey stick shape in the Crowley and Lowery 2000 reconstruction.

Apart from purporting to demonstrate that the modern warming was in excess of the Medieval Warm Period, the principal finding of Crowley and Lowery 2000 was that the timing of the Medieval Warm Period was inconsistent, appearing in different places at different times. As we saw in Chapter 1, this observation would imply that the different warmings were likely to have had different causes and that the Medieval Warm Period was therefore insignificant compared to twentieth century warming. As Crowley put it:

> None of the records between Germany and western China about 100 degrees of longitude contribute significantly to peak [Medieval Warm Period] warming from about 1070–1105.[179]

The evidence to support this claim was restricted to just four proxy series: Polar Urals, the Dunde ice cores, a study of snowfall dates in China by Zhu, and finally QiLian Shan, a Chinese tree ring series. We have already seen that Polar Urals was entirely unreliable, but there turned out to be huge question marks over the other series too.

The snow in China

The Zhu snowfall study was published in the early 1970s. The authors had examined the dates of the last snows each winter during part of the Song dynasty (960–1279), and had concluded that the temperatures around the year 1200 were rather low.[183] However, more than twenty years later Zhang De'er, a researcher from the Chinese Academy of Meteorological Sciences and a sometime co-author of Crowley, attempted to replicate Zhu's paper. To his surprise, Zhang discovered that it contained a major flaw: Zhu had made a mistake in converting the dates in the source records from the Song dynasty lunar calendar into a modern, solar calendar format. When this error was corrected, all of the dates of the final snowfalls each year shifted back towards the start of the year, implying that the climate was relatively warmer. This had the effect of reintroducing the Medieval Warm Period in the record. Zhang also confirmed this finding by means of a survey of taxation records from the same era. From these ancient documents he found that he could determine the distribution of citrus trees in Song dynasty China and he concluded that these had been growing at much higher latitudes than at present. The records suggested, Zhang wrote, that 'the annual mean temperature in the mid-thirteenth century was 0.9°C higher . . . than at present'.[184]

These findings raised some uncomfortable questions about Crowley's own research. How was it that he had ended up using the incorrect Zhu series from 1973 rather than Zhang's more recent corrected version? After all, he knew Zhang well – they had written a paper together. McIntyre decided to probe the issue and wrote to Crowley once again to ask about his reasoning. Unfortunately, in his reply Crowley failed to answer the question and no explanation has ever been forthcoming.

QiLian Shan

The QiLian Shan proxy series was also extremely suspect. As we saw in Chapter 2, one of the important criteria for using a tree in a temperature reconstruction is that its growth should be limited by temperature rather than any other factor. QiLian Shan is, however, located in semi-desert in Western China and a number of authors had reported that the growth of trees in the area was in fact limited by rainfall, as would be expected from a tree in this kind of terrain.

When McIntyre posted this finding on his website, he was strongly criticised by a number of paleoclimatologists. Professional scientists were generally given a pretty hard time by Climate Audit readers, so it was very difficult for them to interact meaningfully when anything they wrote would be pounced on by a mass of more or less irate sceptics. McIntyre therefore set up a post especially for these professionals, indicating that they could leave their comments there and that any responses by the readership would be removed. The results were quite revealing. One scientist, who didn't want to be identified, wrote this of the QiLian Shan series:*

---

*   The comment was posted to the Climate Audit comment thread by Rob Wilson, a paleoclimatologist now working at St Andrews University. He described the author as being 'a friend and colleague'.

> Those in the know, who really know the science, know not
> to use that chronology and know who still use that
> chronology. The work that uses that chronology for a
> temperature reconstruction is less-respected than others.
> Please, do not cast the whole field as deceitful or ignorant
> of this.[185]

This was a remarkable thing to say, given that Crowley and Lowery 2000 was widely cited in paleoclimate studies and at that time was being given a prominent role in IPCC reports, apparently without objections from the paleoclimate community. One can only wonder why this anonymous scientist did not make his feelings known to the IPCC during the review process.

Ice cores

Lonnie Thompson's work was different in many ways to the studies we have looked at so far. His temperature reconstructions were based on ice core records rather than tree rings, an area of study which put a whole new set of issues and uncertainties on the table. Recreating temperatures from ice cores is almost as fraught with difficulties as the tree ring studies. The principle is to take air trapped in the ice cores and to measure the amount of the ^{18}O isotope in it. This isotope *should* be a proxy for temperature. That, at least, was the theory. However, it was a theory whose basis in physics was rather suspect. For example, the relationship between ^{18}O and temperature from tropical ice cores like Dunde, which was taken from Chinese mountain glaciers, was the reverse of the relationship used in the polar ice core studies. In McIntyre's words, the basis for the relationship was entirely statistical – the idea that ^{18}O was a proxy for temperature appeared to have no grounding in physics.[186] This is the kind of thing that sounds alarm bells for statisticians on the lookout for spurious correlations.

Thompson was closely connected with the Hockey Team and he seemed to have taken a leaf out of the books of his colleagues: his data was almost impossible to obtain. McIntyre's long-running correspondence with Thompson and *Science*, the journal that had published some of Thompson's most important findings, had drawn as near to a blank as makes no difference. This was not an insignificant issue as there were several different versions of Thompson's data doing the rounds of the paleoclimate community, making replication extremely hard, if not outright impossible.

One area that could be probed, however, was that of verification statistics – just how well did Thompson's processed data match up against actual temperatures? Thompson had this to say about his reconstruction's statistical performance:

> For the period from 1895 to 1985, the correlation coefficient r* is 0.5 (significant at the 99.9% level). This correlation suggests that [changes in] the Dunde Ice Cap ^{18}O should serve as a good proxy for larger-scale temperature variations.[187]

McIntyre, however, was not going to take his word for it and set about a replication, his calculations coming up with a figure for r of 0.48, just a whisker away from Thompson's 0.5. *But*, and there is usually a but with Hockey Team studies, McIntyre didn't stop there:

> Since Thompson is on the Hockey Team, you have to ask yourself why he only did the correlation from 1895 on. Any bets on what the correlation was for 1851–1895? *Minus* 0.36.[186]

It is therefore safe to conclude that Dunde is not a reliable proxy for temperature: it failed its verification statistics.

Methods
With all of the proxy series supporting Crowley's claims about the Medieval Warm Period at best highly dubious and at worse

---

* See note on verification statistics on page 16.

completely refuted, it is clear that Crowley and Lowery 2000 cannot be much of a confirmation of the Hockey Stick. But data issues aside, there were further surprises from Crowley when it came to his methodological decisions. When regressing the proxy records against the temperature records to establish the mathematical relationship between the two, Crowley had only performed the calculation for 1861–80 and 1920–65; there was a 40-year gap in the middle. This, said Crowley, was because there was a breakdown in the relationship between the two in this period, caused perhaps by carbon dioxide fertilisation. In order to deal with the problem, Crowley had simply spliced in the instrumental temperature record. What is remarkable is that even with the splice in place, the reconstruction *still* failed its verification statistics. Crowley hadn't actually reported them, but in McIntyre's emulation, which appeared to closely match Crowley's results, the R^2 was 0.005 and the Durbin–Watson statistic* was 1.3, well short of the 1.5 required to give any confidence in the reconstruction.

In a follow-up paper, Crowley went even further and replaced all of the proxies from 1870 onwards with instrumental data.[180] The result was rather like Mann's original hockey stick, with a huge, and in Crowley's case, undifferentiated uptick in temperatures in the twentieth century. Crowley had his hockey stick, and from there on there was no doubt that it would be seen again and again and again – as indeed it was.

Esper 2002

Jan Esper, whom we have already met in Chapter 9, was a young Swiss paleoclimatologist employed by the Swiss Federal Research Institute. His most important contribution to paleoclimate was his 2002 paper in Science.[188] Although not usually considered to be a part of the Hockey Team, his co-authors, Cook and Schweingruber, were pretty much core members.

---

*   See page 244.

Although Esper and his colleagues had concluded that there had been a large-scale Medieval Warm Period, at least in the Northern Hemisphere, it is far from clear that their conclusions were any more robust than those of Mann or Jones or any of the others who had reached the opposite conclusion. Replicating Esper's paper proved to be fraught with difficulty for McIntyre. While Esper had given the names of the sites used, this was as far as he went in citing data. Esper, like so many paleoclimatologists didn't archive his data. Without either a copy of the data as used or an exact citation of a dataset in one of the tree ring databases, it was nearly impossible for an outsider to work out what had been done. McIntyre had therefore been trying to get hold of Esper's original data directly, but it had been a long hard struggle. His first email request was sent as far back as May 2004, and 18 months later he was still knocking at Esper's door. In the face of this intransigence, in September 2005, McIntyre felt he had reached the end of his tether and that he had no choice but to take the matter up with *Science*.

The response from *Science* was quick in coming but was rather odd, in that the editors told McIntyre that correspondence between them should be treated as confidential and could not be posted publicly. One wonders whether they felt a little embarrassment at their inability to enforce their own data archiving policies. However, they did attach 13 of the 14 chronologies used in Esper's paper, but for some reason the Mongolia series was not included. Esper had also omitted some methodological details which McIntyre had requested. Another letter went out to *Science*, itemising the missing information and at the same time, Benny Peiser, a British social scientist who published an influential email newsletter on climate change, wrote a public letter to *Science* in support of McIntyre's request and calling on his readers to petition the journal in support.

Six months later, the correspondence was still dragging on; the methodological details were still missing, the Mongolia chronology was still withheld, as was measurement data for four of the sites. These four included Mongolia and Polar Urals, although the latter was delivered shortly afterwards. The other two were a couple of foxtail sites, and McIntyre had heard on the grapevine that the original author, Lisa Graumlich, had lost the data in an office move – a situation that was eerily reminiscent of Tom Crowley's move from Texas A&M to Duke, and which only reinforced the wisdom of McIntyre's calls for the archiving of all data at the time of publication.

McIntyre was unfailingly polite, which suggests considerable powers of self control, and it is rather remarkable to observe how he worded his *39th* email to *Science*.

> Perhaps it's simply that [Esper] hasn't considered your requests important enough to respond to. If that is the case, perhaps you could write to him in firmer tones than you have done so far.[189]

Cherrypicking again

One of the methodological problems which was intriguing McIntyre was the possibility that Esper had cherrypicked his data. Esper referred in the paper to not using all of the data from certain series, but without explaining how he had decided which data to retain, or why certain data was deemed unsuitable. This was a crucial methodological decision and it was therefore necessary to understand it in order to replicate Esper's findings. In his response to McIntyre, Esper had referred to some remarks he had made in a later paper, saying that these explained the data removals.[190] However when McIntyre examined this purported explanation, what Esper said seemed actually to confirm his worst fears – that Esper was merely cherrypicking hockey stick shaped series:

> Before venturing into the subject of sample depth and chronology quality, we state from the beginning, 'more is always better'. However . . . this does not mean that one could not improve a chronology by reducing the number

of series used if the purpose of removing samples is to enhance a desired signal. The ability to pick and choose which samples to use is an advantage unique to dendroclimatology.[190]

. . . which is a statement to send a shudder down the back of any reputable scientist.

In the same paper, Esper had also shown that paleoclimatologists didn't only cherrypick those sites they felt best met their purposes, but also, when they collected the raw data in the field, they were cherrypicking the 'best' trees too.

It is important to know that at least in distinct periods subsets of trees deviate from common trends recorded in a particular site. Such biased series represent a characteristic feature in the process of chronology building. Leaving these trees in the pool of series to calculate a mean site curve would result in a biased chronology as well. However if the variance between the majorities of trees in a site is common, the biased individual series can be excluded from the further investigation steps. This is generally done even if the reasons for uncommon growth reactions are unknown.[190]

Esper argued that he had taken these steps to avoid getting a biased chronology. To some readers, however, they might sound much more like a way of *obtaining one*. After all, the object of the exercise was to discover what signal was in the tree rings, not to choose a subsection of the rings that gave a 'desired signal'.

Shortly after this correspondence, *Science* stopped responding to McIntyre's emails, but, undeterred, he set about doing what analysis he could. He had 13 of the 14 chronologies, and he could at least see what they looked like.

Of the 13, two were bristlecones, which obviously over-represented this type of tree in the reconstruction. Of course, they both had markedly hockey stick shaped curves. There were also all sorts of oddities among the other series. For example, there were a couple that were extremely short, only lasting a few hundred years. Even more surprising was the inclusion of the Polar Urals update, the first sighting of this record in a major climate reconstruction. How Esper had arrived at his chosen set of proxies is not clear, but the impact of using the Polar Urals, rather than substituting Yamal as so many of his paleoclimate peers had done, would have been to leave the reconstruction with a pronounced Medieval Warm Period. It is possible that he had then had to introduce some bristlecone series into the datebase in order to drag medieval temperatures back down again.

Some months later, McIntyre was able to attempt a replication of Esper's paper, when he got hold of a version of the Mongolia data that had been published as part of an entirely different study. Esper had still not yielded up any new information on his methodology, so McIntyre decided to take the simple approach by averaging the data series and rescaling them to match Esper's original result.*

The results were intriguing. For most of the series, Esper's reconstruction and McIntyre's emulation of it matched fairly well, until it came to the twentieth century. At this point, McIntyre's reconstructed temperatures headed downwards, while Esper's kept on rising. This was presumably due to some methodological decision Esper had made, but if this was the case, then it strongly suggested that his results were not robust. The conclusions could not reasonably be based on an undisclosed methodological decision that made the results so different to the series mean.

Rutherford et al

We have touched upon Scott Rutherford's paper earlier in the book, insofar as it was used to try to rebut McIntyre's own papers.**[93]

---

*　Rescaling means adjusting the mean and standard deviation of one series to make them the same as another.

**　See page 179.

We noted also that its authors were all core Hockey Team members – apart from Rutherford, there was Mann, Bradley, Hughes, Briffa, Osborn and Jones, so this could not be seen as an independent confirmation of the Hockey Stick. The arguments in Rutherford et al were the basis of Mann's ripostes to McIntyre's *Nature* submission, and some of what follows will therefore be familiar from earlier in the story.

Rutherford's paper attempted to demonstrate the reliability of the Hockey Stick by means of two new reconstructions. The first of these took a roster of proxies and used a completely different methodology to Mann, called RegEM. As we have seen, Mann had argued that the similarity of Rutherford's results to his own demonstrated that the Hockey Stick was correct. The problem with this approach was that McIntyre was able to show that the proxy roster used by Rutherford was identical to the one used in MBH98, including Mann's faulty PC1. Merely processing the same biased dataset through a different algorithm hardly demonstrated Mann's point. The second reconstruction took a different approach, eliminating the PC analysis step from the reconstruction, with the proxy series going straight into the RegEM calculation. You may remember that the rationale for using PC analysis was to stop certain types of proxies being overrepresented in the reconstruction. Eliminating PC analysis therefore left the proxy database with a huge preponderance of North American bristlecone pines, so again, Rutherford was not demonstrating anything significant, other than that there were many different ways in which a biased reconstruction could be created.

Rutherford's paper had originally been submitted to the *Journal of Climate* in 2003, but there was an unusually long delay until it finally appeared in 2005. While there was no firm evidence, it is likely that this delay was caused by the publication of McIntyre's 2003 paper in *Energy and Environment* with its description of the litany of errors in Mann's dataset. This impacted directly upon Rutherford's paper, which used the MBH98 PC1 unchanged – short centring and all. In fact, as McIntyre looked into Rutherford's work, he was able to explain a great deal of the mystery that had surrounded the pcproxy.txt file of proxy data that he had originally received from Rutherford back in 2003 when he started his investigations into MBH98.* It looked very much as if the version of pcproxy.txt that Rutherford had sent him had been originally prepared for Rutherford's own paper. In preparing these figures, he seemed to have introduced errors into the database – the same errors that had alerted McIntyre to the possibility that there were serious problems with MBH98. Amusingly, McIntyre discovered similar mistakes in Rutherford's collation of *instrumental* temperature records for the same paper.

So, when McIntyre's *Energy and Environment* paper had hit the presses, it looked as if Rutherford been forced to clean up the database and rewrite his paper, hence the delays in getting it published. McIntyre had since come to the conclusion that the version of the data he was originally sent was actually not the one Mann used in MBH98, although it was hard to be certain because of the number of different versions of the dataset that had been issued by the Hockey Team.

While the datasets in Rutherford et al were therefore just as flawed as MBH98, Rutherford had managed to avoid any discussion of the criticisms McIntyre had made in MM03. In normal circumstances it would have been the job of the journal editor and the peer reviewers to require the authors to address these issues, but unfortunately the editor of *Journal of Climate*, Andrew Weaver, was a fierce opponent of McIntyre and McKitrick. He had gone on the record as saying that McIntyre and McKitrick's first paper should never have been published and stating that it was 'dangerous' to give equal space to both sides in a scientific dispute.[191] It was therefore unlikely that he would to be responsive. Nevertheless, the attempt had to be made, and McIntyre wrote an email to the journal pointing out the failure to address the findings set out in MM03, and going on to suggest to

---

*   See page 74.

Weaver that he was probably going to have to ask Rutherford and his colleagues to certify that they had made 'full, true and plain disclosure'.

A little later, a reply came through from Weaver, indicating that he *had* received an assurance to this effect from Rutherford. However, Weaver had also said that the paper would go ahead in its current form, without discussion of the McIntyre papers. Whether failing to address McIntyre's criticisms of the dataset constituted 'full, true and plain' disclosure is, of course, debatable.

There was another disclosure issue with Rutherford's paper too. One of the problems with creating a reconstruction that closely tracked the results of MBH98 was that Rutherford's reconstruction was necessarily going to fail its R^2, just as Mann's had done. Like Mann, Rutherford had chosen to discuss only the RE statistic. However, Rutherford did make a better fist of trying to explain why, going to the slightly absurd lengths of presenting some theoretical cases where the R^2 would have given the wrong answer. While it was true that in the scenarios Rutherford discussed R^2 would not have given a reliable answer, these kinds of situation were well known, and statistical measures had been developed to deal with them. That is why the statistical authorities all recommend the use of a suite of verification statistics.

There were other issues too: splices of instrumental data, truncations, adhoc changes to standard methodologies. When all of these issues are put together it becomes clear that Rutherford et al was neither an independent confirmation of the Hockey Stick nor a study on which great reliance could be placed.

Moberg

Another paper that made a big impact on publication came from a group of researchers led by Anders Moberg of the University of Stockholm. The paper, Moberg et al 2005, was an exciting development for McIntyre because Moberg's group was genuinely independent of the Hockey Team[192] – a first for multiproxy studies – although it should be said that Moberg had worked with Hughes in the past. The paper was announced to the world by *Nature*, who, for their own reasons, illustrated it with a picture of the Hockey Stick rather than Moberg's own work. The reasons for this error became clear when the actual paper was examined. Moberg's reconstruction found a clear Medieval Warm Period and Little Ice Age. However the reconstruction still suggested that current temperatures were unprecedented, a position that was achieved by truncating the proxy records, which often showed declining temperatures, and, just as Mann had done with the original Hockey Stick, by overlaying the end of the reconstruction with the upward-trending instrumental record.*

Data

Moberg's approach to temperature reconstruction was rather different to previous researchers. Noting one of the problems with tree ring reconstructions, namely that they were thought to miss longer term trends in temperature, Moberg decided to use mainly non-tree ring records such as lake sediments and speleothems instead. While these 'low-resolution' proxies could not distinguish between the temperatures of individual years in the way that tree rings could, they did have the advantage of extending much further back into the past, and the Moberg reconstruction was to be a 2000-year history. It was hoped that 'low resolution' proxies would pick up longer-term climatic changes that were being missed by the tree ring proxies. However, his reconstruction would also use tree ring data in the modern period, combining the two datasets to give the best of both worlds.

The limited involvement of the Hockey Team was immediately apparent in the paper, in that there were clear data citations given for most (but not all) of the proxy series. That is to say that Moberg had provided hypertext links to the actual version of each series used. This was a huge step forward for temperature reconstructions, but nevertheless was still less than adequate

---

*   See page 33.

because of the exceptions: of the eleven low resolution series, two were not archived, and in fact were 'grey'* versions.

Because he was trying to create such a long reconstruction, the problem of a lack of suitable data was just as pertinent for Moberg and his team as it had been for earlier researchers. There were a few surprises though. Firstly Moberg didn't directly use Thompson's ice core records, whose problems we have already discussed, although he did use a series known as the Yang composite which included some Thompson data. The other low frequency proxies included series based on speleothems, ice melt records and forminafera, which are fossilised shells of plankton. By now you will probably be able to have a reasonable guess at the tree ring proxies used. There were no fewer than three bristlecone series, including one which was not even a current version, and of the other tree ring series, our old friends Yamal and Tornetrask were there as well as a couple of unfamiliar Siberian series, Taimyr and Indigirka.

Series 8, an analysis of the ^{18}O isotope of oxygen in a Norwegian stalagmite proved to be a bugbear for McIntyre. When he wrote to Moberg to ask about the data series, which had been derived from grey data, he was told that it had been obtained from the original author, Lauritzen. However, Moberg's version of the data ended in 1938, whereas Lauritzen's original article only extended to the end of the nineteenth century. McIntyre's analysis of the two articles suggested that Moberg may have made an error somewhere in his data processing and applied the wrong values to the wrong dates by some 80-odd years. In answer to his enquiry, Moberg suggested that McIntyre approach Lauritzen for the original data. Lauritzen, however, refused to comply, saying that the figures were unpublished. Given that they appeared to have been used in Lauritzen's paper as well as Moberg's this seemed implausible.

Indigirka soon turned into another paleoclimate farce too, with Moberg explaining that he'd got the data from someone else,

---

*   See page 269.

and they'd had it from someone else again. Perhaps, he suggested to McIntyre, the original Russian authors might supply the figures. When the request was passed on to the Russians, the reply soon came back that 'the series developers do not want to disseminate it. They say this series will be re-calculated soon to reject some errors in it'. Another dead-end. The irritation started to show in McIntyre's postings:

> If the series developers did not want to disseminate it, you'd think that allowing it to be used in a multiproxy study in *Nature* is pretty strange way of not disseminating it. Secondly, if the series developers now change the series, what good is the new version in understanding Moberg et al 2005? So if I want to actually look at the data, I now need to get into the same old war: write a materials complaint to *Nature* and fight with [them] for 12 months. And I'm sure that this stuff is ear-marked for [the IPCC's Fourth Assessment Report]. What a goofy way of running a scientific community. Then people get mad at me for being hard to get along with.

There was nothing for it but to start the whole tedious process of a materials complaint. Unlike the earlier complaint about Mann's data and code, and contrary to McIntyre's expectations, this time there was actually a relatively speedy outcome. It turned out that Moberg had been using Lauritzen's data without permission, and *Nature* now required him to produce a corrigendum and provide the data.

Glob. bulloides

As soon as Moberg et al was published, McIntyre had set to work, as he always did, to find which series were driving the shape of the reconstruction. Experience had shown him that in most temperature reconstructions, the 'hockey stick-ness' was driven by just a few of the proxies, with the others all representing noise and cancelling each other out in the final reckoning. Moberg's reconstruction was slightly different in that most of the series were actually tending to produce a pronounced Medieval Warm Period, with just a few others pulling its peak down to the level of the modern warming.

The Moberg proxy series had been calibrated against local instrumental records, in order to check that the proxy was in fact responding to temperature. . . except for two of them. One of these, the Arabian Sea, *Glob. bulloides** series, turned out to be the only proxy that had modern values higher than those of the medieval warming. In other words it was one of the ones that was suggesting that modern temperatures were unprecedented. However, when McIntyre started to examine the scientific literature on *Glob. bulloides*, its inclusion in the reconstruction started to look very strange indeed. *Glob. bulloides* turned out to be a subpolar species of forminafera, which proliferated in the Arabian Sea only when there was an upwelling of cold water from the bottom of the ocean. So at first sight, proliferation of *Glob. bulloides* would suggest *lower* temperatures and not higher ones, as was implied in Moberg's paper. Moberg had explained that he had included the *Glob. bulloides* series in order to give a better geographical spread to the proxy data, and that the series was only indirectly related to temperature. What he seemed to be saying was that the cold water upwelling was evidence of warming elsewhere. But for increased prevalence of a subpolar species to be evidence of higher temperatures was rather bizarre even for the eccentric world of paleoclimate. And as Soon and Baliunas knew to their cost,** the paleoclimate community in general and Mann in particular were highly critical of scientists who did not demonstrate a direct relationship between each proxy and temperature. Fortunately for Moberg, there was no similar outcry over his use of *Glob. bulloides*. Why should Mann complain so

---

*   *Globivalvulina bulloides*.
**   See page 56.

vociferously about Soon and Baliunas but remain silent about Moberg?

Apart from *Glob bulloides*, the rest of the twentieth century warming pattern in Moberg's reconstruction seemed to be driven by just two other series: Yang's composite, which, as we have seen, owed its shape to Thompson's undisclosed ice core data, and the Agassiz ice melt record, this latter being the other series which hadn't been calibrated against temperature.

Although McIntyre was able to understand the main factors behind the shape of the reconstruction, without the correct data it was not going to be possible to emulate it properly and the process was slowed up considerably because, as always, the source code was unavailable. However, with the information he had, it appeared that Moberg's statistical handling of the reconstruction had also been inadequate, failing many of the key tests and lacking convincing confidence interval calculations.

Osborn and Briffa 2006

After the publication of McIntyre's papers in *Energy and Environment* and *Geophysical Research Letters*, the Hockey Team responded with a new series of papers to back their argument that even if McIntyre was right about short-centred PCs (and of course they weren't accepting that he was) the independent confirmations still suggested that Mann's conclusions were right. Osborn and Briffa's 2006 paper published in *Science* was one of the most prominent of these, a contribution to the debate by the European wing of the Hockey Team.

When the paper was published towards the start of 2006 it immediately attracted significant media attention, with a BBC article entitled 'Climate warmest for millennium' trumpeting its findings.[193] McIntyre, meanwhile, published his own review on Climate Audit. He was less than impressed with what he found.

Proxies
Of the 14 proxies used in Osborn and Briffa, two were cores from

bristlecones or foxtails. Among the other members of the roster were Yamal and Tornetrask, which we have just seen are also unreliable, together with some of Thompson's unarchived ice core studies. But even the inadequacies of Osborn and Briffa's proxy roster was as nothing compared to the next surprise. McIntyre discovered that one of their proxies was a PC series, and it used Mann's discredited short centring algorithm.

Still picking cherries

So, apart from poor quality proxies and biased PC analysis, what were the main problems with Osborn and Briffa? The question that hangs over the paper, like so much of the rest of the paleoclimate field is that of cherrypicking. While they had eventually used only 14 proxies, they had actually started out with a much larger database, whittling this down to the final 14 by eliminating any that were not genuinely responding to temperature. To do this, they had matched up the tree ring widths to the instrumental temperatures in the local area (the gridcell), and had then withheld from the calibration all those series where there was not a reasonable correlation between the two. The problem was that while they said that they had applied this test to some of the proxies, there was no mention of whether they had applied it to the PC series. They merely asserted that there was a correlation. McIntyre ran a check on the data to make sure, and found that of the six series in the PC calculation only one had a positive correlation with temperature.

McIntyre wasn't the only one who noticed problems with the Osborn and Briffa paper. Gerd Bürger (see below), of the Free University of Berlin, submitted a comment to *Science*, picking up on the cherrypicking issue and expanding on it.[194] Bürger explained that since Osborn and Briffa had screened a large number of proxies for those with a positive correlation to temperature, they risked ending up with some whose correlation was only a matter of chance – in other words, just because the ring widths went up in line with twentieth century temperatures, it didn't mean that they always tracked each other. It was quite possible that the hockey stick shape Osborn and Briffa achieved had only come about because they had effectively selected only series with a twentieth century uptick. To explain this a little, if you have several series of random data and you take an average, they will all cancel out to nothing – the peaks in one series will tend to offset the troughs in another. If you take enough random series the average will be a straight line centred on zero. On the other hand, if you take a set of random series and select only those with twentieth century upticks, you will get a hockey stick. This is because in the twentieth century, the upticks are all in synch and add together, giving a strong uptick in the average. The earlier periods are all still random numbers, so they continue to cancel out, giving the long flat shaft. This effect was just as pertinent to Mann's papers as to Osborn and Briffa.

One of McIntyre's readers, statistician David Stockwell, had published a short article demonstrating the effect.[195] Stockwell simply took several red noise series, calibrated them against the actual temperature records, and then averaged the result. The answer: a hockey stick. As Stockwell concluded:

> All the salient aspects of past climate usually associated with millennial reconstructions are essentially already encoded into the methodology, so that a 'hockey-stick' shape is inevitable on any data resembling natural . . . series.[195]

Stockwell's observations were quite distinct from McIntyre's earlier experiments on getting hockey sticks from red noise.* McIntyre had demonstrated that the PC algorithm had a tendency to produce hockey sticks from red noise. What Stockwell was demonstrating on the other hand was that the calibration process could produce hockey sticks from red noise in its own right. So now it was clear that not one but two separate steps of Mann's paper had an inbuilt tendency to produce hockey sticks from nothing.

---

*   See page 114.

Bürger was quite explicit about what this meant for Osborn and Briffa's results:

> [T]he reported anomalous warmth of the 20th century is at least partly based on a circularity of the method, and similar results could be obtained for any proxies, even random-based proxies.[194]

Bürger emphasised that it was necessary to adopt the most stringent statistical testing to deal with this issue. He was able to show that when the effect of the proxy selection procedure was factored in to the significance levels for the reconstruction, Osborn and Briffa's hockey stick suddenly lost its statistical significance. Osborn and Briffa seemed to recognise the validity of this argument, saying 'we agree with Bürger that the selection process should be simulated as part of the significance testing process in this and related work'.[196] However, they tried to minimise the impact of Bürger's findings by claiming that this screening had only a small effect on their results since they had eventually used 'almost all' of a small number of proxy series available. There were, they claimed, only 16 suitable ones available and they claimed that this meant that the significance levels they had demonstrated were highly unlikely to have been achieved by chance. McIntyre found this statement extremely surprising, since he was personally aware of a multitude of other suitable series that they could have used, but hadn't. Osborn and Briffa had even listed Rutherford et al 2005 as one of the papers whose proxies were covered by the statement of 'almost all', but this latter paper included all of Briffa's network of 300 or more tree ring series. Why didn't they use all those?

McIntyre charitably wondered if perhaps Osborn and Briffa actually meant 'proxy series that went back to the year 1000'? But even then, there were still far more than 14 or 16 series – at least double, he thought. And it wasn't just the tree ring records. What about the ice cores? Why didn't they use Mount Logan? Perhaps because it had a divergence problem – twentieth century values were falling rather than rising. Or what about Rein's offshore Peru (strong medieval warmth) or Mangini's speleothems (medieval warmth again)?

McIntyre summarised his thought to his readers:

> [W]hen Osborn and Briffa say that the universe from which they've selected can be represented by selecting 14 of 16 [proxy series], this is completely absurd. There has probably been cherrypicking from at least 3 times that population. But aside from all that, the active ingredients in the 20th century anomaly remain the same old whores: bristlecones, foxtails, Yamal. They keep trotting them out in new costumes, but really it's time to get them off the street.[197]

. . . and all the others

We have looked at several of the independent confirmations of the Hockey Stick in detail and have seen that they leave much to be desired. Little is added by making the same criticisms against the others. However, for the record, these studies (and the issues affecting them) are Hegerl et al (cherrypicking, data not archived), D'Arrigo et al (divergence problems, data not archived) and Mann and Jones 2003 (bristlecones, short-centred PCs).

It's also worth summarising what we have learned about the commonality of proxies in the various studies used by the IPCC (see Table 10.1). Is it possible to maintain that these studies are in any way independent of the Hockey Stick when the same proxies, most of which are known to be flawed, turn up again and again and again?

Before we move on to the next chapter, there are a couple of other studies that are worth looking at briefly because they come from researchers who were genuinely independent of the Hockey Team. Their findings are important because they throw some more light on the lack of robustness of the IPCC's temperature reconstructions.

TABLE 10.1: Commonality of proxies in temperature reconstructions

Proxy series	MBH98 & MBH99	Rutherford	Jones 98	Crowley 00	Briffa 00	Esper 02	Mann, Jones 03	Moberg	Osborn, Briffa	D'Arrigo
Polar Urals	x	x	x	x	x	x	x	x	x	x
Tornetrask	x	x	x	x	x	x	x	x	x	x
Jacoby Mongolia	x	x			x	x	x	x	x	x
Jacoby treeline	x	x	x	x	x					x
Bristlecones	x	x		x		x	x	x	x	
Dunde/Yang	x	x		x		x	x	x	x	
Greenland δ^{18}O	x	x	x	x			x		x	x
Jasper			x	x	x	x			x	x
Taimyr					x	x	x	x	x	x
European docs	x	x	x						x	
CETR	x	x	x	x						

Adapted from the Wegman report.[15]

Loehle

Just because paleoclimate researchers select proxies from a short menu of hockey stick shaped tree ring series, doesn't mean that this is the only way to do a temperature reconstruction. We have seen how tree ring studies are fraught with problems – the apparent inability to pick up longer term trends, the apparently inverted U-shape response to temperature, the problems with calibration and carbon dioxide fertilisation, and so on. In the light of all these issues, one of McIntyre's readers, an ecologist called Craig Loehle, decided to see what would happen if you didn't use tree rings at all. His study was not so much an attempt to create a new temperature reconstruction as a somewhat simplistic attempt to see what the patterns in non-tree ring data looked like. His idea was that by keeping everything very simple, much rancorous debate would be avoided. On this score at least, he was to be sorely disappointed.

We have seen how paleoclimate studies were criticised, by sceptics at least, for failing to describe their selection criteria for proxies, the suspicion being that this permitted cherrypicking of hockey stick shaped series. Loehle was keen to avoid criticism on this score and therefore adopted the innovative, for paleoclimate research, approach of laying out his selection criteria in black and white: he chose all available non-tree ring series that were at least 2000 years long, had been calibrated to temperature and had at least one measurement per century over the length of the record. This gave him only 18 series, although as we have seen this is still a larger number than in some paleoclimate reconstructions. Demanding that the series be calibrated to temperature would deal, he hoped, with the criticisms that had been directed at Soon and Baliunas, and which should also have been directed at Mann.

Loehle's methods were rather simple when compared to the intricacies of Mann's papers – smooth the series to remove noise, subtract the mean to put them all on the same baseline and take the average.

Although his index ended in 1980, missing much of the modern warming, his results showed that non-tree ring proxies suggested a Medieval Warm Period that had been a real, global phenomenon, with temperatures as much as 0.3°C above those prevalent in the modern period. He was also able to demonstrate a certain amount of rigour in the results he had calculated by showing that the story was largely the same if you removed a series from the database or if you took a random subsample of the proxy database. While Loehle was careful not to read too much into his results, his study did seem to suggest that the hockey stick shape of the other paleoclimate studies might be, in part, an artefact of the use of tree rings rather than a genuine reflection of climatic history.

The paper was initially submitted to *Geophysical Research Letters*, where it was rejected out of hand on the grounds that the journal was no longer interested in publishing temperature reconstructions. This was slightly surprising given the political and scientific importance of the subject. However, taking the hint, Loehle next sent the manuscript to *Energy and Environment*, where the reception was much more favourable.

The publication of Loehle's paper was met with a storm of criticism and not only from scientists working in the paleoclimate mainstream.[198] Despite the fact that Loehle's conclusions reinforced their preconceptions, many Climate Audit readers also joined in with the pulling apart of the paper, happy to point out any flaws they could find. Both sides pointed out that Loehle's data was not available at the point of publication, an oversight which was careless, given the criticisms that had been regularly launched at the paleoclimate mainstream on this score. While this was remedied in fairly short order, a moral victory was probably given away unnecessarily on this point.

The more substantive criticisms of the paper related to the absence of any estimates of errors. Loehle said that he couldn't see how these could be calculated, but the Climate Audit readership, which included some fairly high-powered academics, was able to demonstrate that it was in fact possible, if less than obvious.

There were other errors too. For example, it was pointed out that some of the data series presented figures on a regular basis while others had irregular gaps, and it was suggested that this could lead to a distortion of the results. A dating error was also revealed.

In the face of all of these issues, Loehle issued a correction.[199] His main findings still stood, but he now had a much firmer basis on which to stand his conclusions. While some commenters had been quite dismissive of his work, in many ways, the story of his paper is a model of how science should work in the twenty-first century. The review was open and vigorous; even exciting. The data was available for anyone interested to look at (shortly after publication at least). The author responded publicly to questions and criticisms. The paper was not perfect, but what was extraordinary about the affair was the speed with which the community, both professional scientists and interested amateurs, were able to identify the errors, determine how to resolve the difficult issues, and force a correction. The original paper was published in November 2007, the corrigendum in January 2008.* It is remarkable to compare this rapid turnaround to the other paleoclimate studies, where on occasion there would be a delay of years even before the study data was made available.

Bürger and Cubasch

As Hans von Storch once pointed out, McIntyre and McKitrick's papers on the Hockey Stick made one very important change to the culture of climatology. Before MM03, criticism of Mann's work was not possible. Anyone who stuck their heads above the parapet would have been shot down by a barrage of criticism, as Soon and Baliunas had discovered to their cost. In the years after the appearance of MM03 and particularly MM05(GRL), it became considerably easier to publish studies critical of MBH98, although it is fair to say that it was sometimes still a struggle.

In Chapter 8, we looked at von Storch and Zorita, one of the papers critical of Mann's work. Another set of authors adopted a rather different approach in a series of papers published in the years following MM05(GRL). Gerd Bürger, whom we have just met as the author of a comment on Osborn and Briffa, together with a colleague called Ulrich Cubasch, fired their first shots at MBH98 in a 2005 paper called 'Are multiproxy climate reconstructions robust?'[200] The title was somewhat misleading, because it focused very much on MBH98, and in particular the effect of key methodological decisions on the shape of the reconstruction. Bürger and Cubasch identified six decisions that they felt were key

---

* Readers should note that Loehle's paper was published too late to be considered in the IPCC report, the story of which is told in the next chapter. I have included it here for the sake of tidiness of the narrative.

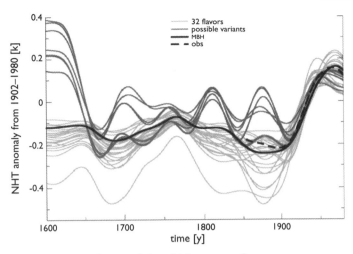

FIGURE 10.4: Some of the 64 flavours of temperature reconstruction

The figure shows just 32 of the different possible temperature reconstructions. Reproduced from Bürger and Cubasch.[200]

to Mann's methodology. The different combinations of these decisions meant that in the final reckoning there were 64 different ways in which the reconstruction could have been put together, none of which could be intrinsically preferred over the others. In fact McIntyre was able to point out several other important choices that Bürger and Cubasch had overlooked, so there were actually more than 64 possible combinations, but the point remained the same: if you couldn't choose one combination over the others, how could you possibly decide which of the mish-mash of different reconstructions was the correct one (see Figure 10.4).*

Bürger and Cubasch made other criticisms as well. For example, they pointed out that it was not legitimate to extrapolate

---

* Remarkably, during the Barton hearings, Gerald North said that he thought all of these 64 reconstructions looked like hockey sticks, although he added that they were 'a bit curved'.[163]

temperatures in the reconstruction beyond the range of the calibration. In other words, you couldn't reconstruct temperatures that were outside the range of the temperatures during calibration. You just didn't know if your model still worked outside that range. This proviso was, as Bürger and Cubasch put it, the basic condition of statistical regression. Unfortunately it was *exactly* what Mann had done, with some proxies being used far outside their calibration range.

There were by now a lot of nails in the coffin of Mann's reconstruction. Would they be enough to bury it once and for all? The IPCC was going to be the final arbiter, and the answer would not be long in coming.

11 The Hockey Stick and the IPCC

IPCC reports are being produced in a very open process under the discipline of science, where honesty and balance are hallmarks of that discipline.

Sir John Houghton

The point is that every single man who was there knows that the story is nonsense, and yet it has never been contradicted. It will never be overtaken now. It is a completely untrue story grown to legend while the men who knew it to be untrue looked on and said nothing.

Josephine Tey
The Daughter of Time

Introduction

In August 2005, McIntyre received an email from the IPCC informing him that he had been nominated to be an expert reviewer of its Fourth Assessment Report (4AR), which was due to be published at the start of 2007. While in theory this was a valuable opportunity to influence the content of the report, McIntyre had no illusions that the IPCC was going to take his criticisms on board wholesale and report that Mann's Hockey Stick was indeed flawed. However, much would be revealed about the attitudes of the IPCC by the way that they reacted to dissenting views, and when the truth finally came out, politicians and IPCC bureaucrats would be hard pushed to carry off a claim that nobody knew about the flaws in Mann's work. Their own policies required IPCC lead authors to collate all of the different opinions on any particular issue and to report any which were technically valid, even where these opinions conflicted with or could not be reconciled to the positions held by the majority. In theory, at least, they couldn't avoid reporting on McIntyre's findings on the Hockey Stick.

IPCC *protocols*

The IPCC review process was promoted as a model of its kind, but in reality it had serious flaws. In fact, so serious were these shortcomings that they would not have passed muster in any other professional setting. Not least of these was the issue of conflict of interest. We saw in Chapter 1 how, for the IPCC's Third Assessment Report, Mann had been appointed the lead author on the paleoclimate chapter.* From this powerful position, he had been responsible for reviewing his own work, forming an opinion on the rest of the scientific corpus and writing the final text. What is more, as a result of his dominant position within the IPCC report-writing process, he reached a position of leadership in the paleoclimate world, a position that was only enhanced by his status as the author of the Hockey Stick studies, which he had himself highlighted in the IPCC Third Assessment Report. As McIntyre explained to the Climate Audit readership, this situation would have been entirely unacceptable in a commercial situation, and in fact would have been entirely *illegal* outside of a banana republic.

> For someone used to processes where prospectuses require qualifying reports from independent geologists, the lack of independence [in the IPCC report-writing process] is simply breathtaking and a recipe for problems, regardless of the reasons initially prompting this strange arrangement.

> Businesses developed checks and balances because other peoples' money was involved, not because businessmen are more virtuous than academics. Back when paleoclimate

---

* See page 39.

research had little implication outside academic seminar rooms, the lack of any adequate control procedures probably didn't matter much. However, now that huge public policy decisions are based, at least in part, on such studies, sophisticated procedural controls need to be developed and imposed.[201]

This conflict of interest had not gone unnoticed, and several commentators in the science policy community had picked up on the subject after Barton had raised it in his letters to the Hockey Team (see Chapter 9).[202,203] One comment had come from von Storch, who sagely observed that to have scientists who already dominated a debate also authoring the key review of that debate was a sure road to trouble; the situation demanded the involvement of scientists who really were independent.[204] However, when the authors for 4AR were announced, it soon became clear that none of these objections had been taken on board: Mann had been replaced as lead author by none other than his British teammate, Briffa. The IPCC had no intention of allowing an independent review of the field.

Nature of the review

To some extent, IPCC reports are much like a listing prospectus for a large company. Appearing every five years or so, they set out our complete understanding of climatology: what is known, what is still not understood and what uncertainties there are. The reports cover paleoclimate studies which look back in time, instrumental records describing the present and climate models which peer into the future. They should be a warts-and-all summary of the state of the science, enabling politicians and the public to reach policy decisions on a rational basis.

With the importance of the subject in mind, readers might therefore expect the IPCC review process to involve a very detailed examination of the scientific papers that inform our understanding of the climate. However, IPCC due diligence turns out to be entirely different to the kind of work that an auditor working on a listing prospectus would peform. Business auditors examine all of the major assumptions made in preparing accounts and prospectuses, often tracing key values back to source documents and reperforming key calculations and estimates. All information is available, everything is open to question. IPCC reviews, on the other hand, consist merely of the assembling of a panel of experts who survey the climatological literature and form an opinion of where the truth lies. Like the NAS panel, they are content to 'wing it'. And, as we will see, the IPCC also takes a different approach to data availability.

In its defence, the IPCC review process is at least rather more open than that of the NAS, in that anyone can contribute comments on the draft reports, and the protocols in place require the author teams to respond to every one. McIntyre may therefore have felt that he could at least try to bring a little more scepticism and rigour to the process, but it soon became apparent that IPCC bureaucrats had other ideas.

Soon after starting work on his review, McIntyre noticed that the draft paleoclimate chapter referred to two unpublished manuscripts – one by Rosanne D'Arrigo, the other by Gabriele Hegerl. The IPCC's own policies referred to reviewers being supplied with 'specific material referenced in the document being reviewed, which is not available in the international published literature' and therefore, almost as a matter of course, McIntyre decided to get hold of the authors' data, assuming, not unreasonably, that this was an entirely appropriate step for an IPCC reviewer to take. In September, therefore, McIntyre wrote to the IPCC Technical Services Unit (TSU) based at UCAR in Boulder, Colorado, to ask them to obtain for him the download locations for the datasets used by Hegerl and D'Arrigo.

By now of course, you will know that data is rarely made available in paleoclimate circles and that official bodies tend to be quite happy to connive in this secrecy. The IPCC turned out to be no exception and a few days after his email was sent McIntyre received a reply from TSU explaining that they would not be able to help him. According to TSU's Martin Manning, the IPCC only assessed published literature and did not want to 'act as a global clearing house' for scientific data. Undeterred, McIntyre wrote back explaining that his request merely required a short email to the authors from TSU staff, and pointed out that if D'Arrigo and Hegerl subsequently refused to release the data then this would (and should) colour the IPCC's opinions on whether the papers should be cited in the final drafts. Unfortunately, Manning rebuffed this idea in an ill-tempered email, which he presumably thought would end the matter.

McIntyre, however, was not yet willing to play along and decided to appeal Manning's decision to Susan Solomon, the chairman of IPCC Working Group I, the subcommittee responsible for the scientific report. His letter reiterated the request and pointed out that it appeared to be well within the stated terms of reference for reviewers. Meanwhile, he also sent a direct request for the data to D'Arrigo and Hegerl.

The almost inevitable refusals were received shortly afterwards. For her part, Hegerl refused to release any data until after her paper was published and, while D'Arrigo declined to correspond with McIntyre directly, one of her co-authors referred him to the editors of the *Journal of Geophysical Research* (JGR), the journal which was to publish *her* paper.

Being referred to the journal should have been the end of the matter: JGR was a publication of the American Geophysical Union (AGU) and therefore had clear data policies, which stated unequivocally that all datasets used must be archived in a public database. On the face of it then, D'Arrigo should not be able to resist a data request for very long. McIntyre therefore penned a long letter to the journal's editors, explaining the inadequacies of the data citations in D'Arrigo's paper and noting the apparent conflict with the journal's policies. He asked politely that the information be made available in a format that would allow him to reproduce the results. Needless to say, the data was never sent.

While this correspondence was ongoing, McIntyre also reverted to Susan Solomon, explaining the refusal from Hegerl and D'Arrigo and asking once more for the IPCC to intervene. Again, this was to no avail, and it now looked as if Solomon was running short of excuses. Her reply was startling: she first declared that the attempt to obtain the data from the AGU was a breach of McIntyre's terms of reference as a reviewer, this document apparently requiring that any material provided as part of the review should not be 'distributed, quoted or cited' without permission. Readers can judge for themselves whether McIntyre had actually done any of these things in asking for the data.

More bizarrely still, Solomon then accused McIntyre of trying to influence editorial decisions at the journal, saying that it was 'inappropriate' for him to use his IPCC status to obtain information from the authors. It seems fairly unlikely that data requests really were 'inappropriate' under IPCC rules, since Manning had told McIntyre to make such a request just a few days earlier. However, with Solomon closing her letter by threatening to remove McIntyre's reviewer status if the line was not toed, it was pretty clear that this enquiry was unlikely to bear any fruit, and McIntyre reluctantly decided to call a halt. Nobody, it seemed, should look too closely at any paleoclimate studies under the IPCC's auspices – like the NAS, they expected reviewers just to 'wing it'.

First Order Draft

And so to the First Order Draft.[205,206] When McIntyre saw the proposed text, he realised that the tone of the paleoclimate chapter was going to be less promotional than it had been when Mann had been lead author – at least as far as the Hockey Stick was concerned. However, as then drafted it was not going to be any kind of a triumph for him and McKitrick. The Hockey Stick was still there and the spaghetti graph was still there, with the Briffa's temperature reconstruction still truncated. On the other hand there was not a mention of the divergence problem and as we will see, the coverage of the Hockey Stick affair was controversial, if not downright scandalous.

However, there was no alternative but to enter a response, and McIntyre and McKitrick started to collect their thoughts. Reviewers' comments had to be entered into a spreadsheet, which would be emailed to TSU in Boulder once it was completed. Manning and his team would then compile all of the contributions from the different reviewers into a consolidated pack, which they would issue to the author team, who would in turn compose a response to each comment. In this way, each reviewer would be unaware of the contributions of the others until after the event.

A surprising number of the comments were trivial – votes of thanks and congratulations and so on, but there were plenty of more substantive contributions. McKitrick had begun his comments by firing a warning across the bows of the chapter authors:

> Bear in mind that a great many observers, especially the most motivated critics of the AR4, will start their reading by turning to the paleoclimate chapter and seeing how the IPCC deals with the hockey stick. I will present my comments on this chapter as helpfully and objectively as I can. But I begin with some exasperation at this first draft. You may not want any advice from me, but for what it's worth, do consider. Chapter 6 is obstinate in its rejection of criticisms of the hockey stick, yet is surprisingly weak on the technical issues at stake. If you truly want to proceed with the chapter in its current form then you will not only be handing the IPCC's traditional critics a large club to beat the AR4 with, but you will alienate those many scientists who have hitherto given the IPCC the benefit of the doubt, but who have followed these issues and are looking for a serious treatment of them, not a brittle, dogmatic dismissal.[205]

There was plenty for McIntyre and McKitrick to do just to try to tone down the promotional tone of the first draft. The authors seemed very keen to sell the paleoclimate chapter to their readers: they had expounded at some length on the allegedly high quality of the proxy records, which were said to reflect environmental change 'in a highly quantitative and well-understood manner' and to have undergone 'comprehensive calibration with . . . instrumental data'. As readers will be aware, these two statements are wrong and debatable respectively: as we have seen, proxy-based temperature reconstructions are fraught with difficulty and there was severe doubt that many, or even any, of the proxies were capturing temperature information at all. The authors had gone on to claim that multiproxy reconstructions provided 'more rigorous estimates than single proxy' ones, a position that was again debatable.

Fortunately, McIntyre's comments on this issue were all acknowledged by the chapter authors and were earmarked for redrafting. On the other hand, McIntyre's attempts to get a fuller explanation of the uncertainties in the proxy studies were rejected. While the uncertainties were *mentioned* in the summary, it appeared that it was considered inappropriate to spell them out in too much detail.

Briffa on the Hockey Stick

When it came to the meat of the chapter, the authors first reviewed the main studies from the Third Assessment Report: the Hockey Stick, Jones et al 1998 and Briffa's truncated tree ring study from 2001, before going on to look at developments since that time. McIntyre's work on the Hockey Stick papers was considered worthy of a paragraph of its own, but as a summary of the state of the debate, it was shocking:

> McIntyre and McKitrick (2003) produced a Northern
> Hemisphere reconstruction that differs radically from that
> of Mann et al (1999), in indicating a period of significant

warmth in the 15th century, even though they attempted to employ the same method and [proxies]. However they omitted several important proxy series used in the original reconstruction [and their reconstruction failed verification statistics]. The Mann et al (1999) series was subsequently successfully reproduced by by Wahl and Ammann (2004).

Readers will have noted that there is almost *nothing* correct about this paragraph. As McIntyre pointed out in his comments, he and McKitrick had *not* produced a Northern Hemisphere reconstruction (or indeed any reconstruction at all). They had simply shown that Mann's reconstruction was not robust and failed its verification statistics. It is hard to believe that Briffa and his colleagues can actually have been unaware of McIntyre's repeated statements to this effect, or that they can have missed reference to it in McIntyre's original papers. McKitrick added that it sounded almost as if the authors had got their material directly from the pages of RealClimate rather than from the scientific literature, and their wording did rather seem to echo Mann's accusation that the two Canadians had 'censored' key indicators from the proxy network. Clearly somewhat riled, he went on to point out that the missing data was, in large part, the bristlecone pines, which were not valid proxies.

> So it is not that we 'omit' some important proxies and end up with a lousy result, instead we remove some lousy proxies and end up with an important result: the conclusions fall to pieces. The issue, as we have said over and over, is robustness. Mann's conclusions are not robust. They are not statistically robust, nor are they robust to removal of a small network of bristlecone proxies that are widely viewed among dendrochronologists (including Hughes himself in another paper) to be invalid as temperature proxies. What we have shown is not that the 15th century was 'warm', but that Mann's results do not

provide evidence that the late 20th century was climatologically exceptional.

The silence was almost deafening, although Briffa and his colleagues did refer to the next draft (notably, however, without saying that they accepted McKitrick's arguments). In fact this was a consistent response from the author team: *Noted – see edited text, Noted – see edited text*. The implication of course was that the edited text would cover McIntyre's work more fairly, but McKitrick must have wondered if there was anything he or McIntyre could say that would persuade a core member of the Hockey Team to accept their case. His low expectations were to be entirely justified.

The Hockey Team fights back

While McKitrick was attacking the author team's interpretations of the Hockey Stick controversy, Mann was vigorously defending his version of events. His comments forcefully promoted the claims of Wahl and Ammann ('showed the original Mann et al reconstruction to be robust . . .') and those of Rutherford et al ('demonstrate[s] that each of the criticisms raised by McIntyre and McKitrick (2003) are without merit'). At the same time he was also launching direct attacks, saying that McIntyre's papers had been refuted five times* and also doing his best to malign the separate criticisms of von Storch, whose paper, he alleged, included a 'fundamental error'. Jan Esper and David Ritson were also fighting the Team's corner. Esper was trying to downgrade discussion of the Hockey Stick affair, which he felt was dominating proceedings, while Ritson wondered innocently if, since the authors didn't refer to McIntyre's 2005 papers, they should consider removing discussion of his work entirely.

---

* Mann referred to Ammann's two whoppers in *Climatic Change* and GRL, Rutherford et al and the comments on McIntyre's GRL paper by Huybers and von Storch. These are all considered elsewhere in this book and, as we have seen, are all far from constituting refutations.

McIntyre had plenty of ammunition to counter Mann's arguments, which were fairly brazen since Mann must have known that the author team would have read the replies to Huybers and von Storch. McIntyre had also pointed out that the author team were being highly misleading over Ammann's alleged replication of the Hockey Stick. They had referred to the CC paper as 'Wahl and Ammann 2004', the date suggesting that it had been published already. In fact, as we have seen, at that point it was stuck in a kind of publishing limbo, held up by the problems with the GRL comment.* McIntyre also pointed out that Wahl and Amman had *not* reproduced Mann's claims of statistical skill and had *not* reproduced the Hockey Stick. McKitrick was even more forthright, saying of the supposition that the Hockey Stick had been reproduced, 'The last claim is false'. He also pointed out that if the authors felt that failing verification statistics was sufficient grounds for condemning a paper (as suggested by their spurious point about the McIntyre 'reconstruction'), then they should be condemning MBH98 as well, since it had failed its verification statistics with some panache. This presented Briffa and his colleagues with a dilemma. One one hand they had Mann telling them that Ammann's work replicated the Hockey Stick and on the other was McIntyre saying point blank that it didn't. How would they deal with this difference of opinion? As we saw above, they were duty bound to report differences of opinion.

Quite apart from getting pretty much every fact in their paragraph on the Hockey Stick wrong (even though the paragraph ran to only three sentences), the authors also managed to omit some key indicators of their own: such as the existence of two McIntyre and McKitrick papers from 2005, such as the PC analysis arguments and the bristlecones. As an attempt to represent a complex scientific dispute, it was either incompetent or biased. The authors' vague responses to McIntyre and McKitrick's comments did not bode well for a happy resolution.

---

* See Chapter 8.

The NAS *defence*

It was clear from the rest of the paleoclimate chapter that even if the authors could not save Mann directly, they would still attempt to kill off the Medieval Warm Period by adopting the 'NAS defence': they would declare that 'independent' confirmations of Mann's reconstruction meant that his findings were correct, even if his data and methods were invalid.

The independent confirmations that we looked at in Chapter 10 were all advanced in Mann's support, although with a caveat about the commonality of proxies between them. All were on the receiving end of a vigorous McIntyre attempt to shoot them down.

In his review, McIntyre first flagged up Rutherford et al, which had used the same flawed PC methodology and the same flawed proxies as Mann had used in MBH98. Quite illogically, Briffa and his colleagues on the author team declared that this didn't matter because the rest of Rutherford's methodology was different, as if this somehow excused the use of a biased dataset. Briffa either ignored or missed McIntyre's comment on the divergence problem in D'Arrigo et al since no response was forthcoming. When it came to Hegerl et al, which didn't cite data sources properly, Briffa merely replied that it did.* Briffa's own paper, Briffa 2005, was not even available for review and here at least, the author team appeared to accept the point. They also appeared to accept McIntyre's criticisms of Jones et al 1998, but how these comments would be incorporated in the next draft was anyone's guess. Far too many of them were 'noted' although not, apparently, 'accepted'. An answer to this question would have to wait.

Second Order Draft

With his review comments all submitted, McIntyre returned to his research and his blogging. The second draft of the IPCC report was

---

*   McIntyre's point was presumably that the citation was inadequate to allow the actual dataset to be identified, but this subtlety seems to have been missed by the authors.

due to be issued in April 2006, at which point a whole new set of review comments had to be submitted and responded to. McIntyre was keen to understand the author team's thinking as they moved from the first to the second draft and, fortunately, according to IPCC policies the first draft review comments should be made available at the same time as the second draft was published. McIntyre therefore wrote to the IPCC to ask how he could get access to the paleoclimate chapter comments. To his surprise, a short time later a large package was delivered to his home, which turned out to contain the full set of review comments – rather than email over a file or post a disc, the IPCC had printed out the whole document and *posted* it to him. This was incredibly inconvenient as it meant that it was impossible to search for relevant comments in a document of over 200 pages. Quite why TSU should put themselves to so much trouble when they could simply have emailed an electronic file is unclear. They had also rather oddly marked the pages, 'Confidential, Do Not Cite, Quote or Distribute', so there was to be no discussion of the contents either. So much for an open review process.

Late breakers

Shortly before the deadline for the IPCC review, McIntyre had noticed a surge of submissions of new papers to the review. Apart from Ammann's *Climatic Change* paper* there were new findings, apparently hot off the press, from familiar names such as Hegerl et al and D'Arrigo et al (see above); Wahl, Ritson and Ammann was another, and Osborn and Briffa another. Some months later, in early 2006, having seen how the IPCC had ignored their own deadlines in order to include Ammann's *Climatic Change* paper, McIntyre took the trouble to compare the actual publication dates of these 'late breakers' with the IPCC's timetable requirements.

The IPCC guidelines stated that a paper had to be available in draft form by 12 August 2005, published or in press by the end of

---

*   See page 217.

December and in final preprint form by the end of February 2006. A cursory glance at the publication timetables for the late breakers showed that, as McIntyre had suspected, most of them had actually failed to meet the deadlines but had been carried forward into the review regardless.

TABLE 11.1: IPCC papers and their publication progress

Paper	Submitted	Accepted	Published
IPCC *deadline*	*12 Aug 05*	*16 Dec 05*	*27 Feb 06*
Wahl, Ritson and Ammann	3 Oct 05	27 Feb 06	28 Apr 06
Hegerl et al	8 Jul 05	28 Feb 06	20 Apr 06
Wahl and Ammann	10 May 05	28 Feb 06	
Osborn and Briffa	23 Sep 05	17 Jan 06	10 Feb 06

Clearly, at the end of February the paleoclimate community was in some turmoil, as papers were rushed through the peer review process in time to make the IPCC deadline. Of course, according to the letter of IPCC rules, they were already too late, having missed the deadline for journal acceptance by some margin. It may have been thought that if it could be demonstrated that the papers had received journal acceptance by the later *publication* deadline there was at least a small fig leaf to hide behind.

While this might seem like something of a technicality, it is in fact of great importance to the quality of the review. The lead author meeting at which the second draft was written had taken place in New Zealand on 13–15 December 2005. The authors were supposed to have taken into account comments submitted on the first draft, but of course the reviewers of the first draft could not have read these late-breaking papers, none of which had completed peer review by the time comments had to be submitted. In fact, with the acceptance dates being in February, even the authors preparing the second draft would not have had access to the papers in their final form. So in practice, the review failed to meet even the rather low standards set by the IPCC.

The Hegerl substitution

When McIntyre started his second draft review, he was astonished to find that one of the papers cited had *changed*.[207,208] When he had looked at Hegerl et al during the first draft review, he had read a text from *Nature* which, as we saw above, had only been accepted for publication on 28 February 2006, thus making it ineligible for the report. Had it been cleared for inclusion on a nudge and a wink, like all the other late breakers? However, some time between the first and second drafts, the paper had been switched for an entirely *different* Hegerl et al article, this time from the *Journal of Climate*. The latter paper had not even been accepted for publication in *April* 2006 and therefore, on any measure, could not be considered. The reasons for the swap became clear when McIntyre realised that the *Journal of Climate* paper included a new temperature reconstruction, which had not been mentioned in the *Nature* article at all. McIntyre pointed out this illegitimate switch in his second draft review comments, but was rebuffed with the bald statement that the Hegerl submission met the IPCC guidelines. This was apparently not true, as the IPCC guidelines stated clearly that final preprints must be available by late February.[209]

The new spaghetti chart

McIntyre also gave Briffa strong censure for trying to do a 'Michael Mann' and use his status as an IPCC lead author to promote his own work. Between the first and second drafts, Briffa had managed to insert references to his new paper co-authored with Tim Osborn. This paper, as we saw above, had actually failed to meet the IPCC publication deadlines, but here it was, bold as brass, highlighted in a second spaghetti diagram in the new draft. When McIntyre pointed out this inconvenient fact, his comment was rejected, with another bold assertion that its inclusion did not contravene current IPCC policies.

The new spaghetti diagram was designed to demonstrate how the various proxy series that went into temperature reconstructions all told different stories about the Medieval Warm period, and many of McIntyre's 'same old whores' were there – Mann's PC1 (bashfully labelled 'W USA'), Yamal and so on. The only clear message that came from the diagram was that there seemed to be a general uplifting of proxy values in the twentieth century; the rest of the graph looked rather like random data, which in McIntyre's opinion was exactly what it was: red noise. The twentieth century uplift was simply an artefact of having cherrypicked series which showed such an uplift. Briffa didn't see it that way, of course, and rejected all attempts by McIntyre to point out what was wrong. Why did Briffa include Mann's PC1, which was essentially just the bristlecones, trees which were known to be flawed proxies? Briffa merely replied that the purpose of the figure was 'to illustrate in a simple fashion, the variability of numerous records that have been used in published reconstructions of large-scale temperature changes' and 'not to give a very detailed account of the specific limitations in data or interpretation for each'. So according to Briffa, the findings of the NAS panel and the Wegman report – that the statistical treatments used in Mann's papers were wrong – were to be recast as mere 'limitations'. Readers can no doubt assess for themselves whether using incorrect mathematics is a 'limitation' or whether it is in fact just plain incorrect. Briffa said that the carbon dioxide fertilisation issue was complex and that it would be covered in the final draft.

Meanwhile, we might wonder again why Briffa preferred Yamal over the Polar Urals update, when the latter had a better correlation to temperature records, suggesting it was a more reliable proxy? Of course, Yamal had no Medieval Warm Period, and many will assume that this was the reason. Again, Briffa referred to wanting to demonstrate the variability in the records, although this didn't seem to actually address the point at issue.

The Hockey Stick in the second draft
The authors had suggested that they would take on board McIntyre and McKitrick's criticisms of IPCC's treatment of their papers when

they put together the second draft. However, when this document became available in April 2006,* it was clear that they had done no such thing. The best that could be said about their new summary of the Hockey Stick affair was that it was somewhat different. It remained a travesty of the truth:

> McIntyre and McKitrick (2003) reported that they were unable to replicate the results of Mann et al (1998). Wahl and Ammann (accepted) demonstrated that this was due to the omission by McIntyre and McKitrick of several proxy series used by Mann et al (1998). Wahl and Ammann were able to reproduce the original reconstruction closely when all records were included. McIntyre and McKitrick (2005) raised further concerns about the details of the Mann et al (1998) method, principally relating to [verification statistics and PC methodology]. The latter may have some foundation, but it is unclear if it has a marked impact upon the final reconstruction.

So the argument that McIntyre had 'censored' key data still stood, despite the nature of these alleged omissions having been made abundantly clear. Invalid proxies (like the bristlecones), obsolete data (like Twisted Tree, Heartrot Hill) and extrapolations (like Gaspé) *should* have been omitted from the dataset by any diligent researcher. By failing to point out why McIntyre had omitted this data and by refusing to discuss its validity, the authors were handing the victory to their colleagues in the Hockey Team. McKitrick was forthright once again:

---

*   The official archive of the drafts is at the Harvard University Library, whose online retrieval system seems to be carefully designed to make accessing the information as hard as possible. Readers wishing to access the material in a more user-friendly manner may want to use an alternative.[207]

> The opening sentence . . . misrepresents the situation by
> failing to point out that the 'results' of Mann et al were,
> principally, the supposed findings of unprecedented
> robustness and statistical significance. Not only have these
> results NOT been replicated by others, but they have been
> amply disproven, by teams on both sides.[208]

McKitrick pointed out, yet again, that Wahl and Ammann's purported replication failed its R^2 statistic, but the argument appears to have been lost on Briffa, who failed to address it in his response, which concentrated on the benchmarking for the RE. On the latter point, Briffa launched into an explication of his own views on the issue, without making any supporting citations, and on this basis rejected McKitrick's points. The text, he said, gave a balanced view and he said that McIntyre's 0.51 benchmark level was 'somewhat overstated', although he failed to explain how, if McIntyre's published benchmark was only 'somewhat' overstated, he could justify rejecting it entirely in favour of Ammann's still unpublished benchmark of zero.

Broadening their defence, Briffa and his colleagues on the author team tried to rationalise their summary, explaining that they had received diametrically opposite opinions on the Hockey Stick issue and had tried to strike a balance. This once again highlighted a breathtaking lack of due diligence (or perhaps of intellectual honesty). If the authors had received opposite opinions on Wahl and Amman's paper, should they not have investigated the difference themselves? They had only to ask Ammann for his verification statistics and the answer would have become crystal clear – that the Hockey Stick was not a credible reconstruction. As it was, they had been silent on the dispute, and Briffa was to find himself relying on a draft paper that was subsequently shown to be not just wrong, but cynically so, a finding that cast him in a very bad light.

McIntyre also pointed out that the version of Wahl and Amman which had been accepted by *Climatic Change* differed markedly from the version which was considered by the IPCC – as we have seen, the paper eventually included an admission that its verification statistics were a failure, making it useless for assessing historic climate. Once again, Briffa and his colleagues were not interested, failing to address the specific point and brushing McIntyre aside with a statement that his comment was 'rejected' and that the inclusion of each paper was 'allowed under current rules'. In fact the declaration about conformity with current rules was something of a feature of the second draft, with each of McIntyre's objections to the late-breaking papers being rebuffed with this statement or a variation on it. This slightly awkward turn of phrase turned out to be important, although its significance was not discovered until some time later.

Briffa also made a lengthy response to the question of the robustness of the reconstructions to the PC algorithm. Unsurprisingly, he sided with the rest of the Hockey Team, echoing the claims of Mann and Ammann that, provided you retained enough PCs, you still got a hockey stick. It is worth reminding ourselves what this means in terms of the flow of data from proxies to final reconstruction. The bristlecone pines appear, under a standard centring regime, in the PC4, representing just 8% of the variance of the total dataset. What this means is that their hockey stick shape is a rather unimportant pattern in the dataset, as would be expected since bristlecones are a couple of closely related species from a small area of the western USA. However, because they correlate well to temperature in the twentieth century, they dominate the calibration results and hence the reconstruction too. In this way the temperature reconstruction for the whole Northern Hemisphere is made to look like the growth pattern of a few trees in the White Mountains of the USA. This is apparently accepted as reasonable by the IPCC's author team. It had certainly not been viewed as reasonable by Wegman, who had told one of the senators on the House Energy and Commerce Committee in no uncertain terms that adjusting your methodology after the event was not an appropriate way of conducting a statistical experiment:

> Wahl and Ammann [argue that] if one adds enough principal components back into the proxy, one obtains the hockey stick shape again. This is precisely the point of contention. . . . A cardinal rule of statistical inference is that the method of analysis must be decided before looking at the data. The rules and strategy of analysis cannot be changed in order to obtain the desired result. Such a strategy carries no statistical integrity and cannot be used as a basis for drawing sound inferential conclusions.[210]

What of the report's assertion that McIntyre's criticisms of the biased PC methodology 'might have some foundation'? It is hard to see this as a fair assessment of the state of the scientific debate. Two expert panels – the NAS and Wegman – had considered the issue in detail and both had concluded that Mann's methodology was wrong. Von Storch had also said that short centring was wrong. Huybers had agreed. How could Briffa possibly overturn this with a vague statement that 'it might have some foundation'; how could he possibly claim that he had 'struck a balance'?

The rest of the chapter was little better. The spaghetti graph of multiproxy reconstructions still included the Hockey Stick, it still included the Briffa truncation and it still included bristlecones and foxtails in abundance. Briffa's conclusions appeared even more wayward when the subject of the divergence problem was raised. Richard Alley* raised Briffa's failure to mention this question in the second draft, and McIntyre had pointed out the truncation of the results in Briffa's own notorious papers:

> Show the Briffa et al reconstruction through to its end; don't stop in 1960. Then comment and deal with the 'divergence problem' if you need to. Don't cover up the divergence by truncating this graphic. This was done in IPCC TAR; this was misleading.

---

*   See page 233.

The reply from Briffa and his co-authors must have amazed everyone:

> Rejected – though note 'divergence' issue will be discussed, still considered inappropriate to show recent section of Briffa et al. series.

This was an extraordinary answer. It was apparently considered 'inappropriate' for the IPCC to set out the truth, warts and all. McIntyre was outraged at what appeared to be a disgraceful example of political expediency standing in the way of the truth, and when he posted Briffa's words up at Climate Audit, the IPCC added insult to injury, demanding that he remove them immediately. If their authors were going to behave in this way, then it was perhaps better if the general public knew nothing about it.

The lead author meeting

The comments on the second draft were returned to IPCC at the start of June, ready for the lead author meeting on the 25th–30th in Bergen, Norway at which the final draft would be prepared. Shortly after the end of this meeting, there was another extraordinary development, which revealed the reason for the author team's repeated declarations that the inclusion of the late breakers didn't breach 'current rules'. In the days after the closing of the meeting, an email was issued to all IPCC reviewers, including McIntyre, which contained an entirely new set of guidelines for the inclusion of literature in the review.[211] These are reproduced below:

GUIDELINES FOR INCLUSION OF RECENT SCIENTIFIC
LITERATURE IN THE WORKING GROUP I FOURTH
ASSESSMENT REPORT

> We are very grateful to the many reviewers of the second draft of the Working Group I contribution to the IPCC

Fourth Assessment Report for suggestions received on issues of balance and citation of additional scientific literature. To ensure clarity and transparency in determining how such material might be included in the final Working Group I report, the following guidelines will be used by Lead Authors in considering such suggestions.

In preparing the final draft of the IPCC Working Group I report, Lead Authors may include scientific papers published in 2006 where, in their judgment, doing so would advance the goal of achieving a balance of scientific views in addressing reviewer comments.

However, new issues beyond those covered in the second order draft will not be introduced at this stage in the preparation of the report. Reviewers are invited to submit copies of additional papers that are either in-press or published in 2006, along with the chapter and section number to which this material could pertain, via email to ipcc-wg1@al.noaa.gov, not later than July 24, 2006.

In the case of in-press papers, a copy of the final acceptance letter from the journal is requested for our records. All submissions must be received by the TSU not later than July 24, 2006 and incomplete submissions cannot be accepted.

It appeared then that the IPCC was going to allow three extra weeks for new papers to be included, provided that they didn't open any new topics and provided the papers were accepted before the 24 July. Suddenly the significance of Briffa's statements about the late breakers being allowed 'under current rules' become clear. Presented with overwhelming evidence that Wahl and Amman's CC paper, Hegerl et al and Osborn and Briffa should all have been excluded from the review, the IPCC, and perhaps Briffa himself, had simply rewritten the rules to permit their inclusion.

Even then, the IPCC wasn't entirely off the hook, because Wahl and Amman's CC paper relied upon their GRL comment for its statistical arguments, and the GRL comment had been rejected by the journal. In order to maintain the fiction of a March acceptance date they needed the GRL comment to find its way through peer review at some point. The story of how this came about will be told in the next chapter.

The SPM

The Summary for Policymakers (SPM), a short distillation of the whole IPCC report, was due to be released in February 2007. However, even the simple act of scheduling the publication of this document opened up another set of questions over the propriety of the IPCC's procedures.

When the publication date was announced, the IPCC also stated that the main body of the report, on which the SPM was based, was not itself to go into print until three months later. This was a little odd, but when seen in the light of the procedures for making final revisions to the report, it became rather sinister. The procedures read as follows:

> Changes (other than grammatical or minor editorial changes) made after acceptance by the Working Group or the Panel shall be those necessary to ensure consistency with the Summary for Policymakers or the Overview Chapter.[212]

In other words, the IPCC was intending to ensure that the policy document was consistent with the scientific one by making changes to the *scientific* document! As McIntyre commented to his readers:

> Unbelievable. Can you imagine what securities commissions would say if business promoters issued a big promotion and then the promoters made the 'necessary' adjustments to the qualifying reports and financial statements so that they matched the promotion? Words fail me.[213]

As the publication date for the SPM approached, IPCC insiders started to leak the contents to the media, and it looked very much as if there were going to be doom-laden headlines once again. In a widely distributed report from the Associated Press, Andrew Weaver* declared that 'This isn't a smoking gun; climate is a battalion of intergalactic smoking missiles . . .'[214] Meanwhile climatologist Kevin Trenberth advised readers to 'Look for an "iconic statement" – a simple but strong and unequivocal summary – on how global warming is now occurring'.[214] While these statements didn't address the Hockey Stick question directly, neither did they suggest that Briffa was going to rebalance his summary of the Hockey Stick.

A couple of days later Rajendra Pachauri, the head of the IPCC, continued the process of whipping up media attention by telling Reuters that 'There are a lot of signs and evidence in this report which clearly establish not only the fact that climate change is taking place, but also that it really is human activity that is influencing that change'.[215] Unnamed sources told the same reporter that the Fourth Assessment Report would declare that the IPCC 'is at least 90 per cent sure than human activities, led by the burning of fossil fuels, are to blame for global warming over the past 50 years'.

Then, on the last day of January, the news was leaked that the sceptics had all been expecting: the Hockey Stick would survive to fight again another day. This was the unavoidable conclusion of an article in the *Toronto Globe and Mail*, based on yet another IPCC leak of the SPM:

> It concludes the higher temperatures observed during the past 50 years are so dramatically different from anything in the climate record that the last half-century period was likely the hottest in at least the past 1,300 years.[216]

---

*   See page 292.

This quotation was strongly suggestive that the Hockey Stick had been given a reprieve, and sure enough when the SPM made its official appearance the following day, Mann's conclusions were right there at the start of the paleoclimate section:

> Average Northern Hemisphere temperatures during the second half of the 20th century were *very likely* higher than during any other 50-year period in the last 500 years and *likely* the highest in at least the past 1,300 years.[217]

Recent studies, it reported, had shown rather more variability than had been concluded from the Third Assessment Report (in other words, there was more of a Medieval Warm Period apparent than Mann had said six years earlier) but drew 'increased confidence from additional data showing *coherent* behaviour across multiple indicators in different parts of the world' (emphasis added) – in other words the independent confirmations were alleged to have confirmed the Hockey Stick position on the Medieval Warm Period. This was a surprising statement, firstly because as we have seen, the proxies were all behaving in a very *incoherent* fashion, but also because, as McIntyre noted, the word 'coherent' was not used (at least in this context) in either of the drafts of the report. So either the idea of coherence had been inserted in the final draft by Briffa (and had therefore not formed part of the IPCC's much-vaunted review process) or it appeared in the Summary for Policymakers but not in the still-unpublished report itself (in which case it would presumably be one of those issues where the scientific report would have to be changed to bring it into line with the SPM.)

Another striking feature of the SPM was that, while the IPCC seemed to like the conclusions of the Hockey Stick papers, they omitted any mention of the Hockey Stick itself. This shift in emphasis from the Third Assessment was picked up by some watching journalists who wondered whether the change was significant. One of the coordinating lead authors of the paleoclimate chapter, Eystein Jansen, told a Norwegian newspaper that the omission was necessary for reasons of space and also because it was a bit difficult to explain to politicians![218] He went on to explain, however, that the IPCC had performed no specific evaluations of McIntyre's arguments, another claim which seemed bizarre. If McIntyre's claims about the validity (or lack of it) of bristlecone pines as proxy thermometers were correct, then many of the studies which the IPCC was relying on for its position on the Medieval Warm Period were thrown into severe doubt. How then could they explain a failure to specifically assess McIntyre's findings?

The full report

'The Physical Science Basis: Contribution of Working Group I to the Fourth Assessment Report of the Intergovernmental Panel on Climate Change' was released very quietly on 29 April 2007. It was almost as bad as it could have been. Most of McIntyre and McKitrick's comments on the second draft had been ignored, including, incredibly, those in which they had corrected Briffa and his team on what their claims actually were – he was still claiming that the Canadians had created a reconstruction of their own, in the face of their vehement statements to the contrary.

McIntyre's comments that the Hockey Stick and Ammann's purported replication of it had both failed their verification statistics seemed to have had some effect, because Briffa had decided to drop the subject entirely from the text. However, hiding the lack of robustness of the Hockey Stick behind a veil of silence could hardly be said to give a fair reflection of the scientific literature, if indeed that had been the intention of the report. The general thrust of the final text was that McIntyre's criticisms might have some validity but that they were probably not of material importance.

There were some other significant changes too. Briffa had clearly felt he could not credibly pass over the divergence problem and the bristlecones without mentioning them at all, and a lengthy paragraph had been added. Of course, having ignored McIntyre and McKitrick's comments on these issues in the first draft and then, in the second round having referred them to the final draft, this new text was all Briffa's own and was entirely unreviewed.

The carbon dioxide adjustment that wasn't

While the new paragraph did address the subject of possible carbon dioxide fertilisation, Briffa's treatment of it was brief and insubstantial. His new position was that Mann had adjusted for possible carbon dioxide fertilisation in MBH99 and, since this latter paper still had a hockey stick shape, the possibility that the bristlecones contained a non-climatic signal could be discounted.

With Briffa having introduced this new argument only in the final report, it was not possible for McIntyre or McKitrick to point out the holes in Briffa's case, and there was no doubt that more than one aspect of his position could be disputed. Firstly, arguing that Mann had adjusted for carbon dioxide fertilisation didn't address the effect of carbon dioxide fertilisation on all the other reconstructions in the spaghetti graphs, which used bristlecones without any adjustments. These other studies must have been tainted just as much as the Hockey Stick.

The second flaw, however, was more intriguing. The claim that Mann had adjusted for carbon dioxide fertilisation had been around for some time, and McIntyre had looked into it previously. He had prepared a graph of the two Hockey Stick papers side by side and had calculated the difference between them in the period of overlap from 1400 to 1980. It turned out that in the overlap period the results presented in the two papers were *identical* (see Figure 11.1).[155] So either Mann had not adjusted the final result, or his adjustment calculation amounted to a convoluted way of getting back to where he started from.

Dealing with divergence

To return to the IPCC report, the divergence problem was addressed in rather clearer terms than were the bristlecones, although Briffa

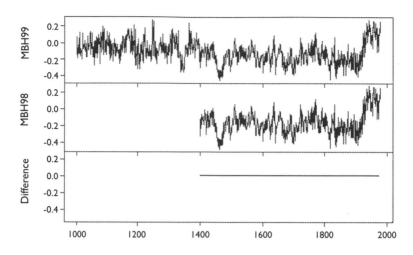

FIGURE 11.1: Was a correction made to MBH99?

had claimed, without a supporting citation, that the problem was limited to 'some northern, high latitude regions'. This was a surprising position for him to take because it appeared to contradict his statement in 1998 that the divergence problem was an issue which affected the whole of the northern hemisphere.* However, he had at least done something about his infamous truncation of the problem. Noting the excision of the data, he said that it had been done while 'implicitly assuming that the "divergence" was a uniquely recent phenomenon'. He went on, however, to note that certain 'others' (Hockey Team members don't like to mention McIntyre by name) had argued for a breakdown in the relationship between tree rings and temperature. If these 'others' were correct, he went on, 'this would imply a similar limit on the potential to reconstruct possible warm periods in earlier times'. In other words the IPCC's claim that modern temperatures were unprecedented might be resting on a scientific method that was incapable of detecting warmings in the climate.

---

*   See page 64.

This was an amazing admission: it meant that most of the paleoclimate chapter was in danger of having to be thrown out as worthless conjecture. It was clearly hugely important, with major implications for politicians, and yet this vital fact had *still* not found its way into the SPM. How different would the latter document have looked if its key finding had read as follows?

> Average Northern Hemisphere temperatures during the second half of the twentieth century may have been higher than during any other 50-year period in the last 500 years. However, there is evidence that the data and methods used reconstruct past climates may be incapable of detecting warming episodes in earlier centuries and this position is therefore subject to considerable uncertainty.

12 The IPCC Aftermath

> At last the secret is out, as it always must come in the end;
> The delicious story is ripe to tell to the intimate friend;
> Over the tea-cups and in the square the tongue has its desire;
> Still waters run deep, my dear, there's never smoke without fire.

W.H. Auden

At last the secret is out

A little is revealed

In Chapter 8 we saw how Ammann and Wahl's CC paper went forward into the IPCC review process in rather peculiar circumstances, and also how the paper relied on the GRL comment in order to establish its RE benchmark of zero. But the GRL comment had then been rejected twice by the journal, leaving the CC paper with no justification of its claims of statistical significance.

By early 2007, and a year after it had been accepted for publication, the CC paper was nowhere to be seen. As the publication date for the final IPCC report loomed, this left the IPCC and *Climatic Change* with a problem. McIntyre observed:

> I'm intrigued as to what the final [CC paper] will look like. They have an intriguing choice: the inclusion of a reference to this article in [the IPCC's Fourth Assessment Report] was premised on [it] being 'in press', which would prohibit them from re-working [it] to deal with the [rejection of the GRL comment]. But the article needs to be reworked since it will look pretty silly to describe [the GRL comment] as 'under review' over 18 months after it has been rejected.[219]

The new comment

In the background, however, much had been happening. In September 2007, and with the IPCC report published, the CC paper suddenly appeared, preceded in the same journal by *another* paper by the same authors. Remarkably, in view of his apparently having been misled by Ammann over the rejection of the GRL comment, it nevertheless looked as if Stephen Schneider, the editor of *Climatic Change*, had decided to allow Wahl and Ammann to rewrite their rejected GRL comment and to submit it to *Climatic Change* instead. All reference to the rejected GRL comment in the CC paper could be replaced by reference to the new paper.

One advantage of this approach was that it allowed the CC paper to retain its original acceptance date, and hence justify its inclusion in the IPCC review. It did leave the IPCC with the embarrassing problem that a paper that was allegedly accepted in March 2006 relied upon another paper that even the journal itself said was only received in August that year, (and in reality, it was even later than that). And this *mattered* because unless the CC paper had been accepted by the journal before the IPCC deadline, it should not have been accepted for inclusion in the Fourth Assessment Report. But, as we've seen, the IPCC needed the CC paper so that they could claim that McIntyre's work had been rebutted. So despite the inconsistency being pointed out to them (see Chapter 11), they had waved the objections aside as irrelevant.

With identical authorship and a maze of cross-references between them, it was extremely difficult to discern how the arguments in the two *Climatic Change* papers relied on each other. Ammann and Wahl's claim of statistical significance for the CC paper was based on their having established an RE benchmark of zero. This claim was repeated at several places in the main text of the paper and each time they referred to an appendix and to the new paper. In the appendix, however, there was further discussion of the issue and a further citation of the new paper, but no justification of it. Mystifyingly, however, the new paper referred back to the CC paper, citing its use of a benchmark of zero. There were *some* details of what they had done to establish the benchmark, but the calculations themselves were apparently to be found in the online Supplementary Information (SI). This presented a problem for McIntyre, because there was no trace to be seen of the SI. In fact, as far as McIntyre could tell, even the peer reviewers had not been given access to it. McIntyre wrote to Ammann one more time to request the benchmarking data and code, and once again Ammann refused to hand it over, this time in terms which left no room for doubt. His reply was remarkable:

> Under such circumstances, why would I even bother answering your questions, isn't that just lost time?[220]

The appearance of the benchmark

Again, everything fell silent. For the next year nothing more was heard of the two papers. McIntyre continued to press from his blog for release of Ammann's SI and with the IPCC report now published, politicians were able to take advantage of the political space it had created. With Ammann and Wahl's alleged refutation of McIntyre in print and in the IPCC report, naysayers could safely be ignored and the policy agenda advanced.

Then in August 2008, and entirely unannounced, the Supplementary Information suddenly appeared on Caspar Ammann's website, some three years after that first press release announcing the alleged refutation of McIntyre's work. With it, and a godsend to McIntyre, was the code used to establish the benchmark for the RE statistic. With no more than a few days work, McIntyre was able to establish exactly what had been done. What he found was stupefying.

It turned out that Ammann had calculated almost *exactly the same figure* as McIntyre. The number he had arrived at was 0.52, just a whisker away from McIntyre's own 0.54. As we saw, however, Ammann had reported in the paper that it was sufficient only to score above zero.

So how had Ammann reconciled the two numbers – 0.52 and zero? A benchmark of 0.52 must have represented a big problem for him and Wahl. It was entirely inadequate for their purposes because Mann's Hockey Stick had a verification RE of 0.48, leaving it tantalisingly just below the calculated benchmark. Unless they could somehow work things so that the benchmark came down a little, the Hockey Stick would be declared 'statistically insignificant', with all the disastrous ramifications that would have on political and climatological careers around the world. However, achieving this feat and getting the Hockey Stick up above the benchmark level would have been tricky. Remember, a thousand runs of random data were pushed through the statistical sausage machine. In other words, a thousand attempts were made to reconstruct the temperatures of the past using random data instead of proxies. The RE number – the correlation with the actual temperature records – was recorded for each. Then all the runs were sorted in order of RE value, the best runs having the highest RE and the worst the lowest. Ammann needed to show that the Hockey Stick RE was right up there with the best simulations – in the top one percent. It was no good simply removing runs which had a higher score than the Hockey Stick because this would be seen as statistical malpractice. It would be like crossing out all the questions you got wrong in a test, then claiming a perfect score.

The method Ammann and Wahl had chosen to achieve their objective turned out to be only a little less obvious, but no more legitimate. They said that each of the 1000 runs should be examined to compare the RE score it achieved in the verification period to the one it got in the calibration. Their argument was that if it did too well in the verification compared to calibration, it should be rejected. They proposed therefore that the ratio of the RE scores in the calibration and verification periods – which I will refer to as the CV ratio – should be used as a test. This wasn't a test of whether the Hockey Stick methodology was correct but of whether the random data they were going to compare the Hockey Stick to was 'suitable'. They were essentially going to filter out some of the random data. This meant, of course, that it was no longer actually random, although Wahl and Ammann didn't say so.

Wahl and Ammann next argued that if one of the runs had a CV score that indicated it was unsuitable, rather than being thrown out entirely, its RE score should be changed arbitrarily to −9999. This meant that it remained in the list of runs, but was pushed to the bottom, making the Hockey Stick appear relatively more significant.

This extraordinary set of methodological steps was entirely unknown to statistics or to any other branch of science – it appeared to have been contrived entirely for the purposes of saving the Hockey Stick. But in order to do this, they needed to set the correct level for the CV ratio. If it were set too low, say around 0.5, not enough of the high scoring runs would be knocked off the top of the list and the Hockey Stick would not appear in the top 1%. Set it too high and the Hockey Stick itself would fail the CV ratio test and would have its RE score of 0.813 reset to −9999, indicating a total failure. In the event, Wahl and Ammann chose, apparently arbitrarily, to set the CV test level to be 0.75 and noted with satisfaction that under the terms of their new methodology, Mann's results were significant.

The new ratio was a bizarre contrivance, but it provided at least some kind of a fig leaf for the Hockey Team and *Climatic Change* to hide behind. With this new, and pretty much arbitrary, step in place, Wahl and Ammann were able to reject several of the runs which stood between the Hockey Stick and what they saw as its rightful place as the gold standard for climate reconstructions. That the statistical foundations on which they had built this paleoclimate castle were a swamp of ad hoc and arbitrary methodological steps was, to the Team, apparently an irrelevance. For political and public consumption, the Hockey Stick still lived, ready to guide political and economic decision-making for years to come.

Review comments again

The IPCC's sleight of hand over the eligibility of Ammann's paper for the Fourth Assessment Report was now out in the open, but there were to be some more strange revelations about the conduct of the review. Shortly before the publication of the final report, McIntyre had requested from IPCC's Technical Support Unit (TSU) the full set of review comments for the second draft. Having experienced the farce of being sent a paper copy by the IPCC for the first draft comments, he did wonder what Martin Manning and his TSU colleagues would get up to this time round and he was certainly not disappointed.

According to IPCC policies, once the final report was published, the second draft review comments would be placed in a public archive. Manning's initial response to McIntyre's request for the review comments was that TSU had been 'setting up the arrangements', but that the review comments were now available from George Clark, the curator of the Environmental Science and Public Policy Archives at Harvard University.[221] This seemed simple enough, but when McIntyre submitted a request to Clark for a copy, he received an extraordinary reply.

> [P]lease let me know your desired time to visit (no later than one week prior) so that I can make sure the materials will be ready for you. I will be away from the office June 21–July 5, so the materials will not be available during that date range.[221]

In other words it appeared that Clark was expecting him to travel from Toronto to Boston if he wanted to look at the comments. Wearily, McIntyre set about the trying to grind down the bureaucrats before they ground him down. He duly composed a letter to the secretary of the IPCC, Renate Christ, complaining about the suggestion that he would have to travel to Boston to see the comments. Eventually, a reply was sent by Manning denying that TSU had said that this was necessary – it was of course Clark at Harvard who said that – and that Clark had confirmed that he was 'very willing' to copy and send what McIntyre required.

Clark had in the meantime indicated the terms on which he would perform the photocopying:

> I can provide a photocopy of up to 100 pages for research purposes only (not republication) for our interlibrary loan fee of $34 plus 40 cents per page [there were a total of 1834 pages of comments]. Copyright of the material resides with its authors. It may be possible for you to hire a research assistant locally to look over the materials if that would be helpful in selecting materials of most interest. I can recommend someone if you like.[221]

McIntyre's complaint dragged on for months, with the IPCC seemingly moving heaven and earth to avoid giving him a convenient searchable digital copy. It wasn't until the end of June that they relented a little, Manning sending an email indicating that he was now willing to make the comments available, provided that they were not redistributed to anyone. It was not at all clear whether Manning meant that McIntyre could not quote from the text or merely that he was not to distribute the document to anyone else. Neither could McIntyre understand from where Manning thought he derived the authority to add this restriction – review comments were, after all, meant to be kept in an open archive, so they could hardly be said to be confidential.

The next day, McIntyre received a large package from the IPCC, but given the restrictions that Manning had placed on its use, McIntyre chose to leave it unopened: opening it might have implied acceptance of Manning's conditions. The correspondence continued to flow both ways, with Manning resorting to some fairly bizarre justifications for restricting the use of the comments, but eventually he backed down further and agreed to make an electronic copy available. Even so, he wanted to retain the restrictions on their use. An open review process could only be achieved he said, if the review comments could not be selectively quoted.

> We would not be promoting a transparent and open process, nor would we be acting responsibly to our authors and many expert reviewers, if there were no restraint on others selectively editing and redistributing review materials.[221]

With those watching the process now doubled up in laughter at Manning's intellectual contortions, and a barrage of freedom of information requests threatening to force disclosure in the near future, the end was clearly not far off. A few days later the IPCC capitulated and the comments were finally posted up at UCAR's website.

Some things remain secret

The contents of the second draft comments document have mostly been discussed in Chapter 11, but there were other intriguing aspects to this part of the story that need to be recounted. When he had looked over the contributions of his fellow reviewers, McIntyre had realised that, despite all the submissions and rejections and resubmissions of Ammann and Wahl's two articles, neither of his two opponents appeared to have made a comment on the IPCC paleoclimate chapter. This had seemed so unlikely as to be simply unbelievable, but at the time, there was little McIntyre was able do about it and he had set the issue aside.

Then in 2008, fully two years later, when McIntyre was going over the second draft review comments, he found himself considering one of the answers Briffa had given to a comment left by McKitrick. McKitrick had been discussing the benchmarking of the RE statistic and the shenanigans over the submission dates of the CC paper, and had received a lengthy response from Briffa rejecting his points. However, to his surprise, McIntyre noticed that the arguments used by Briffa were identical to those used by

Ammann in his new CC comment.[220] This was rather extraordinary, because these arguments had not appeared anywhere in the literature at the time of the IPCC review; Ammann's comment on MM05(GRL) hadn't been resurrected until August 2006, months after the deadline for IPCC submissions. This could mean only one thing: Ammann and Wahl had been allowed to make contributions to the IPCC report in secret and entirely outwith the normal review process.

A new member of the audit team

As McIntyre's examination of the IPCC review process continued, more and more oddities came to light. For example, when the review comments were posted up on the UCAR website, it was discovered that, although the contributions of the expert reviewers and the official government reviewers had been made available, there was no sign of the review editors' comments. Review editors were a kind of umpire for the review process – their job was to ensure that where there were disputes, the lead authors didn't simply enforce their own point of view, but fairly represented both sides of the argument.

> Review Editors will assist the Working Group/Task Force Bureaux in identifying reviewers for the expert review process, ensure that all substantive expert and government review comments are afforded appropriate consideration, advise lead authors on how to handle contentious/ controversial issues and ensure genuine controversies are reflected adequately in the text of the Report.

> Review editors will need to ensure that where significant differences of opinion on scientific issues remain, such differences are described in an annex to the Report. Review editors must submit a written report to the Working Group Sessions or the Panel . . .[222]

Clearly, the treatment of the Hockey Stick affair was hugely controversial, and yet the final draft had sided pretty clearly with Mann and his team. What then had been the contribution of the review editors? Had they tried to have McIntyre's objections incorporated and yet been ignored, or had the checks and balances for which they were responsible been worthless?

David Holland, a British Climate Audit reader, took it upon himself to investigate. The review editors for the paleoclimate chapter were John Mitchell, the chief scientist of the UK's Meteorological Office, and Jean Jouzel of France. Holland decided to approach Mitchell for the information and sent a request off to the Met Office. This went unanswered for several weeks, and Holland eventually decided to put in a request under the UK's Freedom of Information Act. At this point, a reply was received from Mitchell, but it was evasive: the sign-off by review editors, Mitchell explained, was only available from Manning's TSU, which was in the process of being closed down, its task complete. Fortunately this appeared to be Mitchell excusing himself for not replying to Holland earlier: the Freedom of Information request did its job and at the end of January 2007, Holland took possession of the full set of review editor comments.

The review editor comments were based around a standard form letter, although this letter was slightly different for each of the IPCC working groups. These letters were remarkably brief, amounting to just a single paragraph. Working Group II, which looks at the impacts of climate change, required review editors to confirm that review comments had been appropriately considered by the authors 'in accordance with IPCC procedures'. Working Group I, which oversaw the scientific papers, and therefore the paleoclimate chapter, had a slightly but significantly different sign-off, omitting the last few words used by Working Group II. In other words, they seemed to be neglecting to get confirmation that the authors had dealt with the review in accordance with IPCC procedures. And there was another oddity in Working Group I. One of the review editors had chosen not to use the form letter but had submitted his own report: none other than Mitchell, the review editor of the paleoclimate chapter.

So, how had Mitchell dealt with the Hockey Stick controversy? How had he ensured that scientifically valid comments were properly covered in the report. His comments were as follows:

> . . . I can confirm that the authors have in my view dealt with reviewers' comments to the extent that can reasonably be expected. There will inevitably remain some disagreement on how they have dealt with reconstructions of the last 1000 years and there is further work to be done here in the future, but in my judgment, the authors have made a reasonable assessment of the evidence they have to hand. . . . This has gone some way towards reconciliation but I sense not everyone is entirely happy.
>
> With these caveats I am happy to sign off the chapter. . .[223]

And as far a substantive comment went, this was the totality of Mitchell's output as a review editor. He clearly recognised that there was a dispute, but he seemed to have taken it upon himself to consider whether the authors' assessment was reasonable, rather than whether they had reported both sides of the argument as their mandate required them to do. He certainly did not offer an opinion as to whether the authors had complied with IPCC procedures.

Jouzel's report was even more scant:

> . . . I can confirm that all substantive expert and government review comments have been afforded appropriate consideration by the writing team in accordance with IPCC procedures.[223]

McIntyre and Holland were both amazed at how perfunctory these reviews seemed to have been. The level of attention that appeared to have been given to the process made a mockery of the IPCC's claims about the thoroughness of their procedures. Holland therefore decided to make sure that there was really nothing else to uncover, sending off another request for supplementary and working papers related to the review.

The reply from Mitchell was again rather surprising. Firstly, it seemed that there had been *no* supplementary papers submitted with the review, but the real surprise was Mitchell's revelation that he had *destroyed* his working papers. There was, he said, no requirement to retain them. This appeared to conflict directly with IPCC policies, which stated that review comments would be kept in an open archive for a minimum of five years.

Holland persisted, submitting a series of email requests covering all the possible ways in which IPCC might be trying to evade handing over the information. He first asked for all of the emails Mitchell had sent in his capacity as an IPCC review editor, but received the extraordinary reply that there had been none, apart from a few emails to IPCC colleagues prompted by Holland's initial request. So despite having been in the position of review editor for several years and despite there being a major difference in scientific opinion in Mitchell's chapter, he had not actually managed to exchange any emails with the authors.

Holland made a further request, which was again stonewalled, and he therefore decided to take a slightly different tack, asking when Mitchell had destroyed his emails and working papers and requesting that the Met Office retrieve them from their backups and archives. This new approach seemed to have the desired effect, since there was a sudden change in tune from the scientists. Met Office officials were mistaken, it seems, in advising Holland that Mitchell's emails had been destroyed. Mitchell's work had allegedly been performed in a *personal* capacity and as such, all of the emails relating to his work were therefore not disclosable. As Holland persevered, requesting details of expense claims and information about Mitchell's holidays, the Met Office changed tack once again, invoking a derogation from the Freedom of Information Act that permitted information to be withheld where its release would affect

British relations with an international organisation, by which they presumably meant the IPCC.

Meanwhile, Holland had also been pursuing Briffa for information about the IPCC process and was meeting with a similar wall of obfuscation, not the least of which was Briffa's failure to reply to Holland's emails. Holland had, however, managed to get hold of some of Mitchell's correspondence, among which was an email from Briffa in which he had told Mitchell that he would make a brief reply to Holland 'when he got round to it'. Tiring of the delays, Holland took the legal route once more, requesting the data under the stronger terms of the Environmental Information Regulations (EIR), and this time also asking for copies of any correspondence between Ammann and Briffa on the subject of the Fourth Assessment Report, a request designed to probe the issue of whether Ammann had been providing Briffa with unpublished findings. He promptly received an acknowledgement from Briffa's employers at the University of East Anglia, which confirmed that his request was being considered under the terms of EIR.

Then a month later, he received a refusal under the Freedom of Information Act. The grounds given were partly cost and partly because it was alleged that the correspondence was confidential. This was another astonishing claim, given that the IPCC review process was allegedly open and transparent. The rejection letter explained that 'the persons and organisations [i.e. Ammann] giving this information to us . . . believe it to be confidential and would expect [it] to be treated as such'.[224] As the IPCC was the only organisation that can conceivably have been in a position to ask for confidentiality, this claim strongly suggested that IPCC bureaucrats had sought to undermine the openness that governments had built into the operating procedures for the review. Either way there was to be no light shone on Mitchell's work or on Ammann's review comments from this corner.*

---

*   At the time of writing, David Holland continues to press Briffa and Mitchell to release the papers. The two scientists are still steadfastly refusing.

13 Update the Proxies!

Nature is often hidden, sometimes overcome, seldom extinguished.

Francis Bacon

Mann's comments

The IPCC's conclusion that the divergence problem was restricted to a few proxies in a few geographical areas did not appear to have a firm grounding in the scientific literature. Part of the difficulty in assessing the situation more rigorously was that, incredibly and despite all the billions of dollars poured into global warming research, virtually none of the proxy records had been updated since 1980. We have seen a couple of exceptions, albeit problematic ones, in the shape of the Polar Urals and Gaspé updates. Even the original MBH98 paper only included proxy records up to 1980, lagging nearly twenty years behind the publication date. By the time of the Fourth Assessment Report in 2007, the proxies had seemingly still made no further progress, with the record apparently remaining stuck at 1980. Even the scale on the IPCC's spaghetti graph only ran up to the year 2000. If they had shown the scale right up to date, the proxy records would have ended nearly 30 years short of the date of the report. But by truncating the scale at the end of the millennium this was reduced to 20 years, helping to obscure the failure to update the record and at the same time avoiding any discussion of the lack of a rise in temperature since the end of the millennium.

Early on in the blog war between Climate Audit and RealClimate, Michael Mann had posted up a partial explanation of why there had been no updating of proxy records for nearly 25 years. Responding to a question on the divergence problem he said:

> Most reconstructions only extend through about 1980 because the vast majority of tree-ring, coral, and ice core records currently available in the public domain do not extend into the most recent decades. While paleoclimatologists are attempting to update many important proxy records to the present, this is a costly, and labor-intensive activity, often requiring expensive field campaigns that involve traveling with heavy equipment to difficult-to-reach locations (such as high-elevation or remote polar sites). For historical reasons, many of the important records were obtained in the 1970s and 1980s and have yet to be updated.[225]

Mann's response clearly covered all the different proxy types, but it is fair to say that the vast majority of these are tree rings. While McIntyre had not collected any tree ring samples himself, his career in the mining industry had given him an easy familiarity with out-of-the-way places, and his gut feel was that Mann was making much more of the difficulties than was justified. Indeed there were hints in the literature that it was considerably easier to collect tree ring samples than Mann would have his RealClimate readers believe. In one of the classic bristlecone studies, Lamarche had written:

> D.A.G. [Graybill] and M.R.R. [Rose] collected tree ring samples at 3325 m on Mount Jefferson, Toquima Range, Nevada on 11 August 1981. D.A.G. and M.R.R. collected samples from 13 trees at Campito Mountain (3400 m) and from 15 trees at Sheep Mountain (3500 m) on 31 October 1983.[226]

So if it was possible to collect samples from 28 trees on two different mountains on a single day in the 1980s, it can hardly have been very difficult to do the same in the twenty-first century, especially now that funding was pouring into climatology. For sure there were some proxies such as the Himalayan ice cores that would require more work and cost to update, but the majority should be entirely routine.

McIntyre had amused himself by ridiculing Mann's claims that these sites were difficult to reach, quoting extensively from a guidebook to one of the bristlecone sites, Almagre Mountain, an area which was much frequented by hikers and cyclists. As he explained:

> To get to these sites from UCAR headquarters in Boulder, a scientist would not merely have to go 15 miles [south-west] of Colorado Springs and go at least several miles along a road where they would have to be on guard for hikers and beware of scenic views, they would, in addition, have to go all the way from Boulder to Colorado Springs. While lattes would doubtless be available to UCAR scientists in Colorado Springs, special arrangements would be required for latte service at [Almagre], though perhaps a local outfitting company would be equal to the challenge. Clearly updating these proxies is only for the brave of heart and would require a massive expansion of present paleoclimate budgets. No wonder paleoclimate scientists have been unable to update these records since Graybill's heroic expedition in 1983.[227]

Testing the Starbucks hypothesis

In 2007 McIntyre started to look anew at the issue of the divergence problem. Over the previous three years, several of his critics had accused him of attacking other people's results rather than actually doing original work of his own. Now, with the NAS and IPCC reports behind him, he thought he saw an opportunity to disprove Mann's assertions about the difficulties of collecting tree ring samples, shoot down the arguments of his critics, and at the same time make an important contribution to the climate debate: he would update the proxies himself. With nearly twenty-five years of new growth on many of the trees, there was a wonderful opportunity to answer the question that had plagued dendroclimatology for years – were the tree rings actually capturing any temperature information at all? A whole new verification – an out-of-sample test – could be performed on the last 25 years simply by collecting a few tree ring samples.

By happy coincidence, McIntyre had a sister living in Colorado Springs, close to many of the key bristlecone sites, and so, in the summer of 2007, he arranged to pay her a visit. McIntyre's idea was to try to find some of the trees that had been sampled by Graybill back in the 1980s. Finding the exact trees was going to be no easy task because, while their positions had been mapped and the trees would normally also be marked in some way, finding these markers was going to take a lot of doing. In order to share the load and to make more of a party of the trip, he therefore arranged to link up with a Climate Audit reader, called Pete Holzmann, who lived locally and had indicated that he was willing to help out. Having obtained a permit from the US Forest Service, the two men hired an off-road vehicle and armed themselves with supplies and sampling equipment. Their tongue-in-cheek idea was to test the hypothesis that it was possible to have coffee at Starbucks in the morning, sample some tree rings during the day and still be home in time for supper. Much amusement was had posing for photos outside the local Starbucks before McIntyre and Holzmann, together with their wives, headed for their chosen site: Almagre mountain.

Getting to Almagre was just as straightforward as they had expected, and they were able to extract several cores on day one. Finding the original trees that Graybill had sampled back in the 1980s was slightly more difficult though. Modern dendrochronologists use GPS to record the exact position of each tree they sample, but Graybill's work predated this technology. There was no alternative but to examine each tree in turn, trying

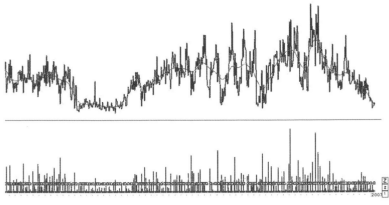

FIGURE 13.1: Analysis of rings of Almagre tree 84–55

to find the small plaques with which Graybill had marked the trees. Fortunately, after three days of hunting high and low, Holzmann finally located a group of marked trees, their identities confirmed in several places by checking their appearance to photos that Graybill had taken during his research. By the time they returned to Colorado Springs, the Climate Auditors had managed to take 64 cores from 36 trees at five different locations. They had found 17 Graybill trees and had resampled 8 of these. Mission accomplished.

Safely back in Canada, McIntyre posted up a report of his trip for his Climate Audit readers. The samples were immediately sent off to a professional dendrochronology lab for analysis, and the results trickled out over the next weeks and months. By October, McIntyre was able to report the first analysis from one of the original Graybill trees, and the figures were just as he had expected: tree ring widths, in this tree at least, had been declining for all of the 1980s and 1990s and into the twenty-first century. They had completely failed to capture the late twentieth-century temperature spurt that was shown in the instrumental record (see Figure 13.1).

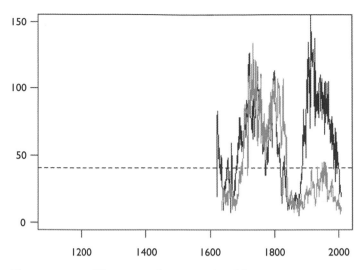

FIGURE 13.2: Two cores from tree 84–56

Of course, this was just a single tree and nothing could be concluded from a sample of one, but it certainly whetted the appetites of the Climate Audit readers. As further results came in though, it became rather more intriguing. On one particular tree, number 84–56, Holzmann had decided to extract two separate cores – one from the west and one from the southwest – and to McIntyre's astonishment, it turned out that the ring measurements from these two samples were entirely different. Figure 13.2 shows the ring analysis of the two cores.

The explanation for this discrepancy turned out to be surprisingly obvious. When sizing up the tree in question, Holzmann had noticed that the tree wasn't in fact circular in cross-section – it was a distinct oval, and this distortion appeared to have been driven by the fact that the tree was stripbarked. Stripbarking, you will recall,* is a form of die-back of the bark on one side of the tree, which can cause a growth spurt on the opposite side in future

---

*   See page 172.

years, as the tree compensates for the loss of part of its bark.

What this suggested, then, was that the growth spurt could have been caused, not by temperature changes as proposed by Mann, not by carbon dioxide fertilisation, as Graybill had mooted all those years ago, but by the process of the tree compensating for stripbarking. Graybill had reported that he had actually *sought out* stripbarked trees, thinking these would show the effects of carbon dioxide fertilisation better. Was it possible that he had inadvertently documented an entirely different effect? The implications of McIntyre's finding would be immense, if repeated across the other trees in the sample. Remember, the hockey stick shape of Mann's PC1 was derived almost entirely from Graybill bristlecone pines. If the growth spurts that Graybill had found were due to stripbarking rather than temperature rises, the Hockey Stick would no longer be credible. Another, perhaps decisive, nail would be hammered into the coffin of Mann's paper.*

The Ababneh thesis

As if this were not enough, at around the same time an entirely new source of evidence about the origins of the growth spurt in bristlecone pines unexpectedly came to light. McIntyre had been aware that one of Malcolm Hughes' students had been studying bristlecone pines and he had heard that this student had updated the Sheep Mountain chronology in 2002. Unfortunately, like so many other proxy updates, it had never appeared in print, an omission that was surprising in view of the importance of the site and the paucity of updated proxy records. Once again, McIntyre's mining background coloured his judgement of why this should be – in his experience, delayed results usually indicated that the results didn't give the 'required' answer. He had followed this hunch up several times, attempting to contact the student in question, whose name was Linah Ababneh. Unfortunately she

---

*   McIntyre presented the results of his trip at the AGU Fall Meeting in 2007, but has yet to publish the findings in a journal.

failed to reply to his emails and McIntyre had set this inquiry aside for another day.

Then one day, while trawling through Hughes' website to see if there was anything new to read, he chanced upon a web page dedicated to the PhD theses of students in Hughes' department. And there, at the top of a list he noticed the newly published thesis of Linah Ababneh.[228] It was breathtaking.

Ababneh's project had involved an investigation of the differences between the stripbark and wholebark trees, looking specifically at sites visited by Graybill and Idso. She had visited two sites, Sheep Mountain and Patriarch Strip, sampling both types of tree in both sites. What her thesis showed was that during the 1980s and 1990s there was no sign of the surge in ring widths that should have accompanied the rise in instrumental temperatures, a clear confirmation of the divergence problem. But the implications were even more serious than that. When you compared her Sheep Mountain results to Graybill's figures for the same site and to the earlier Lamarche study, it was immediately apparent that her new results didn't even confirm the existence of the growth spurt that the two earlier researchers had reported. Figure 13.3 shows Ababneh's Sheep Mountain update alongside Graybill's results for the same location. There is no hockey stick shape shown in new figures, and remember, Mann's PC1 was dominated by Sheep Mountain.* The implications were of huge importance to climate science and to the political world.

What was still more amazing was that, while one of Ababneh's supervisors was Hughes, her thesis contained *not a mention* of the divergence problem. Hughes did not even seem to have required her to attempt to reconcile her findings to Graybill's, an amazing omission given his intimate knowledge both of Graybill's work and its importance in the Hockey Stick papers. And all the while, Hughes was still publishing papers that used the old Graybill version of the data (with the hockey stick shape).* To McIntyre

---

*   See page 121.

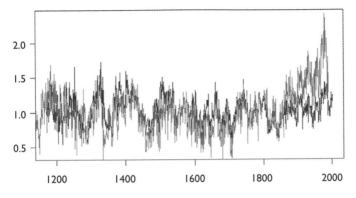

FIGURE 13.3: The Ababneh update to Sheep Mountain

Black: Ababneh update; grey: Graybill

and his readers, this was simply unbelievable. Why would he not use his own student's more-up-to date data, or at the very least, discuss it in his publications?

The significance of these findings is hard to overestimate: not only had Mann used incorrect statistics to force his temperature reconstruction into the hockey stick shape of the bristlecone pines, but it now appeared as though that shape was merely an artefact of stripbarking. Nor was MBH98 the only paper where Sheep Mountain was a key proxy. It was used in MBH99 and Mann and Jones 2003 as well. From these papers it had also found its way into Rutherford et al, which used Mann's PC1, and to Hegerl et al 2006 and Osborn and Briffa 2006, both of which used the Mann and Jones PC1.

Ababneh had indicated in her thesis that her underlying data would be archived in due course. However, by the time McIntyre noticed her thesis, there was still no sign that she had done this and so he decided to contact her directly. Since the completion of her doctorate, Ababneh had left the University of Arizona and had

---

*   See, for example, Hughes and Salzer 2007, which includes the Graybill version of Sheep Mountain rather than the Ababneh update.[229]

moved to a new base at William and Mary College in Virginia. Fortunately McIntyre was able to find an email address for her and sent off a brief message enquiring after the data and asking a few questions about her work. There was no reply, and further enquiries at the University of Arizona determined that Ababneh's data was no longer held there. So, in an eerie echo of the farce of the Gaspé series, the data for an updated tree series, which was found to have lost its former hockey stick shape, had not been archived and had then promptly been lost.

Shortly afterwards, a Climate Audit reader managed to make contact with Ababneh by telephone. Obliged to respond, Ababneh said, somewhat implausibly, that her attorney had advised her that McIntyre's approach had been 'improper' and that she should not supply the data.[230] When McIntyre tried again to make email contact, his messages were returned undelivered, presumably due to his IP address having been blocked.*

Without Ababneh's data, it was tricky, but not impossible, to make further progress. Hans Erren,** who was still involved in the sceptic efforts, digitised the graphical representations of the data in Ababneh's thesis, and while this wasn't quite as good as having the actual data, it did allow McIntyre to make a powerful case. For example, when he recreated the Mann and Jones PC1 using the updated Sheep Mountain chronology, he found that the Hockey Stick disappeared entirely. The effect is shown in Figure 13.4. The top chart includes the original Graybill data, processed through a Mannian short-centred algorithm, which produces a pronounced hockey stick shape. The lower chart shows what happens when the Graybill numbers are swapped for figures derived from the Ababneh thesis and a standard centring algorithm is used: the hockey stick shape disappears.

---

*   McIntyre suggested he thought that Ababneh was a pawn whose actions should be viewed sympathetically: 'I really don't want to discuss Linah Ababneh's decisions or motives any more; she's young and trying to make her way in the climate community.'[230]

** See page 55.

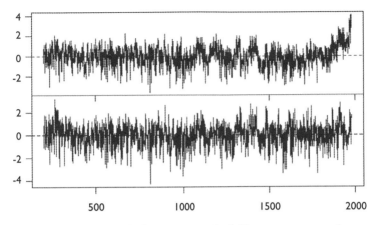

FIGURE 13.4: Mann & Jones PC1 with different versions of Sheep Mountain

Top: Using Graybill version and short centring.
Bottom: Using Ababneh update and standard centring.

The results now seemed unarguable. Evidence from multiple sources was now demonstrating unequivocally that the bristlecone pines, which gave the Hockey Stick its attention-grabbing shape, were not in fact capturing temperature information at all. Surely now, after so many years of work, after so many flaws had been uncovered and demonstrated to the satisfaction of all but its most dogmatic supporters, surely now the Hockey Stick was broken once and for all.

14 A New Hockey Stick

When you're in a hole, stop digging.

Denis Healey

By 2007, there was little more for McIntyre and McKitrick to discover about the Hockey Stick. Certain areas remained a mystery – the confidence interval calculations and the basis on which Mann had retained PCs, for example – but without the rest of Mann's code, these were unlikely ever to be understood. Climatology is a big subject, however, and there were other areas into which McIntyre would delve from time to time, often with amazing results.

One particularly fruitful subject was the quality of the instrumental temperature records and the code used to process the weather station data into headline figures. With the help of a rag-tag group of occasional climate auditors, McIntyre started asking probing questions about every aspect of the temperature records. As far back as 2005, he had unearthed a significant and embarrassing error in the adjustments the UK's Climatic Research Unit had been making to its sea surface temperatures. In 2007, another error was found, this time in James Hansen's NASA surface temperature records. This discovery led to a reassessment of how hot the 1990s were in relation to the rest of the instrumental record for the USA. Hansen had previously claimed that the 1990s were hotter than anything seen previously, but with the correction in place, it emerged that it had in fact been hotter in the 1930s. The findings were seen by sceptics as something of a coup for McIntyre, as they led to another round of newspaper headlines around the world.[231-233] Shortly afterwards, all the media attention forced Hansen to issue a correction and to finally release his long-withheld computer code. This quickly caused another embarrassment when observers noted that the code was both extraordinarily labyrinthine and also very badly documented. Eyebrows were raised by participants on both sides of the global warming debate.

Meanwhile, another close associate of the Climate Audit website, former TV weatherman Anthony Watts, began a volunteer effort to survey the hundreds of weather stations that were the basis of the land record for US temperatures. The poor quality of the siting and maintenance of the majority of the stations again raised questions over how reliable the instrumental record really was.

By 2007, Climate Audit had become a hub for global warming sceptics, where news and opinion were exchanged alongside discussion of the results of the many research projects undertaken by McIntyre and the readers. But it was more than just that. Guest writers, among them several professional scientists, were posting articles on subjects as diverse as hurricanes, polar ice records and climate models. Even some prominent climatologists started to risk the opprobium of the research community by posting comments and engaging in the debate at Climate Audit. Meanwhile, the recognition of McIntyre's work by many professional scientists continued to develop and he now received regular invitations to lecture at seminars around the world.

With so much expertise now assembled around him, McIntyre's humble website had itself been transformed into something resembling a series of scientific seminars. Despite its unstructured nature, which critics might characterise as 'haphazard', McIntyre's online community was demonstrating a remarkable ability to get to the bottom of difficult scientific questions. In 2008, the Climate Auditors had a chance to show just what they could do.

On the morning of 2 September 2008, McIntyre posted a terse notice, informing his readers that there were media reports of a new

Mann temperature reconstruction, which was shortly to be published in the *Proceedings of the National Academy of Sciences* (PNAS).[234] Over the next few hours some of the details of what was in the study started to creep out. It was soon obvious that this was going to be a serious attempt to breathe life back into the Hockey Stick. Despite all the evidence that proxy records were not reaching new levels in the twentieth century, Mann was apparently still maintaining that modern temperatures were unprecedented: the new study was another hockey stick, this time suggesting that modern temperatures were the highest they had been in 1300 years. Mann was certainly not one to throw in the towel. However, knowing that most proxies were not hockey stick shaped, McIntyre and his readers could be fairly certain that the shape of the new study was either erroneous or contrived. They just had to uncover how it had been done.

From a public relations point of view, the big selling point of the new paper was to be a claim that Mann had been able to get a hockey stick without using tree rings. This was obviously important in view of the controversy over the reliability of tree ring proxies in general and the bristlecones in particular. As Mann explained in the news release:

> Ten years ago, we could not simply eliminate all the tree-ring data from our network because we did not have enough other proxy climate records to piece together a reliable global record.[235]

This was a curious statement. For the previous ten years, Mann had maintained that the Hockey Stick was reliable, and moreover that it was 'robust to the presence/absence of dendroclimatic indicators'.[14] He had even maintained this stance in the face of strong evidence, in the form of the CENSORED directory, that this was not so and that he knew it was not so.

The media were ready and waiting for the new paper. The BBC was first out of the blocks, reporting that the 'Climate "hockey stick" is revived', closely followed by Canada.com, who reported that 'Past decade warmest in 1,300 years' and the *Christian Science Monitor* whose headline read 'A gnarlier "hockey stick", the same message'.[236–238] *Nature* was of course in the forefront of the media barrage, predicting that the new paper would silence Mann's critics.[239]

While the media had received their press releases in advance, the Climate Auditors were caught unawares, but within minutes of McIntyre posting up his report, sceptical readers were swinging into action. At first there were only the news reports to look at, but by mid-morning the paper itself had been posted online and, by eleven a.m., McIntyre was linking to some early comments on the article by one of his regular correspondents, a Czech theoretical physicist. Just minutes later, Demetris Koutsoyiannis, professor of hydrology at the University of Athens, also posted some observations on the shape of the reconstruction.

One hopeful sign in the paper was a statement indicating that Mann had taken on board the findings of the NAS panel:

> We were guided in this work by the suggestions of a recent National Research Council report [i.e. the NAS report] concerning an expanded dataset, updated data, complementary strategies for analysis, and the use of thoroughly tested statistical methods.[234]

This was encouraging as far as it went, although it rather alarmingly didn't mention the panel's warnings about the unsuitability of bristlecone pines for use in temperature reconstructions. Would Mann dare to use bristlecones in the new paper? Only time would tell.

To everyone's surprise the data and code appeared to have been published alongside the new paper, in an apparent triumph for McIntyre's ceaseless efforts to improve standards in this area. Climate Audit readers busied themselves with downloading everything they could. Interesting details were picked up and posted with bewildering speed. Some of Mann's new reconstructions seemed to have much larger Medieval Warm Periods than MBH98. The R^2 was still not being used. McIntyre pointed out that Mann was still using a method that fished for proxies with correlations to temperature, with all the attendant risks of spurious correlation and the creation of a hockey stick entirely from noise.

Craig Loehle, the ecologist we met in Chapter 10, noticed something odd in the smoothing of the instrumental records. His observation was picked up within the hour by a retired professor of statistics. Somehow, Mann had managed to get a forty-year smooth of the instrumental record to extend up to the present day. By rights it should not have been possible to get past the 1960s, because the smoothing formula replaces the raw value for a data point with a weighted average of values prior to and after it. By evening time, another statistician had worked out the extraordinary story of how the instrumental record had been artificially extended by padding the data for many years into the future, the future values created by repeating the most recent real values in reverse order.

Several readers raised the subject of the new paper's peer review. At the end of the paper, was the following acknowledgement:

> We are indebted to G. North and G. Hegerl for their valuable insight, suggestions, and comments and to L. Thompson for presiding over the review process for this paper.[234]

This suggested strongly that Gerry North and Gabriele Hegerl had been the peer reviewers of the paper, something that might have been considered rather unfortunate in view of Wegman's criticisms of the paleoclimate community being too insular. The absence of a statistician among the peer reviewers also seemed inappropriate. However, with North having headed the NAS panel, there was at least an expectation that he would bring his experience to bear and that many of the problems with MBH98 would be avoided in the new paper.

Unfortunately, within days of the paper's publication, McIntyre was reporting that most of the 'usual suspects' were included in the proxy database: Briffa's Tornetrask record had been used instead of Grudd's update, Yamal replaced the Polar Urals update once again. And as everyone feared, the bristlecones were there too in the shape of Sheep Mountain, the obsolete hockey stick version used instead of Ababneh's update. This was remarkable when one realises that Hughes, who had supervised Ababneh's thesis, was a co-author on the new paper. He must have known that Ababneh's work failed to replicate Graybill's. The inclusion of the series was even more extraordinary when one realises that it flew in the face of the NAS panel's recommendation against the use of bristlecones, and the fact that North failed to spot this fatal flaw during his peer review.

There were some new data series as well, and McIntyre started to post graphs of these for his readers to view. They were devoured, several people noticing that few if any of them showed any modern warming. The sole exception among the first batch was a group of four lake sediment series from Finland, known as the Tiljander proxies. Readers delved and dug, locating the original paper in which the series had been described and also the PhD thesis of its author, Mia Tiljander. It turned out that the twentieth century uptick in Tiljander's proxies was caused by artificial disturbance of the sediment caused by ditch digging rather than anything climatic. Mann had acknowledged this fact, but then, extraordinarily, rather than reject the series, he had purported to demonstrate that the disturbance didn't matter. The way he had done this was to perform a sensitivity analysis, showing that you still got a hockey stick without the Tiljander proxies.

Great care is needed when reading scientific papers, particularly in the field of paleoclimate, and this was one of the occasions when one could have come away with an entirely wrong impression if the closest attention had not been paid. The big selling point of Mann's new paper was that you could get a hockey stick shape without tree rings. However, this claim turned out to rest on a circular argument. Mann had shown that the Tiljander proxies were valid by removing them from the database and showing that you still got a hockey stick. However, when he did this test, the hockey stick shape of the final reconstruction came from the bristlecones. Then he argued that he could remove the tree ring proxies (including the bristlecones) and still get a hockey stick – and of course he could, because in this case the hockey stick shape came from the Tiljander proxies. His arguments therefore rested on having two sets of flawed proxies in the database, but only removing one at a time. He could then argue that he still got a hockey stick either way. As McIntyre said, you had to watch the pea under the thimble.

As the readers dug on, another of the sceptics' favourite proxies reappeared. Briffa's tree ring density series turned up like a bent penny in Mann's new paper. Once again the inconvenient divergence had been truncated at 1960. This time though there was a remarkable new twist to the story. The inclusion of the Briffa series would have presented Mann with a problem because he needed data that ran right up to the present day. However, Mann had a trick up his sleeve. This involved a new approach to the problem of filling in gaps in the proxy series. We have seen how, where there was a gap in a series in MBH98, Mann had extended the final value of the series up to the year 1980. In the case of Gaspé he had copied the value from the year 1404 into earlier years, extending the start date of the series backward to the year 1400. In the new paper, however, he adopted a more sophisticated methodology.

> The RegEM algorithm of [Tapio Schneider] was used to estimate missing values for proxy series terminating before the 1995 calibration interval endpoint, based on their mutual covariance with the other available proxy data over the full 1850–1995 calibration interval.[234]

In other words the missing data had been infilled using a mathematical algorithm, which looked at the other series and calculated a likely value for the missing data. While the question of infilling data in this way is fraught with difficulty at the best of times, the effect in the case of the Briffa series was remarkable. Here, there was no missing data anyway, or at least there wouldn't have been if the inconvenient downward trending twentieth century hadn't have been deleted. However, with the truncation in place, the RegEM algorithm infilled the gap it had created with a new, upward trending set of data points. The downtick had become an uptick. This procedure had passed peer review. Climate Audit readers were speechless.

Another bombshell the same day was the discovery that any comments on the paper had to be submitted within three months of publication. After that, no contributions would be accepted. If McIntyre and McKitrick were going to submit a critique of the new Mann paper, they and the Climate Audit readers were going to have to pull out all the stops.

Over the next month information poured in. A breakthrough was made when one reader, who had been studying Tiljander's Finnish lake sediment study, noticed that Mann had made an extraordinary error. He had misinterpreted the way the sediment responded to temperature changes. Tiljander and her colleagues had explained in their original study that a high x-ray density measurement in the sediment was indicative of lower temperatures, as higher snowfall led to a bigger spring melt, increased erosion and so to a higher mineral content in the water.[240] But as we saw above, twentieth century ditch digging had also caused higher mineral content in the water. Mann's new algorithm, however, had picked up the high x-ray density this had caused and matched it up against global temperatures. Finding a good correlation, it had therefore interpreted high x-ray density as evidence of *higher* global temperature. So not only did ditch digging cause the uptick rather than climate, but Mann's interpretation of the series was also upside-down.

In fact, Tiljander had noted evidence of a strong Medieval Warm Period in the record, something that Mann had presumably missed because he was looking at it upside-down. Finnish TV had even broadcast a documentary on Finnish climate history, including the Tiljander series as evidence of medieval warmth. The point was somewhat moot because, as we have seen, the twentieth century uptick in the series (or from Tiljander's perspective, downtick) was due to ditch digging rather than climatic factors, but nevertheless it was an embarrassing error for Mann and one that threatened to undermine his non-tree ring reconstructions, which appeared to be strongly influenced by the Tiljander proxies.

Some strange screening of the proxy series was also uncovered. Mann had reported that he had rejected proxy series that didn't have a significant correlation to temperatures in their local gridcell – there were thresholds set for each type of proxy. However, it was actually slightly more complicated than that. If the series failed the test of correlation to the local gridcell, Mann gave them another chance by testing the correlation to the next nearest gridcell. Then, one of McIntyre's professional statistician readers, who was trawling through Mann's code, stumbled across yet another oddity. On some of the reconstructions it turned out that Mann had used a different screening procedure, simply removing any series that had a negative correlation to temperature. However, whichever way you looked at it, these screenings would essentially only give series with upward trending twentieth century records. As we saw on page 300, if you add such series together, you will get a hockey stick, simply because the twentieth century portions add together to give the blade, while the ups and downs in the earlier sections cancel out to give the long flat handle. Mann should have taken this into account in calculating his new benchmark significance levels, but it appeared that he had not done so.

It went on . . . and on. Among the new proxy series, readers were astonished to see one that was apparently a documentary record of temperatures in East Africa dating back to 1400. If true this would have overturned everything known about the history of the continent, but it was quickly discovered that Mann had inadvertently swapped the latitude and longitude, and the series should have been located in Spain. So in an amusing echo of the 'rain in Maine falling in the Seine', it looked as though 'the rain in Spain was falling mainly in the plains' of Kenya. It became positively farcical when it was discovered a short time later that the proxy wasn't a documentary record at all – it really *was* a rainfall record.[241] It was doubly surprising when one notes that this series had apparently passed the test of correlation with its gridcell temperature record. This test seemed less than plausible if one could test it against the wrong gridcell and still have it pass.

In the event, someone on Mann's team was clearly reading Climate Audit because the error was corrected and a revised reconstruction was posted to Mann's Supplementary Information site. The correction led to a change of 0.5°C in the eighteenth century. As McIntyre noted, Mann had claimed at the NAS hearings that he knew temperatures right back to the fifteenth century to an accuracy of 0.2°C, a position that now looked increasingly untenable.

There was so much ammunition that it was hard for McIntyre and McKitrick to know where to start when they sat down to write their formal comment on Mann's new paper.[242] This excess was a problem because as well as demanding that comments be received within three months of the publication of the original article, PNAS also imposed a maximum of 250 words. However, the restrictions did at least force the two men to concentrate only on the most significant issues. It was eventually decided that they would discuss confidence intervals briefly, pointing out that using conventional statistical methods they could show that Mann's uncertainty bounds were infinitely large prior to 1800 – in other words that his new reconstruction was of no use prior to that date. Apart from this, they would restrict themselves to listing the most important of the remaining issues – the calibration process producing hockey sticks from red noise, the strange screening process, proxies incorporating thermometer data, the upside-down Tiljander proxies, and the use of proxies which were not responding to temperature, including the bristlecones. A short comment was duly composed and submitted to the journal. At the end of 2008, the comment was accepted for publication, the journal indicating that Mann and his co-authors would be invited to submit a reply.

Mann's reply appeared at the start of February 2009. McIntyre described it as 'amusing'. With as little space to develop his arguments as McIntyre, Mann appeared to be content to brush most of the Canadians' critique aside.[243] He did, however, manage to include his usual robust rejections of any criticism:

> McIntyre and McKitrick raise no valid issues regarding our paper. . .
> McIntyre and McKitrick's claim . . . is unsupported in peer-reviewed literature and reflects an unfamiliarity with the concept of screening regression/validation . . .
> The claim that upside down data were used is bizarre . . .
> McIntyre and McKitrick misrepresent [the NAS] report . . .
> In summary their criticisms have no merit . . .

Mann's claim that McIntyre had misrepresented the NAS panel was also very surprising, since the report had been unequivocal: 'strip-bark samples should be avoided for temperature reconstructions'.[153] Mann appeared to be relying on Wahl and Ammann's *Climatic Change* paper to support this position. Ammann had made a fairly weak attempt to defend the use of bristlecones, arguing that Mann's approach did not require proxies to correlate with their local temperature at all, the idea being that trees could capture temperature signals from far away by means of teleconnections.* How Mann thought that this argument was tenable in a paper where proxies were specifically screened for correlation to local temperatures is unclear. With so few words to put his case though, any meaningful explanation was impossible.

---

*   See page 47.

Mann's approach to the upside-down Tiljander proxy took a similar line. He appeared to be arguing that it didn't matter because the upside-down proxy correlated to temperature and that was all his algorithm cared about. If this was his case then it was strange indeed, because without any physical mechanism to justify the proxy shape, all Mann had was a correlation. As McIntyre had observed right back at the start of the story, just because you could get a correlation between interest rate futures and temperature didn't mean that you could reconstruct historic temperatures from data extracted from a Bloomberg terminal.*

It was the same for Mann's claim about McIntyre's comments on screening. Despite Mann's assertion that McIntyre's claim was unsupported in the literature, this was the essence of Bürger and Cubasch's paper.** Those of a suspicious frame of mind could also note that Mann didn't *actually* say that McIntyre's claims were wrong, although again, it is difficult to be sure, since the word count was so short.***

---

* The use of the Tiljander proxies in an inverted orientation was repeated in a later paper by Kaufman et al.,[244] its author team including many familiar names from this story, such as Bradley, Briffa, Overpeck, Ammann and Otto-Bleisner (but not Mann). Speaking of this paper, one prominent expert in the field, in no way a sceptic, made the following observation: 'Proxies have been included selectively, they have been digested, manipulated, filtered, and combined, for example, data collected from Finland in the past by my own colleagues has even been turned upside down such that the warm periods become cold and vice versa'.[245] Subsequent to this error being pointed out, Kaufman agreed that he had made a mistake and issued a corrigendum. Mann has yet to follow suit.

** See page 299.

*** At time of writing, McIntyre continues his research into Mann's new hockey stick, in particular the methodologies used.

15 The Meaning of the Hockey Stick

When later generations learn about climate science, they will classify the beginning of the twenty-first century as an embarrassing chapter in the history of science. They will wonder about our time and use it as a warning of how the core values and criteria of science were allowed little by little to be forgotten, as the actual research topic of climate change turned into a political and social playground.[245]

Atte Korhola, Professor of Environmental Change
University of Helsinki

The Hockey Stick and the independent confirmations of it appear to be fatally flawed. We have seen two expert panels agree that Mann's short centring methodology was wrong. We have learned that bristlecones are in near-universal use in millennial temperature reconstructions even though there is widespread agreement that they are not reliable temperature proxies.

Does it matter that the Hockey Stick was wrong? Does it matter if all the millennial temperature reconstructions are wrong? Or even if we can never know what the temperature of the past was? What does the Hockey Stick affair tell us about the IPCC and the way that professional climatologists operate? Does any of what you have just read *matter*?

Relying on peer review

The Hockey Stick was a peer-reviewed paper, published in one of the world's most prestigious scientific journals. It passed another allegedly much more detailed review on its way to the position of prominence it attained in the IPCC's Third Assessment Report. How was it that so many leading climatologists failed to notice its many flaws? What were these panels of experts thinking of? Before we can answer these questions, we need to understand a little about the peer review system: how it evolved and how it actually works in practice.

History of peer review

Peer review is as old as science. There are traces of scientists' work being reviewed by their fellows in sources as far back as the writings of the great Arab physicians of the Middle Ages; European examples are known from the earliest days of the Enlightenment. It will surprise many readers, however, to learn that peer review was only rarely used in mainstream Western scientific publication before the middle of the twentieth century, and that some of the greatest works of Western science made their way into the literature without a peer reviewer's imprimatur. Notable examples abound and they include the great works of Albert Einstein from 1905 and Watson and Crick's paper on the structure of DNA from 1953. Einstein's *Annalen der Physik* papers were reviewed by the journal's editors, Max Planck and Wilhelm Wien, but were given the nod on their say-so alone. John Maddox, the editor of *Nature* stated that no review of the paper was necessary because it was self-evidently correct.* This was very much the way that journals operated at the time, with reviewers called in to provide input on the suitability of a paper for publication only as and when the editors felt it necessary.

Since that time, procedures have tightened up considerably, so that nowadays, nearly all papers are reviewed, but the fact that peer review used to happen only when required tells us something rather profound about what it was designed to do. Before the growth in its use in the second half of the twentieth century, it was *assumed* that scientific papers were suitable for publication, based on the review of

---

*   Watson and Crick's paper was actually handled by Maddox's Predecessor, Jack Brimble.

the journal editor. Peer review was a process that dealt with the exceptions – those papers where the editor required specialist input or a second opinion in order to decide if the paper should go forward. Sign-off by peer reviewers did not and does not automatically make a scientific paper correct, a point made eloquently by Mann in his RealClimate posting ahead of McIntyre's two papers in 2005: 'a necessary but not sufficient condition' was the way he put it.[97] In fact, it is fairly clear that even this is going too far: if Einstein, Watson and Crick, and nearly everyone else before 1950 could all get by without peer review, then it seems fairly clear that in terms of discovering the truth, it is not even a necessary condition.

So what is peer review for then?

What then does peer review do for society? The answer seems to be that it achieves very little for society. In fact, most of the benefits of the process seem to attain to the journals rather than to society at large. Journals are all seeking the best, the most significant scientific papers. They want articles that are important and perhaps newsworthy too. This kind of paper will sell journals and will bring in the subscriptions, so one of the principal objectives of peer review is to gauge how important a paper is. As far as this end is concerned, peer review probably functions quite satisfactorily. However, journals also want to avoid publishing papers containing errors and so a second objective for a peer reviewer is to identify errors. Here, there can be little doubt that peer review is not up to the job.

Nobody really knows how many scientists perform their reviews carefully, how many merely skim the papers and how many just give them the nod, but attempts have been made to provide an answer to this question. Medical journals have been at the forefront of research into the efficacy of peer review as a way of identifying scientific error and their conclusions have been largely unfavourable. Richard Smith, a former editor of the *British Medical Journal* (BMJ) has been one of the most prominent critics of peer review. As he puts it:

> We have little evidence on the effectiveness of peer review, but we have considerable evidence of its defects. In addition to being poor at detecting gross defects and almost useless for detecting fraud, it is slow, expensive, profligate of academic time, highly subjective, something of a lottery, prone to bias and easily abused.[246]

Smith's successor at the BMJ, Fiona Godlee, seems to share his less than fulsome opinions of the efficacy of peer review. She and her colleagues performed a trial in which eight errors were inserted into a genuine manuscript, which was then sent out to 420 reviewers. Of the 221 who responded, nobody spotted more than five of the mistakes, the typical reviewer spotted only two and a sixth of the respondents missed all eight.[247]

The reasons for these failures become clear when we consider the nature of a peer review. A peer review normally consists only of reading a scientific manuscript through. It does not involve obtaining the data, reviewing the code or reperforming calculations. Peer review is not due diligence in the way a business auditor would understand the term, and there is no pretence by journals that it is. It works as if an auditor read the company's annual report but did not actually examine any of the underlying transactions and estimates. He might be able to offer an opinion on the appropriateness of the company's stated policy on providing for obsolete inventory but he would be unable to comment on whether that policy had been applied in practice or, if it had, whether it had been applied correctly. In fact, even a traditional business audit only claims to give 'reasonable assurance' that a set of accounts are not materially misstated, so it is hard to see how the considerably more cursory checks of a peer review provide any comfort to the reader at all.

Yet despite this, politicians and the public seem somehow to believe that the fact that a paper has passed peer review means that it is correct. There appears to be a striking disconnect between the scientific community and the politicians who rely on their findings to inform important policy decisions. As McKitrick put it in a paper he co-authored with Bruce McCullough of Philadelphia's Drexel University, 'some government staff are surprised to find out that peer review does not involve checking data and calculations, while some academics are surprised that anyone thought it did'.[248]

The supposedly more rigorous process of expert panel review of single issues is little better. As we have seen, expert panels are easily packed with scientists of the 'correct' opinions, dissenters' views can be ignored or suppressed, and reports can be biased to give the answer that is required. We saw in Chapter 9 how McIntyre's attempts to get someone with statistical expertise installed on the NAS panel was sidestepped by means of appointing people closely associated with the Hockey Team, who then signally failed to ask the pertinent questions. Statistician Jim Zidek, writing in the *Journal of the Royal Statistical Society* (JRSS), has noted with bewilderment the absence of statistical expertise among IPCC experts, an oversight that is particularly damning in the area of paleoclimate, where signal processing and statistics are the main areas of contention.[249] How can it be right that Mann, who has freely admitted to not being a statistician, was felt to be an appropriate overseer of the panel's paleoclimate deliberations?

Assembling even the most illustrious panel around a table and 'winging it' through the relevant scientific literature is surely unequal to the task of discovering the truth, particularly through the fog of a scientific debate: after one set of peer reviewers have reviewed a paper during the course of its publication at the journal, very little is added by having another set of experts read it through again. The findings of expert panels are often themselves peer reviewed, adding yet more spurious authority to the results, but with no more likelihood of establishing the truth of a scientific question. In the case of the IPCC reports, national science academies have also weighed in, issuing statements of support for the findings and further protecting the 'consensus' from any challenge. But on what basis do they do this? At best, yet another layer of peer review.

The inadequacies of peer review as the basis for finding fraud and error in scientific papers appear clear and well established in the scientific literature. Both sides of the global warming debate apparently agree that peer review is 'a necessary but not sufficient condition'. Yet peer review is the only oversight there is of the validity of the scientific case for catastrophic manmade global warming and on this flimsy basis governments make far-reaching policy decisions that affect everyone and will continue to affect our children for decades into the future.

Sound science

The Hockey Stick affair is not the first scandal in which important scientific papers underpinning government policy positions have been found to be non-replicable – McCullough and McKitrick review a litany of sorry cases from several different fields – but it does underline the need for a more solid basis on which political decision-making should be based. That basis is *replication*. Centuries of scientific endeavour have shown that truth emerges only from repeated experimentation and falsification of theories, a process that only begins after publication and can continue for months or years or decades thereafter. Only through actually reproducing the findings of a scientific paper can other researchers be certain that those findings are correct.

In the early history of European science, publication of scientific findings in a journal was usually adequate to allow other researchers to replicate them. However, as science has advanced, the techniques used have become steadily more complicated and consequently more difficult to explain. The advent of computers has allowed scientists to add further layers of complexity to their work and to handle much larger datasets, to the extent that a journal article can now, in most cases, no longer be considered a definitive record of a scientific result. There is simply insufficient space in the pages of a print journal to explain what exactly has been done. This has produced a rather profound change in the purpose of a scientific paper. As geophysicist Jon Claerbout puts it, in a world where powerful computers and vast datasets dominate scientific research, the paper 'is not the scholarship itself, it is merely advertising of the scholarship'.* The *actual* scholarship is the data and code used to generate the figures presented in the paper and which underpin its claims to uniqueness.

In passing we should note the implications of Claerbout's observations for the assessment for our conclusions in the last section: by using only peer review to assess the climate science literature, the policymaking community is implicitly expecting that a read-through of a *partial* account of the research performed will be sufficient to identify any errors or other problems with the paper. This is simply not credible.

With a full explanation of methodology now often not possible from the text of a paper, replication can usually only be performed if the data and code are available. This is a major change from a hundred years ago, but in the twenty-first century it should be a trivial problem to address. In some specialisms it is just that. We have seen, however, how almost every attempt to obtain data from climatologists is met by a wall of evasion and obfuscation, with journals and funding bodies either unable or unwilling to assist. This is, of course, unethical and unacceptable, particularly for publicly funded scientists. The public has paid for nearly all of this data to be collated and has a right to see it distributed and reused.

As the treatment of the Loehle paper shows,** for scientists to open themselves up to criticism by allowing open review and full data access is a profoundly uncomfortable process, but the public is not paying scientists to have comfortable lives; they are paying for rapid advances in science. If data is available, doubts over exactly where the researcher has started from fall away. If computer code is made public too, then the task of replication becomes simpler

---

* Although originally writing about computer science, Claerbout's point applies just as much to climatology and other fields of study. The actual words quoted are a distillation of Claerbout's ideas due to Buckheit and Donoho.[250]

** See page 303.

still and all doubts about the methodology are removed. The debate moves on from foolish and long-winded arguments about what was done (we still have no idea exactly how Mann calculated his confidence intervals) onto the real scientific meat of whether what was done was correct. As we look back over McIntyre's work on the Hockey Stick, we see that much of his time was wasted on trying to uncover from the obscure wording of Mann's papers exactly what procedures had been used. Again, we can only state that this is entirely unacceptable for publicly funded science and is unforgiveable in an area of such enormous policy importance.

As well as helping scientists to find errors more quickly, replication has other benefits that are not insignificant. David Goodstein of the California Insitute of Technology has commented that the possibility that someone will try to replicate a piece of work is a powerful disincentive to cheating – in other words, it can help to prevent scientific fraud.[251] Goodstein also notes that, in reality, very few scientific papers are ever subject to an attempt to replicate them. It is clear from Stephen Schneider's surprise when asked to obtain the data behind one of Mann's papers that this criticism extends into the field of climatology.* In a world where pressure from funding agencies and the demands of university careers mean that academics have to publish or perish, precious few resources are free to replicate the work of others. In years gone by, some of the time of PhD students might have been devoted to replicating the work of rival labs, but few students would accept such a menial task in the modern world: they have their own publication records to worry about. It is unforgiveable, therefore, that in paleoclimate circles, the few attempts that have been made at replication have been blocked by all of the parties in a position to do something about it.

Medical science is far ahead of the physical sciences in the area of replication. Doug Altman, of Cancer Research UK's Medical Statistics group, has commented that archiving of data

---

*   See page 160.

should be mandatory and that a failure to retain data should be treated as research misconduct.[252] The introduction of this kind of regime to climatology could have nothing but a salutary effect on its rather tarnished reputation. Other subject areas, however, have found simpler and less confrontational ways to deal with the problem. In areas such as econometrics, which have long suffered from politicisation and fraud, several journals have adopted clear and rigorous policies on archiving of data. At publications such as the *American Economic Review, Econometrica* and the *Journal of Money, Credit and Banking*, a manuscript that is submitted for publication will simply not be accepted unless data and fully functional code are available. In other words, if the data and code are not public then the journals will not even *consider* the article for publication, except in very rare circumstances. This is simple, fair and transparent and works without any dissent. It also avoids any rancorous disagreements between journal and author after the event.

Physical science journals are, by and large, far behind the econometricians on this score. While most have adopted one pious policy or another, giving the appearance of transparency on data and code, as we have seen in the unfolding of this story, there has been a near-complete failure to enforce these rules. This failure simply stores up potential problems for the editors: if an author refuses to release his data, the journal is left with an enforcement problem from which it is very difficult to extricate themselves. Their sole potential sanction is to withdraw the paper, but this then merely opens them up to the possibility of expensive lawsuits. It is hardly surprising that in practice such drastic steps are never taken.

The failure of climatology journals to enact strict policies or enforce weaker ones represents a serious failure in the system of assurance that taxpayer-funded science is rigorous and reliable. Funding bodies claim that they rely on journals to ensure data availability. Journals want a quiet life and will not face down the academics who are their lifeblood. Will *Nature* now go back to Mann and threaten to withdraw his paper if he doesn't produce the code for his confidence interval calculations? It is unlikely in the extreme. Until politicians and journals enforce the sharing of data, the public can gain little assurance that there is any real need for the financial sacrifices they are being asked to accept.

Taking steps to assist the process of replication will do much to improve the conduct of climatology and to ensure that its findings are solidly based, but in the case of papers of pivotal importance politicians must also go further. Where a paper like the Hockey Stick appears to be central to a set of policy demands or to the shaping of public opinion, it is not credible for policymakers to stand back and wait for the scientific community to test the veracity of the findings over the years following publication. Replication and falsification are of little use if they happen after policy decisions have been made. The next lesson of the Hockey Stick affair is that if governments are truly to have assurance that climate science is a sound basis for decision-making, they will have to set up a formal process for replicating key papers, one in which the oversight role is peformed by scientists who are genuinely independent and who have no financial interest in the outcome.

The Hockey Stick and the global warming hypothesis

The case that global warming is happening, is manmade and will be catastrophic does not rely on the paleoclimate studies alone and we therefore need to understand the other strands of the argument. Once we are clear on how the whole global warming hypothesis stacks up, we can assess the effect of eliminating the Hockey Stick. Before we do that, we first need to be clear on what is meant by a hypothesis.

The classic explanation of the scientific method that was outlined by the Austrian philosopher, Karl Popper, in the years before the Second World War, describes the formulation of a hypothesis – an idea to be tested – and the performance of experiments that seek to falsify it. As each attempt at falsification is rejected, confidence in the hypothesis grows until it is accepted, for the moment, as the truth.

The global warming hypothesis is, in very simple terms, that man-made emissions of carbon dioxide into the atmosphere are causing the Earth to heat up – the case made by Arrhenius more than 100 years ago. Since Arrhenius's time, this simple idea has become vastly more sophisticated, particularly over the last couple of decades, as scientists have learned more and more about the factors that affect the climate system and how they interact. Vast computer models now incorporate an array of different inputs to the climate system (known as 'forcings') – sunlight, carbon dioxide, emissions from volcanos and so on, together with the feedbacks such as clouds and rainfall. As the models grow and grow, they become steadily more complex but, it is hoped, in the process these vast artificial worlds will become more realistic representations of the world's climate than the simple models that have gone before.

It is important to recognise one important fact about the climate models: they are hypotheses. Newer and more sophisticated and perhaps better hypotheses, but hypotheses all the same. It is extraordinarily common to hear qualified scientists to talk about 'the evidence from the models', as if evidence could be derived from anything other than the real world. The distinction is not a trivial one. It is terribly easy for scientists to fool themselves that what emerges from their models is a reflection of reality rather than of the assumptions they have fed into them. Garbage in, garbage out is a constant concern for climate modellers.

The question of whether the models are reliable is not particularly relevant to deciding if the Hockey Stick affair is important or not, so let us leave that question aside for the moment without getting into the details. For the purposes of the hypothesis that man's activities are causing the world to warm and will make it warm still further in future, we can assume that the models represent the best current understanding of the climate system. We must then test that hypothesis against evidence from the real world.

So what evidence can we look at that will affirm or contradict the hypothesis? We can, of course, observe that surface temperatures rose at the end of the twentieth century: good evidence in favour of the hypothesis. We could of course observe the stalling of the rise since the millennium, apparently contradicting it. We could also look to see if clouds are behaving in the way that the models suggest they should, or if changes to the major weather patterns are consistent.

Where do the Hockey Stick and the other temperature reconstructions fit in here? Reconstructions of past temperatures say nothing about whether current temperatures are rising. We have seen that most of the proxies terminate before 1980, shortly after the models suggest that the largest effects of global warming should kick in; inexplicably there have been few attempts to update them since. The reconstructions tell us nothing about whether or by how much temperatures might continue to rise. In fact, a reliable temperature reconstruction (if such a thing were possible) would only tell us if the current temperatures were unprecedented or not and, when you consider the implications of this, it is clear that hockey sticks are peripheral to the case for manmade global warming. It is entirely possible that, if it were not for carbon dioxide heating the climate system up, temperatures would actually be rather low in historic terms. In other words, temperatures could be entirely *precedented* and the global warming hypothesis could still be correct. Similarly, in a world that is still emerging from the last ice age, temperatures should be rising and we should expect them to be higher than they were before. So temperatures can be unprecedented and the global warming hypothesis could still be wrong. Unprecedented temperatures are persuasive but far from conclusive.

If, having seen the evidence presented here, we believe that the temperature reconstructions are not reliable, where does that leave us? In terms of the case for drastic action, the argument has changed just slightly from 'temperatures are rising in line with the models and are now unprecedented' to 'temperatures are rising in line with the models'. Again, sceptics will want to note the twenty-first century stall of the warming, but let's leave that aside for now.

Whether you feel that either of these arguments represents a *compelling* case for drastic political action is largely a matter of opinion, but it is probably fair to say that fewer people will be convinced by the latter argument than by the former. Those who favour the so-called precautionary principle will always want to avoid future costs and will choose drastic action regardless. My own view is that this is unreasonable; I prefer to consider both costs and benefits of any possible actions, but people will differ on these issues.

To this extent then, the Hockey Stick affair is not the beginning and end of the global warming story: it can cause the arguments to be framed differently but it will not decide the outcome. In fact, this position was agreed very early on in the debate, when McIntyre and two Hockey Team members, Rasmus Benestad and William Connolley, all posted comments on the blog of a climate policy writer, Roger Pielke Jnr, coming to similar conclusions on the Hockey Stick's scientific importance – namely that it wasn't very important at all.[253] In fact, even the IPCC agreed with the position that paleoclimate reconstructions were not particularly important to the scientific case for manmade global warming. This being the case, McIntyre had suggested that they delete the whole paleoclimate chapter from the draft Fourth Assessment Report, but the powers within the bureaucracy did not take him up on the idea.

So why then *have* you just read a whole book about this particular scientific paper? Why has the debate over the Hockey Stick been so drawn out and so heated? Why does the Hockey Stick matter so much to the IPCC?

As we will see in the next section, the chief importance of the Hockey Stick lies not in that it is central to the case for manmade global warming, but in the fact that the IPCC promoted it *as if it were*.

The Hockey Stick and the IPCC

In that same Pielke comment thread, McIntyre had put forward the view that the Hockey Stick's centrality was not so much to the scientific debate, as to the political one. In other words, its importance lay in its use as a promotional tool: a single compelling graphic that could be used to persuade the public and policymakers of the strength of the case for dramatic action. We saw right back at the start of the story just how the Hockey Stick has been used to promote the idea of manmade global warming.* It was the Hockey Stick that was behind Sir John Houghton at the launch of TAR, it was the Hockey Stick that was behind the constant claims that modern temperatures are unprecedented. Even now, years after it has been shown to be flawed, it still appears in school textbooks and government and environmentalist literature. This being the case, it is hard to disagree with the BBC's opinion that 'it is hard to overestimate how influential this study has been'. There can be little doubt that it would have been much harder to sell the idea of manmade global warming if the Third Assessment Report had been illustrated with, for example, Briffa's reconstruction. Even with the divergence effect erased from the record, as shown in Figure 15, the rhetorical effect is considerably weaker than that of Mann's reconstruction. Of course, this effect could have been reversed to some extent by overlaying the end of the record with the instrumental record, as Mann did in the original Hockey Stick paper, but whether this would have been a fair representation of the proxy records is another question.

The IPCC's sale of a dud to governments, which governments then sold on to their citizens, has a much wider significance than its narrow relevance to the scientific debate. Hans von Storch explained some of these issues in comments he made shortly after the publication of the IPCC report.

> The debate about the hockey stick is most significant when it comes to the culture of our science. Posting the hockey stick as key evidence in the [Summary for Policymakers] and Synthesis Report of the IPCC was simply stupid and evidence for what [biologist Dennis] Bray calls post-

---

* See page 39.

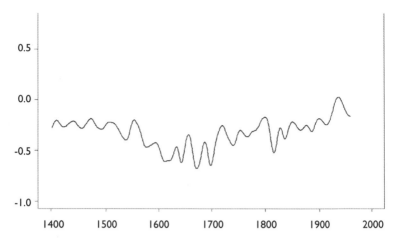

FIGURE 15.1: Briffa's reconstruction from 2001

sensible science – as science which is encroached [upon] by moral entrepreneurship. Or post-normal science. We have more cases of this type of claim-making, which is usually a mix of 'good' political intentions and personal drive for the limelight. Have we, as a community, become better in rejecting such claims? I am afraid we have not.[254]

So in addition to what it tells us about the efficacy of peer review, the Hockey Stick tells us about the culture of climate science and in particular the IPCC, and it tells us about the culture of science in general. Its effect in these areas should give us great cause for concern.

As a discipline, climatologists seem to have got themselves into a rut. Procedures that would be seen as shocking in other areas of scientific study appear to be routine when climate is being studied – use of inappropriate data, cherrypicking, truncations and extrapolations and reliance on ad-hoc and untested methodologies are just the start of it. Climatologists appear to have become fixated on the idea of catastrophic manmade global warming and react in horror to any questioning of their findings, with frantic appeals to a spurious consensus that is worthless in establishing the truth. They appear unable to break out of the mindset that 'the science is settled', an odd position for a discipline that is still very much in its infancy. It has yet to reach the state of maturity in which it can welcome questioning from outsiders and take it in its stride.

A surprising number of climatologists seem to believe that their work should not be subject to scrutiny by others and particularly not by outsiders. The reaction of the profession to criticism of the Hockey Stick was, almost universally, to circle the wagons and to attack the mere suggestion that Mann's work was less than perfect. Requests for data and code appear to be routinely rejected. This is not the behaviour of the scientist but of the political activist. When data was withheld by one of their number, climatologists sat silent or jeered from the sidelines. Likewise, not one of the great, learned societies took a stand on the issue. Yet when Joe Barton wrote to Mann, Bradley and Hughes asking for the MBH98 code there was uproar, with the American Association for the Advancement of Science, the National Academy of Sciences and the European Geophysical Union all fulminating against Barton's political 'interference'. The idea that they should have condemned Mann for withholding his code does not seem to have occurred to them. Nor do they seem to have pondered the idea that Barton was only getting involved in the first place because the scientific community – academics, journals, funding agencies and learned societies – had all signally failed to police the problem themselves.

The Hockey Stick was, from the very beginning, a test of the IPCC's impartiality. It was clear before the publication of Mann's paper that the Medieval Warm Period was a major problem for those who argued that man's activities were having an adverse effect on the climate. The public would simply not be convinced of the case for drastic action if temperatures appeared to have been warmer a few hundred years ago. The arrival of the Hockey Stick and its startling rise to prominence should have given the IPCC pause for thought on these grounds alone. Their handling of the paper represented an opportunity for IPCC officials to demonstrate that their organisation could be an 'honest broker' between environmentalists and sceptics, a chance for it to show that its procedures were fair and balanced and that it could be a reliable source of advice to politicians. On each count, its failure was complete and catastrophic. At each step along the path to the report, IPCC insiders made certain that any criticisms of the Hockey Stick were quashed and that doubts about the veracity of the other paleoclimate reconstructions were swept aside. The panel was stacked against the critics, rules were bent and broken and criticisms were ignored or brushed off; all with apparent impunity and without a word of protest from anyone in the climatological or the wider scientific community.

The fact that the IPCC promoted a Hockey Stick that was not central to the scientific debate simply because it was a good sales tool, and then defended it in the face of all criticism shows us that it is not a disinterested participant in the debate. It has chosen to be a advocate rather than a judge. It has an agenda. How then, can those who are undecided on the global warming issue accept anything it says as an unbiased judgement on the facts rather than a statement of a political position? They can no longer be sure.

Quite apart from what the Hockey Stick tells us about the positioning of the IPCC in the global warming debate, the panel's need for a sales tool also suggests something important about the overall case for manmade global warming. None of the corruption and bias and flouting of rules we have seen in the course of this story would have been necessary if there is, as we are led to believe, a watertight case that mankind is having a potentially catastrophic effect on the climate. What the Hockey Stick affair suggests is that the case for global warming, far from being settled is actually weak and unconvincing.

The implications for policymakers are stark. They have granted an effective monopoly on scientific advice to an organisation that has proven itself to be corrupt, biased and beset by conflicts of interest. Their advisers on the global warming issue are essentially a law unto themselves, the only oversight of their actions and findings provided by volunteers like McIntyre and his ragtag band of sceptic supporters. There is no conceivable way that politicians can justify this failing to their electorates. They have no choice but to start again.

Was the Hockey Stick an isolated problem?

Critics of the arguments I have put forward in this final chapter might well ask whether the promotion of the Hockey Stick was a one-off. Perhaps, they might suggest, these problems are only only found in the area of paleoclimate. It is unfair, they might say, to write off the IPCC as wholly biased based on the problems with just one fairly peripheral area of their remit. Should we not give them the benefit of the doubt?

So let us finish then by looking at another remarkable episode in the story of the Fourth Assessment Report, unrelated to the contentious area of paleoclimate – the way that clouds affect the climate.

Of itself, the direct warming produced by manmade emissions of carbon dioxide would not be enough to trouble mankind. The potential effects are, it seems, only catastrophic because of feedbacks – other effects caused by the initial warming. For example, it is hypothesised that the small direct warming caused by carbon dioxide will cause some melting of the ice caps. As the white cover to the poles reduces in size, it is said, less and less heat will be reflected back into space, and so there will be a further warming which would cause further melting and further warming and so on. These so-called 'positive feedbacks' are what makes manmade global warming so dangerous.

However, as well as positive feedbacks there are also negative feedbacks and the most important of these is the influence of certain types of clouds. The particular kind of clouds with which we are concerned are low-level or 'boundary layer' clouds.

In 2009, McIntyre reported on a paper written in 2006 by a French researcher, Sandrine Bony.[255] Bony et al was a review paper, surveying recent developments in scientific understanding of climate feedbacks including, of course, the critical effects of water vapour and clouds.[255] On the subject of boundary layer clouds, Bony and her co-authors had this to say:

> Boundary layer clouds have a strongly negative [feedback effect] . . . and cover a very large fraction of the area of the Tropics . . . Understanding how they may change in a perturbed climate therefore constitutes a vital part of the cloud feedback problem.[255]

Unequivocally then, boundary layer clouds cool the planet, and strongly so. How then was this knowledge conveyed to the public and policy-makers in the Fourth Assessment Report? Chapter 8 of the report was authored by, amongst others, Sandrine Bony herself, and we should therefore note in passing that this represented another example of IPCC authors reviewing their own work. Certainly, much of the relevant section had been lifted almost word for word from Bony's paper: in the following extract, parts of the text of the Fourth Assessment Report are identical. The differences are, however, significant.

> Boundary-layer clouds have a strong impact . . . and cover a large fraction of the global ocean . . . . Understanding how they may change in a perturbed climate is thus a vital part of the cloud feedback problem . . .[256]

So suddenly, the strongly *negative* feedback noted by Bony in her *Journal of Climate* paper became only a 'strong impact'.

Later in the same paper, Bony had noted the findings of two earlier researchers, Klein and Hartmann, who had observed a correlation between cloud cover and temperature stability in the tropics. This, Bony reported, 'leads to a substantial increase in low cloud cover in a warmer climate . . . and produces a strong negative feedback'. So once again, there was an unequivocal case being made that the feedback from boundary layer clouds is both strong and negative – tending to cool the Earth rather than warm it. However, by the time this statement found its way through to the Fourth Assessment Report it was once again much more ambiguous. The correlation between low-level clouds and temperature stability had 'led to the suggestion that a global climate warming might be associated with an increased low-level cloud cover, which would produce a negative cloud feedback'. In other words, there was now only a suggestion rather than a firm conclusion and it appeared to relate to an effect that was negative, but not apparently strongly so.

Another instance was the reporting of a disagreement over whether clouds in a warmer world would reflect more heat back into space (a higher 'albedo', in the jargon). Bony et al had reported two papers finding in favour of such an effect, and three against. Yet once again, by the time the IPCC came to pronounce upon the issue, things had changed radically, with all mention of a possible cooling effect removed, leaving only those papers arguing against it.

Who knows what other instances there are of arguments contrary to the IPCC consensus disappearing into the ether, of doubts suppressed and questions ignored? That so many strange happenings have been uncovered by the handful of sceptics actively researching the subject would suggest that these problems are just the tip of the iceberg. It is clear that it would be foolish in the extreme to give the IPCC the benefit of the doubt. Their record is too poor, the stakes too high.

16 The Beginning of the End?

You can't fool all of the people all of the time.

PT Barnum

Shortly before calling a halt to the seemingly endless series of corrections and revisions to this book, there were some dramatic developments on one of the many threads of the paleoclimate story. I will attempt to cover these briefly since they are likely to prove important.

Back in Chapter 10, we saw how an update to the Polar Urals series had eliminated its hockey stick shape, jeopardising its use in subsequent paleoclimate studies. Keith Briffa had then come up with a new chronology from the nearby location of Yamal. The Yamal data had been collected by a pair of Russian scientists, Hantemirov and Shiyatov, and had been published in 2002.[257] In their version of the series, Yamal had shown little by way of a twentieth century trend. Strangely though, Briffa's version, which had made it into print even before that of the Russians, was somewhat different.[147] While it tracked the Russians' version for most of the length of the record, Briffa's version had a sharp uptick at the end of the twentieth century – another hockey stick. As we have seen, after its first appearance in *Quaternary Science Reviews*, Briffa's version of Yamal was seized upon by climatologists, appearing again and again in temperature reconstructions; it was virtually ubiquitous in the field: it contributed to the reconstructions in Mann and Jones 2003, Jones and Mann 2004, Moberg et al 2005, D'Arrigo et al 2006, Osborn and Briffa 2006 and Hegerl et al 2007, among others.

When McIntyre started to look at the Osborn and Briffa paper in 2006, he quickly ran into the problem of the Yamal chronology: he needed to understand exactly how the difference between the Briffa and Hantemirov versions of Yamal had arisen. McIntyre therefore wrote to the Englishman asking for the original tree ring measurements involved. When Briffa refused, McIntyre wrote to *Science*, who had published the new paper, pointing out that, since it was now six years since Briffa had originally published his version of the chronology, there could be no reason for withholding the underlying data. After some deliberation, the editors at *Science* declined the request, deciding that Briffa did not have to publish anything more, as he had merely re-used data from an earlier study. McIntyre should, they advised, approach the author of the earlier study, that author being, of course, Briffa himself. Wearily, McIntyre wrote to Briffa again, this time in his capacity as author of the original study in *Quaternary Science Reviews* and once again, as he had expected, the request was refused.

That was how the investigation of the Yamal series stood for the next two years until, in July 2008, a new Briffa paper appeared in the pages of the *Philosophical Transactions of the Royal Society B*, the Royal Society's journal for the biological sciences.[258] The new paper discussed five Eurasian tree ring datasets, which, in fairly standard Hockey Team fashion, were unarchived and therefore not susceptible to detailed analysis. Among these five were Yamal and the equally notorious Tornetrask chronology. McIntyre observed that the only series with a strikingly anomalous twentieth century was Yamal. It was frustrating therefore that he had still not managed to obtain Briffa's measurement data. It appeared that he was going to hit another dead end. However, in the comments to his Climate Audit article on the new paper, a possible way forward presented itself. A reader pointed out that the Royal Society had what appeared to be a clear and robust policy on data availability:*

---

*   The reader in question was in fact the author of this book.

> As a condition of acceptance authors agree to honour any reasonable request by other researchers for materials, methods, or data necessary to verify the conclusion of the article ... Supplementary data up to 10 Mb is placed on the Society's website free of charge and is publicly accessible. Large datasets must be deposited in a recognised public domain database by the author prior to submission. The accession number should be provided for inclusion in the published article.[259]

Having had his requests rejected by every other journal he had approached, McIntyre had no great expectations that the Royal Society would be any different, but it seemed worth another attempt and he duly sent off an email pointing out that Briffa had failed to meet the Society's requirement of archiving his data prior to submission and that the editors had failed to check that Briffa had done so.[260] The reply, to McIntyre's surprise, was very encouraging:

> We take matters like this very seriously and I am sorry that this was not picked up in the publishing process.[261]

Was the Royal Society, in a striking contrast to every other journal in the field, about to enforce its own data availability policy? Had Briffa made a fatal mistake?

Summer gave way to autumn and as October drew to a close, McIntyre had heard nothing more from the Royal Society. However, in response to some further enquiries, the journal sent him some more encouraging news – Briffa *would* be producing most of his data, although not immediately. Most of it would be available by the end of the year, with the remainder to follow in early 2009.

The first batch of data appeared on schedule in the dying days of 2008 and it was something of a disappointment. The Yamal data, as might have been expected, was to be archived with the second batch, so there would be a further delay before the real action could start. Meanwhile, however, McIntyre could begin to look at what Briffa had done elsewhere. It was not to be plain sailing. For a start, Briffa had archived data in an obsolete data format, last used in the era of punch cards. This was inconvenient but it was not an insurmountable problem – with a little work, McIntyre was able to move ahead with his analysis. Unfortunately, Briffa had also thrown a rather larger spanner in the works: while he had archived the tree ring measurements, he had not supplied any metadata to go with it – in other words there was no information about where the measurements had come from: there was only a tree number and the measurements that went with it. However, McIntyre was well used to this kind of behaviour from climatologists and he had some techniques at hand for filling in some of the gaps. Climate Audit postings on the findings followed in fairly short order, some of which were quite intriguing. There was, however, no smoking gun.

There followed a long hiatus, with no word on the remaining data from the Royal Society or from Briffa. McIntyre would occasionally visit Briffa's web page at the CRU website to see if anything new had appeared, but to no avail. Eventually, in late September 2009, a reader pointed out to McIntyre that the remaining data was now available. It had been quietly posted to Briffa's webpage, without announcement or indeed the courtesy of an email to McIntyre. It was nearly ten years since the initial publication of Yamal and three years since McIntyre had requested the measurement data from Briffa. Now at last some of the questions could be answered.

When McIntyre started to look at the numbers it was clear that there were going to be the usual problems with a lack of metadata, but there was also much more than this. In typical climate science fashion, just scratching at the surface of the Briffa archive raised as many questions as it answered. Why did Briffa only have half the number of cores covering the Medieval Warm Period that the Russian had reported? And why were there so few tree ring cores in Briffa's twentieth century? There were only 12 trees contributing to Briffa's estimate of the average ring width for the Yamal area in 1988, an amazingly small number in what should have been the part of the record when it was easiest to obtain data. By 1990 the count was only ten, dropping still further to just five in 1995. Without an explanation of how the selection of this sample of the available data had been performed, the suspicion of 'cherrypicking' would linger over the study, although it is true to say that Hantemirov also had very few cores in the equivalent period, so it is possible that this selection had been due to the Russians and not Briffa.

The lack of twentieth century data was still more remarkable when the Yamal chronology was compared to the Polar Urals series to which it was now apparently preferred. The ten or twelve cores used in Yamal was around half the number used in Polar Urals, which should presumably therefore have been considered the more reliable. Why then had climatologists almost all preferred to use Yamal? Could it be because it had a hockey stick shape?

None of these questions was likely to be answered without knowing which trees came from which locations. Hantemirov had made it clear in his paper that the data had been collected over a wide area – Yamal was an expanse of river valleys rather than a single location. Knowing exactly which trees came from where might well throw some light onto the question of why Briffa's reconstruction had a hockey stick shape but Hantemirov's didn't.

As so often in McIntyre's work, the clue that unlocked the mystery came from a rather unexpected source. At the same time as archiving the Yamal data, Briffa had recorded the numbers for another site discussed in his Royal Society paper: Taimyr. Like Yamal, Taimyr had also emerged in Briffa's *Quaternary Science Reviews* paper in 2000. However, in the Royal Society paper, Briffa had made major changes, merging Taimyr with another site, Bol'shoi Avam, located no less than 400 km away. While the original Taimyr site had something of a divergence problem, with narrowing ring widths implying cooler temperatures, the new composite site of Avam–Taimyr had a rather warmer twentieth century and a cooler Medieval Warm Period. The effect of this curious blending of datasets was therefore, as so often with paleoclimate adjustments, to produce a warming trend. However, this was not the only thing about the series that was interesting McIntyre. What was odd about Avam–Taimyr was that the series seemed to have more tree cores recorded than had been reported in the two papers on which it was based. So it looked as if something else had been merged in as well. But what?

With no metadata archived for Avam–Taimyr either, McIntyre had another puzzle to occupy him, but in fact the results were quick to emerge. The Avam data was collected in 2003, but Taimyr only had numbers going up to 1996. Also, the Taimyr trees were older, with dates going back to the ninth century. It was therefore possible for McIntyre to make a tentative split of the data by dividing the cores into those finishing after 2000 and those finishing before. This proved to be a good first cut, but the approach assigned 107 cores to Avam, which was more than reported in the original paper. This seemed to confirm McIntyre's impression that there was *something else* in the dataset.

At the same time, McIntyre's rough cut approach assigned 103 cores to Taimyr, a number which meant that there were still over 100 cores still unallocated. The only way to resolve this conundrum was by a brute force technique of comparing the tree identification numbers in the dataset to tree ring data in the archives. In this way, McIntyre was finally able to work out the provenance of at least some of the data.

Forty-two of the cores turned out to be from a location called Balschaya Kamenka, some 400 km from Taimyr. The data had been collected by the Swiss researcher, Fritz Schweingruber. The fact that the use of Schweingruber's data had not been reported by Briffa was odd in itself, but what intrigued McIntyre was why Briffa had used Balschaya Kamenka and not any of the other Schweingruber sites in the area. Several of these were much closer to Taimyr – Aykali River was one example, and another, Novaja Rieja, was almost next door.

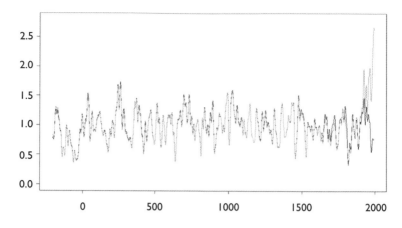

FIGURE 16.1: The Yamal sensitivity test

Grey line: Briffa; black line: McIntyre.

By this point McIntyre knew that Briffa's version of Yamal was very short of twentieth century data, having used just a selection of the available cores, although the grounds on which this selection had been made were not clear. It was also obvious that there was a great deal of alternative data available from the region, Briffa having been happy to supplement Taimyr with data from other locations such as Avam and Balschaya Kamenka. Why then had he not supplemented *Yamal* in a similar way, in order to bring the number of cores up to an acceptable level?

The reasoning behind Briffa's subsample selection may have been a mystery, but with the other information McIntyre had gleaned, it was still possible to perform some tests on its validity. This could be done by performing a simple sensitivity test, replacing the twelve cores that Briffa had used for the modern sections of Yamal with some of the other available data. Sure enough, there was a suitable Schweingruber series called Khadyta River close by to Yamal, and with 34 cores, it represented a much more reliable basis for reconstructing temperatures.

McIntyre therefore prepared a revised dataset, replacing Briffa's selected 12 cores with the 34 from Khadyta River. The revised chronology was simply staggering (see Figure 16.1). The sharp uptick in the series at the end of the twentieth century had vanished, leaving a twentieth century apparently without a significant trend. The blade of the Yamal hockey stick, used in so many of those temperature reconstructions that the IPCC said validated Mann's work, was gone.

At time of writing, the problems with the Yamal series have only been public for a matter of a few weeks. Briffa has made a public response, showing that he can get similar results with an expanded dataset, but without making a substantive defence of the Yamal data. It seems that Briffa was working with the 12 Yamal series in order to compare his standardisation methodology to Shiyatov's, although as others have pointed out, Briffa's preferred method normally requires much larger numbers of samples. It is not beyond the realms of possibility that the assessment of the series will change in the coming weeks and months as people on both sides of the global warming argument study McIntyre's case. For now though, it appears that the tree ring approach to temperature reconstructions lies in tatters.

17 The CRU Hack

Extraordinary claims demand extraordinary evidence.

Carl Sagan

In mid-November 2009 someone accessed the servers of CRU, the Climatic Research Unit at the University of East Anglia (UEA), and extracted vast quantities of data. A matter of a few days later, an archive of a thousand emails together with the long-sought data and code for the HADCRUT temperature index were posted on a web server located in Russia. Pointers to the leaked information were posted on the websites of some prominent sceptics and by the following day the climate blogosphere was in uproar as a series of embarrassing revelations about the conduct of prominent scientists was made public. Over the next few days the story was picked up by the mainstream media until there was a avalanche of outrage and scandal. The immediate reaction was that CRU had been the victim of a hacker, although at time of writing opinion appears to be gravitating towards the culprit being an insider. At time of writing, CRU director, Phil Jones has stepped down pending an investigation into the emails and Mann's employer, Penn State University, have launched an inquiry into his conduct.

Many of the emails leaked are extremely illuminating for our story, and I will attempt to cover here what has been discovered in the few days since the leak. At the time of writing Phil Jones has verified that the unit's systems were compromised, but the volume of data released means that it has not been possible to verify that all the leaked information is genuine. This caveat needs to be borne in mind while reading the rest of this chapter.

Each extract below is headed by an indication of who is

writing, the date and lastly the file number in the archive of leaked emails. These are currently circulating widely on the internet, although there is no obvious permanent repository as yet, so I do not provide a reference. Readers should be able to locate an electronic copy without difficulty.

The Soon and Baliunas paper

In Chapter 3* we saw how the publication of the Soon and Baliunas paper led to the resignation of several of the editors of *Climate Research*, the journal that had published the paper. The story of the resignations from the Hockey Team's side starts in March 2003 with Phil Jones notifying the rest of the team of the publication of the Soon paper and advising his colleagues that they would be best to ignore it. Afterwards he turned his attention to the editorial staff of the journal. The Soon paper had been handled by Chris de Freitas of the University of Auckland in New Zealand, overseen by editor-in-chief, Hans von Storch.

> Jones: 10 March 2003: 1062618881
> The responsible [editor] for this [paper is] a well-known skeptic in NZ. He has let a few papers through by [sceptics Pat] Michaels and [William] Gray in the past. I've had words with Hans von Storch about this, but got nowhere.

Later though, Jones changed his mind about a policy of ignoring the publication.

> Jones: 11 March 2003: 1062618881
> Writing this I am becoming more convinced we should do something – even if this is just to state once and for all what we mean by the [Little Ice Age] and [Medieval Warm Period]. I think the skeptics will use this paper to their own ends and it will set paleo back a number of years if it goes

---

*   See page 33.

unchallenged. I will be emailing the journal to tell them I'm having nothing more to do with it until they rid themselves of this troublesome editor. A CRU person is on the editorial board, but papers get dealt with by the editor assigned by Hans von Storch.

Mann clearly felt the same way:

> Mann: 11 March 2003: 1047388489
> The Soon & Baliunas paper couldn't have cleared a 'legitimate' peer review process anywhere. That leaves only one possibility–that the peer-review process at *Climate Research* has been hijacked by a few skeptics on the editorial board. And it isn't just de Frietas, unfortunately I think this group also includes a member of my own department . . .
>
> The skeptics appear to have staged a 'coup' at *Climate Research* (it was a mediocre journal to begin with, but now its a mediocre journal with a definite 'purpose') . . .
>
> My guess is that von Storch is actually with them (frankly, he's an odd individual, and I'm not sure he isn't himself somewhat of a skeptic himself) . . .

The problem, Mann said, was that having criticised sceptics for not publishing in the scientific literature, they now appeared to have a journal that would publish their views. He went on to explain what he thought should be done.

> Mann: 11 March 2003: 1047388489
> I think we have to stop considering *Climate Research* as a legitimate peer-reviewed journal. Perhaps we should encourage our colleagues in the climate research community to no longer submit to, or cite papers in, this journal. We would also need to consider what we tell or request of our more reasonable colleagues who currently sit on the editorial board . . . [ellipsis in original]

Tom Wigley seemed to share these views, concerned that de Freitas was giving an easy time to sceptic papers. He said that one sceptic paper that he and a colleague had rejected had been accepted by de Freitas, who had rejected Wigley's subsequent complaint, saying that the three other reviewers had been happy to publish.

The team seem to have concluded that their first step should be to issue a formal complaint to the journal. Meanwhile though, emails were exchanged at a furious pace, opinions about what other steps could be taken bouncing around a long distribution list of interested scientists, including James Hansen, Stephen Schneider and the new IPCC chairman Rajendra Pachauri. There was clearly a strong feeling that the Soon and Baliunas paper was very poor and, in Mann's words, was being used to start a political disinformation campaign.

Later that month, conclusions began to be reached. Mann had this to say:

> Mann: 24 April 2003: 1051202354
>
> I would emphasize that there are indeed, as Tom notes, some unique aspects of this latest assault by the skeptics which are cause for special concern. This latest assault uses a compromised peer-review process as a vehicle for launching a scientific disinformation campaign (often vicious and ad hominem) under the guise of apparently legitimately reviewed science . . . Fortunately, the mainstream media never touched the story (mostly it has appeared in papers owned by Murdoch and his crowd, and dubious fringe on-line outlets). Much like a server which has been compromised as a launching point for computer viruses, I fear that *Climate Research* has become a hopelessly compromised vehicle in the skeptics' (can we find a better word?) disinformation campaign, and some of the discussion that I've seen (e.g. a potential threat of mass resignation among the legitimate members of the *Climate*

Research editorial board) seems, in my opinion, to have some potential merit. This should be justified not on the basis of the publication of science we may not like of course, but based on the evidence (e.g. as provided by Tom [Wigley] and Danny Harvey and I'm sure there is much more) that a legitimate peer-review process has not been followed by at least one particular editor.*

Intriguingly Mann also alluded to 'problems' at *Geophysical Research Letters* where, he said, the sheer volume of papers meant that some by sceptics would slip through the net.

> Mann: 23 April 2003: 1051202354
> While it was easy to make sure that the worst papers . . . didn't see the light of the day at [*Journal of Climate*], it was inevitable that such papers might slip through the cracks at GRL.

This remarkable message makes it clear that *Climate Research* was not the first journal where normal procedures had been undermined by the Hockey Team. We will see later that it was not the last time either. It is also interesting to note Mann's comments in light of the statements made by Andrew Weaver, the editor of Journal of Climate, about the suitability of MM03 for publication.**

Other ideas for action included the possibility of a rebuttal in *Climate Research* or perhaps even a more prominent journal. Alternatively, they thought, a direct approach could be made to the US Office of Science and Technology Policy. But while several members of the community voiced their support, the conversation seemed once again to return to the subject of the editors. Tom Wigley was among the hawks:

---

*   Wigley's evidence has been outlined above. The evidence of Danny Harvey, a professor at the University of Toronto, does not appear in the email archive.

**   See page 230.

Wigley: 24 April 2003: 1051190249

I do not know the best way to handle the specifics of the editing. Hans von Storch is partly to blame – he encourages the publication of crap science 'in order to stimulate debate'. One approach is to go direct to the publishers and point out the fact that their journal is perceived as being a medium for disseminating misinformation under the guise of refereed work. I use the word 'perceived' here, since whether it is true or not is not what the publishers care about – it is how the journal is seen by the community that counts.

I think we could get a large group of highly credentialed scientists to sign such a letter – 50+ people.

Note that I am copying this view only to Mike Hulme* and Phil Jones. Mike's** idea to get editorial board members to resign will probably not work – must get rid of von Storch too, otherwise holes will eventually fill up with people like Legates, Balling, Lindzen, Michaels, Singer, etc.*** I have heard that the publishers are not happy with von Storch, so the above approach might remove that hurdle too.

In June 2003, we catch sight of an email from Chris de Freitas to Otto Kinne, the publisher of *Climate Research*. Kinne has apparently had a written complaint from Mike Hulme about the publication of Soon and Baliunas.

De Freitas: 18 June 2003: 1057944829

I have spent a considerable amount of my time on this matter

---

* Professor of Climate Change at the University of East Anglia. In recent years Hulme has made calls for climate change rhetoric to be toned down,[262] so it is interesting to see that he was apparently central to the plan to oust von Storch, thus suggesting that he has only recently taken up his position as the voice of moderation.

** A later email suggests that Wigley is referring to Mike Hulme rather than Mann.

*** The latter are all prominent climate sceptics.

and had my integrity attacked in the process. I want to emphasize that the people leading this attack are hardly impartial observers. Mike himself refers to 'politics' and political incitement involved. Both [Mike] Hulme and [Clare] Goodess are from the Climatic Research Unit of UEA that is not particularly well known for impartial views on the climate change debate. The CRU has a large stake in climate change research funding as I understand it pays the salaries of most of its staff. I understand too the journalist David Appell* was leaked information to fuel a public attack . . .

Mike Hulme refers to the number of papers I have processed for [*Climate Research*] that 'have been authored by scientists who are well known for their opposition to the notion that humans are significantly altering global climate'. How many can he say he has processed? I suspect the answer is nil. Does this mean he is biased towards scientists 'who are well known for their support for the notion that humans are significantly altering global climate?' Mike Hulme quite clearly has an axe or two to grind, and, it seems, a political agenda. But attacks on me of this sort challenge my professional integrity, not only as a [*Climate Research*] editor, but also as an academic and scientist. Mike Hulme should know that I have never accepted any research money for climate change research, none from any 'side' or lobby or interest group or government or industry. So I have no pipers to pay.

He goes on to defend his conduct in publishing Soon and Baliunas:

De Freitas: 18 June 2003: 1057944829
The criticisms of Soon and Baliunas . . . article raised by Mike Hulme in his 16 June 2003 email to you was not

---

* We met David Appell in a later role as media outlet for the Hockey Team on page 65.

raised by the any of the four referees I used (but is curiously similar to points raised by David Appell!). Keep in mind that referees used were selected in consultation with a paleoclimatologist. Five referees were selected based on the guidance I received. All are reputable paleoclimatologists, respected for their expertise in reconstruction of past climates. None (none at all) were from what Hans [von Storch] and Clare [Goodess] have referred to as 'the other side' or what Hulme refers to as 'people well known for their opposition to the notion that humans are significantly altering global climate'. One of the five referees turned down the request to review explaining he was busy and would not have the time.

A few days later we see Phil Jones circulating a copy of de Freitas' email to some of his colleagues, asking that they keep the contents to themselves.

I don't want to start a discussion of it and I don't want you sending it around to anyone else, but it serves as a warning as to where the debate might go . . . I have learned one thing. This is that the reviewer who said they were too busy was Ray [presumably Bradley] . . . It is clear . . . [that] a negative review was likely to be partly ignored, and the article would still have come out.

He goes on to say, however, that de Freitas will not identify the other four reviewers, but that he thinks that one might be a paleoclimatologist called Anthony Fowler, who is not a known sceptic. So it appears that the Team had strong evidence that at least two of the five invited referees were not sceptics, and indeed one was one of their own members. However, they appeared undeterred.

This appears to be the last of the correspondence relating to *Climate Research*. However, as we have already seen, in July 2003, von Storch and two other *Climate Research* editors resigned from their positions on the journal. Did the Hockey Team act on their plans? At the moment we cannot say for certain, although it certainly appears that they planned to do so.

On McIntyre and McKitrick 2003

Shortly before the release of MM03, Mann was passed details of the paper's release by an unidentified source. This source had apparently been provided the information by a third party. Amusingly, the third party included the following statement:

> Anonymous: October 2003: 1067194064
> Personally, I'd offer that [McIntyre's conclusions that there were problems with the Hockey Stick's robustness] was known by most people who understand Mann's methodology: it can be quite sensitive to the input data in the early centuries. Anyway, there's going to be a lot of noise on this one, and knowing Mann's very thin skin I am afraid he will react strongly, unless he has learned (as I hope he has) from the past . . .

Clearly then, as far back as 2003, the knowledge that the Hockey Stick was flawed was not restricted to just sceptics. Mann decided to circulate the news of the paper's publication to the rest of the Hockey Team:

> Mann: 26 October 2003: 1067194064
> This has been passed along to me by someone whose identity will remain in confidence . . .
>
> My suggested response is: 1) to dismiss this as stunt, appearing in a so-called 'journal' which is already known to have defied standard practices of peer-review. It is clear, for example, that nobody we know has been asked to 'review' this so-called paper. 2) to point out the claim is nonsense since the same basic result has been obtained by numerous

other researchers, using different data, elementary compositing techniques, etc. Who knows what sleight of hand the authors of this thing have pulled. Of course, the usual suspects are going to try to peddle this crap. The important thing is to deny that this has any intellectual credibility whatsoever and, if contacted by any media, to dismiss this for the stunt that it is . . .

A few days later, we see Mann thanking some colleagues who had attacked McIntyre's paper, and Mann repeats his claims about McIntyre's use of a spreadsheet:

> Mann: 29 October 2003: 1067450707
> They didn't use the proxy data available on our public FTP site, which I had pointed them too – instead they used a spreadsheet file that my associate Scott Rutherford had prepared. In this file, most of the early series were overprinted at later years. This resulted in the reconstruction becoming increasingly spurious as one goes further back in time – the estimates prior to 1700 or so were rendered meaningless. There were also some other methodological errors that will be detailed shortly, but this was the big one.

It is interesting that Mann does not at this point claim that McIntyre *requested* a spreadsheet, as he would in his later public pronouncements.*

There is also a draft of Mann's formal rebuttal of MM03, a much more strongly worded document than what was eventually published.** Later still, we see a request from Bradley to the CRU team – Jones, Osborn and Briffa – to publish the rebuttal for the Americans.

---

* See page 65.
** See page 70.

> Bradley: 30 October 2003: 1067532918
> If you are willing, a quick and forceful statement from The Distinguished CRU Boys would help quash further arguments, although here, at least, it is already quite out of control. . .

Briffa was happy to help:

> Briffa: 30 October 2003: 1067542015
> I agree with this idea in principle. Whatever scientific differences and fascination with the nuances of techniques we may/may not share, this whole process represents the most despicable example of slander and downright deliberate perversion of the scientific process, and biased (unverified) work being used to influence public perception and due political process.

going on to say that he was minded to ask *Nature* to write an editorial on the subject. He did however have make a caveat:

> Briffa: 30 October 2003: 1067542015
> Much of the detail in Mike's response though is not sensible (sorry Mike) and is rising to their bait.

Over the next few weeks, the Team worked on the response. By the end of October 2003, they had a revised draft. Osborn commented as follows:

> Osborn: 31 October 2003: 1067596623
> The single worst thing about the whole [McIntyre and McKitrick] saga is not that they did their study, not that they did things wrong (deliberately or by accident), but that neither they nor the journal took the necessary step of investigating whether the difference between their results and yours could be explained simply by some error or set of

errors in their use of the data or in their implementation of your method. If it turns out, as looks likely from Mike's investigation of this, that their results are erroneous, then they and the journal will have wasted countless person-hours of time and caused much damage in the climate policy arena.

As we have seen, McIntyre and McKitrick were unable to do this because Mann had cut off communications with them prior to publication.* Osborn was also concerned about becoming too closely associated with the dispute, firstly because he said the CRU team were not fully aware of the details of MBH98, but also because if there was a subsequent independent assessment of the dispute, they would be unable to be involved, having already tied their colours to Mann's cause. He went on to discuss details of McIntyre's findings that were concerning him:

> Osborn: 31 October 2003: 1067596623
> (a) Mike, you say that many of the trees were eliminated in the data they used. Have you concluded this because they entered 'NA' for 'Not available' in their appendix table? If so, then are you sure that 'NA' means they did not use any data, rather than simply that they didn't replace your data with an alternative (and hence in fact continued to use what Scott had supplied to them)? Or perhaps 'NA' means they couldn't find the PC time series published (of course!), but in fact could find the raw tree-ring chronologies and did their own [PC analysis] of those? How would they know which raw chronologies to use?

The impossibility of working out which proxy series went into which step of the reconstruction was, as we have seen, was a real issue for McIntyre and McKitrick.

---

* See page 63.

Osborn continued to make perspicious observations on what might have happened.

> Osborn: 31 October 2003: 1067596623
> Or did you come to your conclusion by downloading their 'corrected and updated' data matrix and comparing it with yours – I've not had time to do that, but even if I had and I found some differences, I wouldn't know which was right seeing as I've not done any [PC analysis] of western US trees myself? My guess would be that they downloaded raw tree-ring chronologies (possibly the same ones you used) but then applied [PC analysis] only to the period when they all had full data – hence the lack of PCs in the early period (which you got round by doing [PC analysis] on the subset that had earlier data). But this is only a guess, and this is the type of thing that should be checked with them – surely they would respond if asked? . . . And if my guess were right, then your wording of 'eliminated this entire data set' would come in for criticism, even though in practise it might as well have been.

As we saw in Chapter 3, Mann's initial response to the publication of MM03 had included the claim that McIntyre had requested the data in spreadsheet format, a claim that was not substantiated when McIntyre published his correspondence with Mann.* This was Osborn's next concern.

> Osborn: 31 October 2003: 1067596623
> (b) The mention of FTP sites and Excel files is contradicted by their email record on their website, which shows no mention of Excel files (they say an ASCII file was sent) and also no record that they knew the FTP address. This doesn't

---

*   See page 65.

matter really, since the reason for them using a corrupted data file is not relevant – the relevant thing is that it was corrupt and had you been involved in reviewing the paper then it could have been found prior to publication. But they will use the email record if the FTP sites and Excel files are mentioned.

We also saw in the footnote on page 64 that there had also been some suggestions that the paper had not been peer reviewed, and this was Osborn's next point.

> Osborn: 31 October 2003: 1067596623
>
> (c) Not sure if you talk about peer-review in the latest version, but note that they acknowledge input from reviewers and Fred Singer's email says he refereed it – so any statement implying it wasn't reviewed will be met with an easy response from them.*

Osborn next addressed the RE statistic. Mann had apparently suggested including some kind of emulation of the McIntyre and McKitrick results. Then by showing that the emulation had a poor RE score, he could argue that it was less reliable than his own paper.

> Osborn. 31 October 2003: 1067596623
>
> (d) Your quick-look reconstruction excluding many of the tree-ring data, and the verification RE you obtain, is interesting – but again, don't rush into using these in any response. The time series of PC1 you sent is certainly different from your standard one – but on the other hand I'd hardly say you 'get a similar result' to them, the time series look very different (see their Fig. 6d). So the dismal RE applies only to your calculation, not to their

reconstruction. It may turn out that their verification RE is also very negative, but again we cannot assume this in case we're wrong and they easily counter the criticism.

Osborn went on to urge Mann to be much more cautious in his public pronouncements:

> Osborn: 31 October 2003: 1067596623
>
> (e) Claims of their motives for selective censoring or changing of data, or for the study as a whole, may well be true but are hard to prove. They would claim that theirs is an honest attempt at producing a key scientific result. If they made errors in what they did, then maybe they're just completely out of their depth on this, rather than making deliberate errors for the purposes of achieving preferred results.

He closed by outlining a list of other issues Mann referred to in the draft.

> Osborn: 31 October 2003: 1067596623
>
> It is again a bit of a leap of faith to say that these *explain* the different results that they get. Certainly they throw doubt on the validity of their results, but without actually doing the same as them it's not possible to say if they would have replicated your results if they hadn't made these errors. After all, could the infilling of missing values have made much difference to the results obtained, something that they made a good deal of fuss about?
>
> To say they 'used neither the data nor the procedures of MBH98' will also be an easy target for them, since they did use the data that was sent to them and seemed to have used approximately the method too (with some errors that you've identified). This reproduced your results to some extent (certainly not perfectly, but see Fig 6b and 6c).

Then they went further to redo it with the 'corrected and updated' data – but only after first doing approximately what they claimed they did (i.e. the audit).

Mann seemed happy enough with Osborn's contribution. However, his reply contains another tantalising turn of phrase:

> Mann: 31 October 2003: 1067596623
>
> Let's let our supporters in higher places use our scientific response to push the broader case against [McIntyre and McKitrick]. So I look forward to people's attempts to revise the first [paragraph in] particular. I took the liberty of forwarding the previous draft to a handful of our closet* colleagues, just so they would have a sense of approximately what we'll be releasing later today – i.e., a heads up as to how [McIntyre and McKitrick] achieved their result . . .

Quite who these 'supporters in higher places' are is of course a mystery, but it is worryingly suggestive of political interference in the scientific process.

Meanwhile, Osborn and Briffa, like others in the course of this story seem to have been less than impressed with Mann's antics. Osborn wrote to his colleague a couple of weeks later:

> Mann: 12 November 2003: 1068652882
>
> I do wish Mike had not rushed around sending out preliminary and incorrect early responses – the waters are really muddied now. He would have done better to have taken things slowly and worked out a final response before publicising this stuff. Excel files, other files being created early or now deleted is really confusing things!

---

* The word 'closet' here may be a typing error, which should actually read 'closest'.

On McIntyre and McKitrick 2005

When McIntyre's GRL paper was published at the beginning of 2005, it clearly presented the Hockey Team with a problem. While a paper by a retired mining executive in *Energy and Environment* could be brushed aside, something in a prominent journal like GRL was much harder to ignore. In the email archive we can see that Mann was quick into the fray: Steve Mackwell, the GRL editor-in-chief is seen replying to a Mann complaint, which appears to have concerned the fact that he had not been allowed to review and respond to the paper before publication. Mackwell explained that the editor responsible was James Saiers, who, he said, had been fully aware that McIntyre's paper challenged Mann's work and had therefore ordered a particularly thorough review. Mackwell was sympathetic, but stood firm and suggested that Mann respond by means of a published comment.

It appears that this was the point at which Mann discovered that Saiers was the editor of McIntyre's paper. Having taken on board this snippet of information Mann swung into action and instigated an immediate investigation of Saiers' background. Later the same day he reported his findings to his colleagues on the Hockey Team:

> Mann: 20 January 2005: 1106322460
>
> Just a heads up. Apparently, the contrarians now have an 'in' with GRL. This guy Saiers has a prior connection [with] the University of Virginia Dept. of Environmental Sciences* that causes me some unease. I think we now know how the various Douglass et al papers [with Pat] Michaels and [Fred] Singer, the Soon et al paper, and now this one have gotten published in GRL.

Tom Wigley appears to have been absolutely horrified that GRL had published McIntyre.

---

* This may be a reference to the prominent sceptic Pat Michaels, a member of staff at the University of Virginia.

Wigley: 20 January 2005: 1106322460

This is truly awful. GRL has gone downhill rapidly in recent years. I think the decline began before Saiers. I have had some unhelpful dealings with him recently with regard to a paper [a colleague] and I have on glaciers – it was well received by the referees, and so is in the publication pipeline. However, I got the impression that Saiers was trying to keep it from being published.

Proving bad behavior here is very difficult. If you think that Saiers is in the greenhouse skeptics camp, then, if we can find documentary evidence of this, we could go through official AGU channels to get him ousted. Even this would be difficult. How different is the GRL paper from the *Nature* paper? Did the authors counter any of the criticisms?

Mann was grateful for the support.

Mann: 20 January 2005: 1106322460

Thanks Tom,

Yeah, basically this is just a heads up to people that something might be up here. What a shame that would be. It's one thing to lose *Climate Research*. We can't afford to lose GRL. I think it would be useful if people begin to record their experiences [with] both Saiers and potentially Mackwell (I don't know him – he would seem to be complicit [in] what is going on here).

If there is a clear body of evidence that something is amiss, it could be taken through the proper channels. I don't [think] that the entire AGU hierarchy has yet been compromised!

In a later message to Malcolm Hughes, he expands on these concerns:

> Mann: 21 January 2005: 1106322460
> I'm not sure that GRL can be seen as an honest broker in these debates anymore, and it is probably best to do an end run around GRL now where possible. They have published far too many deeply flawed contrarian papers in the past year or so. There is no possible excuse for them publishing all 3 Douglass papers and the Soon et al paper. These were all pure crap.
>
> There appears to be a more fundamental problem [with] GRL now, unfortunately . . .

As we have seen, Saiers was ousted as editor in charge of the McIntyre and McKitrick paper, replaced by Jay Famiglietti.* Later that year, Mann can be seen discussing possible further sceptic papers criticising the Hockey Team:

> Mann: 15 November 2005: 1132094873
> The GRL leak may have been plugged up now [with] new editorial leadership there, but these guys always have *Climate Research* and *Energy and Environment*, and will go there if necessary.

So now there appears to be strong evidence that the Hockey Team sought to undermine the peer review process at at least three journals. It seems likely that they were successful on each occasion.

Getting rid of the Medieval Warm Period

The email archive at CRU shines some light on the infamous 'get rid of the MedievalWarm Period' email. In early 2008, David Holland, the sceptic who had been seeking Caspar Ammann's correspondence with Briffa, wrote to Jonathan Overpeck to inquire if the remarks attributed to him by David Deming were true. Overpeck seems to have been at a loss to know what to do, and

---

* See page 161.

wrote to several of his colleagues, including Mann, Jones and Susan Solomon, to ask for advice.

> Overpeck: 25 March 2008: 1206628118
> I have no memory of emailing [Deming], nor any record of doing so (I need to do an exhaustive search I guess),* nor any memory of him [from that] period. I assume it is possible that I emailed . . . him long ago, and that he's taking the quote out of context, since [I] know I would never have said what he's saying I would have, at least in the context he is implying.

So did Overpeck really make the outrageous statement he is alleged to have done. Perhaps we will never know. Were the Hockey Team really trying to 'get rid of the Medieval Warm Period'? Two more emails from the CRU archive can help colour our views.

In 1999, prior to the Third Assessment Report, there is an email from Briffa in which he says:

> Briffa: 22 September 1999: 0938018124
> I know there is pressure to present a nice tidy story as regards 'apparent unprecedented warming in a thousand years or more in the proxy data' but in reality the situation is not quite so simple.

Who it was that was applying the pressure is unclear, however. Briffa goes on to discuss the difficulties in justifying such a conclusion.

> Briffa: 22 September 1999: 0938018124
> We don't have a lot of proxies that come right up to date and [in] those that do (at least a significant number of tree

---

*   It is notable that Overpeck misspells Deming's name as 'Deeming' throughout this email. It is possible that if he searched his emails for 'Deeming' he would have missed the relevant message'.

proxies) [there are] some unexpected changes in response that do not match the recent warming. I do not think it wise that this issue be ignored in the chapter.

There is also a tantalising email from Mann to several team members concerning an article they were proposing to co-author. Discussing the length of the temperature record they would present he said

> Mann: 4 June 2003: 1054736277
> I think that trying to adopt a timeframe of [2000 years], rather than the usual [1000], addresses a good earlier point that [Overpeck] made [with] regard to the memo, that it would be nice to try to 'contain' the putative [Medieval Warm Period], even if we don't yet have a hemispheric mean reconstruction available that far back. . .

It seems clear then that there was outside pressure on the scientists to 'get rid of the Medieval Warm Period', a pressure that in some cases at least, was not entirely unwelcome. And if future developments turn out to show that Overpeck did not make the statement attributed to him, it seems clear that he at least had indicated to his Hockey Team colleagues that he would be happy to 'contain' evidence of past warming.

On the existence of the Medieval Warm Period

In their email correspondence, several of the scientists were much less gung-ho about the extent of medieval warmth than might have been expected. Back in 1999, Briffa had been saying this:

> Briffa: 22 September 1999: 0938031546
> For the record, I do believe that the proxy data do show unusually warm conditions in recent decades. I am not sure that this unusual warming is so clear in the summer responsive data. I believe that the recent warmth was probably matched about 1000 years ago. I do not believe

that global mean annual temperatures have simply cooled progressively over thousands of years as Mike appears to and I contend that that there is strong evidence for major changes in climate over the Holocene that require explanation and that could represent part of the current or future background variability of our climate.

It is striking how few of these doubts found their way into the Third Assessment Report. By 2003, the doubts were still lingering: Ed Cook pointed out that Ray Bradley viewed the Medieval Warm Period as 'mysterious and very incoherent'. He went on:

> Cook: 29 April 2003: 1051638938
>
> Of course [Bradley] and other members of the MBH camp have a fundamental dislike for the very concept of the [Medieval Warm Period], so I tend to view their evaluations as starting out from a somewhat biased perspective, i.e. the cup is not only 'half-empty'; it is demonstrably 'broken'. I come more from the 'cup half-full' camp when it comes to the [Medieval Warm Period], maybe yes, maybe no, but it is too early to say what it is. Being a natural skeptic, *I guess you might lean more towards the MBH camp, which is fine as long as one is honest and open about evaluating the evidence (I have my doubts about the MBH camp).* We can always politely(?) disagree given the same admittedly equivocal evidence.

The emphasis in the last quotation is added. This is an extraordinary statement for Cook to make about one of the most important scientists working in the field of climatology. Briffa's response indicated that he too was very cautious about the reality of medieval warmth.

> Briffa: 29 April 2003: 1051638938
>
> Can I just say that I am not in the MBH camp – if that be characterized by an unshakable 'belief' one way or the

other, regarding the absolute magnitude of the global [Medieval Warm Period].

He did, however, go on to say that he was inclined to believe in the IPCC assessment of the time, namely that there was 'likely unprecedented recent warmth'.

On Wahl and Ammann

The questions over when exactly Wahl and Ammann's CC paper was submitted and accepted by *Climatic Change* have already been outlined.*

We see in the emails the struggle to justify the acceptance of Wahl and Ammann's CC paper by the IPCC deadline. Wahl has been discussing whether the CC paper will meet the IPCC deadline with Jonathan Overpeck and they have agreed to approach *Climatic Change* editor, Stephen Schneider, for advice. Wahl's email to Schneider is as follows:

> Wahl: 11 February: 1139845689
>
> What I have understood from our conversations before is that if you receive the [manuscript] and move it from 'provisionally accepted' status to 'accepted', then this can be considered in press, in light of [*Climatic Change*] being a journal of record.

To which Schneider responds:

> Schneider: 11 February: 1139845689
>
> Your interpretation is fine – get me the revision soon so I have time to assess your responses in light of reviews in time! Look forward to receiving it, Steve

---

*   See page 167.

In other words, for the purposes of Stephen Schneider, it was sufficient to interpret 'accepted' as 'in press'. This then enabled the CC paper to be accepted into the IPCC review.

We have also seen that the CC paper relied on statistical arguments in the other paper, the comment, which was not even submitted until well after the CC paper had gone forward to the IPCC review. In September 2007, just before the final publication of the CC paper, Jones had clearly noticed that this was likely to be the cause of some criticism and emailed Wahl and Ammann accordingly:

> Jones: 12 September 2007: 1189722851
>
> Gene/Caspar,
> Good to see these two [papers] out. Wahl/Ammann doesn't appear to be in *Climatic Change* online [at] first, but comes up if you search. You likely know that McIntyre will check this one to make sure it hasn't changed since the IPCC close-off date July 2006! . . .
>
> [As for the resurrected comment] – try and change the Received date! Don't give those skeptics something to amuse themselves with.

Soon after this startling statement from Jones, Wahl made the following statement, in which he appears to admit the problems with the submission date for the revised comment.

> Wahl: 12 September 2007: 1189722851
> There were inevitably a few things that needed to be changed in the final version of [the CC paper] . . . I tried to keep all of this to the barest minimum possible, while still providing a good reference structure. I imagine that [McIntyre and McKitrick] will make the biggest issue about the very existence of the [revised comment], and then the referencing of it in [the CC paper]; but that was simply something we could not do without, and indeed [the CC paper] does a good job of contextualizing the whole matter.

Then Wahl seems to suggest that Stephen Schneider, who you may remember has been closely associated with the growth of the global warming phenomenon, had been involved in deciding if the Team could get away with the charade of the revised comment:

> Wahl: 12 September 2007: 1189722851
> Steve Schneider seemed well satisfied with the entire matter, including its intellectual defensibility (sp?) and I think his confidence is warranted. That said, any other thoughts/musings you have are quite welcome.

In Chapter 12,* we saw how, at the start of May 2008, David Holland had made his first Freedom of Information requests to Briffa, seeking background information on how certain decisions on the IPCC chapter had been taken and, later, requesting all the information the University of East Anglia held on the Fourth Assessment Report including Briffa's correspondence with Caspar Ammann. The CRU emails reveal just how problematic these requests were for the Hockey Team. In an email to Mann, Bradley and Ammann on 9 May, Jones said of Holland:

> Jones: 9 May 2008: 1210341221
> You can delete this attachment if you want. Keep this quiet also, but this is the person who is putting in FOI requests for all emails Keith and Tim have written and received re Ch 6 of AR4. We think we've found a way around this.

A way round was certainly needed. In their initial response to Holland, CRU had advised him that the request *must* be handled under the Environmental Information Regulations (EIR). This may have been a mistake on their part. The problem was that there were far fewer exemptions available to them under EIR than under the Freedom of Information Act (FOI). At some point over the next

---

* See page 278.

few weeks a decision seems to have been made to tell Holland that the information he requested was not environmental in nature and that EIR would not apply. This would then allow them to handle the request under the much weaker terms of FOI and so to invoke its exemptions for information provided in confidence and requests that would be too expensive to process.

Meanwhile, after the issues regarding the publication date of Ammann's CC paper became public, Holland had written again, probing the possibility that there had been a submission of review comments out with the normal channels. He asked for any additional comments on the drafts that were not in the online database of review comments, for any additional papers that had been submitted to the review, and also for any correspondence between Briffa and Ammann.

The main aspects of this request were dealt with by Phil Jones. Jones' email, which was addressed to Freedom of Information officer, David Palmer, but was copied also to Briffa, Osborn and senior faculty manager, Michael McGarvie, is a truly remarkable document:

> Jones: 28 May 2008: 1212009215
>
> . . . Keith (or you Dave) could say that . . .
>
> (1) Keith didn't get any additional comments in the drafts other than those supplied by IPCC . . . (2) Keith should say that he didn't get any papers through the IPCC process, either. I was doing a different chapter from Keith and I didn't get any.
>
> What we did get were papers sent to us directly – so not through IPCC asking us to refer to them in the IPCC chapters. If only Holland knew how the process really worked!! Every faculty member [in Briffa's department] and all the post-docs and most PhDs do, but seemingly not Holland.

So . . . Keith should say that he didn't get anything extra that wasn't in the IPCC comments. As for [Holland's request for any correspondence with Ammann] Tim has asked Caspar, but Caspar is one of the worse responders to emails known. I doubt either he emailed Keith or Keith emailed him related to IPCC. I think this will be quite easy to respond to once Keith is back. From looking at these questions and the Climate Audit web site, this all relates to two papers in the journal *Climatic Change*. I know how Keith and Tim got access to these papers and it was nothing to do with IPCC.

So clearly, Ammann had not provided a secret review but, shockingly, Briffa had received a copy of the CC paper and the revised comment directly from others. Unpublished material, unavailable to the external reviewers, had been used to inform the IPCC review. Even worse, Briffa, Palmer and a senior member of faculty staff appear to have been sniggering at Holland's attempts to get at the truth, all the time ignoring their statutory duty to help and assist him, thus flouting the spirit of the law.

The last part of Holland's request, in which he asks for copies of Ammann's correspondence with Briffa, seemed to concern the scientists rather more. Despite the fact that Briffa had got hold of Ammann's new comment from someone else, the CRU team still appeared determined that nothing would be released. The problem was that Ammann's correspondence was not obviously confidential. The UK Information Commissioner's guidelines said however, that in order to determine confidentiality, it was necessary to determine that any release of the information was legally actionable, if necessary by consulting the person affected.[263] To that end, on 27 May 2008, David Palmer, the FOI officer at the University of East Anglia, wrote to Tim Osborn asking if the correspondence between Ammann and CRU was in fact confidential. Osborn took the hint and wrote the same day to Ammann.

> Osborn: 27 May 2008: 1211924186
> Our university has received a request, under the UK
> Freedom of Information law, from someone called David
> Holland for emails or other documents that you may have
> sent to us that discuss any matters related to the IPCC
> assessment process. We are not sure what our university's
> response will be, nor have we even checked whether you
> sent us emails that relate to the IPCC assessment or that we
> retained any that you may have sent. However, it would be
> useful to know your opinion on this matter. In particular,
> we would like to know whether you consider any emails
> that you sent to us as confidential.

Ammann replied that he would need to look through his
correspondence:

> Ammann: 27 May 2008: 1211924186
> Well, I will have to properly answer in a couple days when
> I get a chance digging through emails. I don't recall from
> the top of my head any specifics about IPCC.

Ammann clearly hadn't taken the hint about the confidentiality of
the correspondence and Osborn therefore decided to make the
point slightly clearer:

> Osborn: 30 May 2008: 1212166714
> I don't think it is necessary for you to dig through any
> emails you may have sent us to determine your answer. Our
> question is a more general one, which is whether you
> generally consider emails that you sent us to have been sent
> in confidence. If you do, then we will use this as a reason
> to decline the request.

Ammann replied the same day.

> Ammann: 30 May 2008: 1212156886
>
> In response to your inquiry about my take on the confidentiality of my email communications with you, Keith or Phil, I have to say that the intent of these emails is to reply or communicate with the individuals on the distribution list, and they are not intended for general 'publication'. If I would consider my texts to potentially get wider dissemination then I would probably have written them in a different style. Having said that, as far as I can remember (and I haven't checked in the records, if they even still exist) I have never written an explicit statement on these messages that would label them strictly confidential.

This extraordinary vague response appears to have been enough to convince the authorities at the University of East Anglia that a release of Ammann's correspondence would be legally actionable. This was clearly a very weak position and Jones seems anyway to have wanted to be absolutely certain that nothing was going to be revealed. He was prepared to take further extraordinary steps to do so. This is an extract of an email he sent to Mann at the end of May 2008:

> Mann: 29 May 2008: 1212073451
>
> Mike,
>
> Can you delete any emails you may have had with Keith re AR4? Keith will do likewise. He's not in at the moment – minor family crisis. Can you also email Gene and get him to do the same? I don't have his new email address. We will be getting Caspar to do likewise.

Under UK Freedom of Information (FOI) legislation, deleting information that has been requested under the legislation is a criminal offence. It is not clear, however, whether anything was in fact deleted.

So can we ever know who provided Briffa with a copy of the revised comment in *Climatic Change*? Another set of emails appears to answer the question. In July 2006, in the middle of the review process, we see Briffa thanking Eugene Wahl for something.

> Briffa: 21 July 2006: 1155402164
>
> Gene
> Thanks a lot for this – I need to digest and I will come back to you.
> Thanks again
> Keith

A few hours later, Wahl responds, apologising that there is so much to digest, but also providing the intriguing new detail that he has been preparing a briefing paper on the Hockey Stick affair for 'a person in [Washington] DC who is working on all this with regard to the [Barton Hearings].'* The briefing paper is very interesting but is slightly besides the point at issue here. However, one extract will give both a flavour of the piece and a feel for how well it represented the debate. This covers the question of whether the bristlecones are valid proxies or not. (The annotations and capitals are all Wahl's.)

> Wahl: 21 July 2006: 1155402164
>
> Although there are a number of reasons to keep the bristlecone data in, maybe the most compelling reason they are a NON-ISSUE is that, over the common period of overlap (1450–1980), the reconstruction based on using them from 1400–1980 is very close to the reconstruction based on omitting them from 1450–1980. Since the issues about the bristlecone response to climate are primarily about 1850 onwards, especially 1900 onwards [KEITH – PLEASE LET ME KNOW IF I AM NOT ACCURATE IN

---

* The first hearing had been held two days earlier.

THIS], there is no reason to expect that their behavior during 1400–1449 is in any way anomalous to their behavior from 1450–1850. Thus, THERE IS NO REASON TO THINK THAT THE BRISTLECONES ARE SOMEHOW MAKING THE 1400–1449 SEGMENT OF THE MBH RECONSTRUCTION BE INAPPROPRIATELY SKEWED.

Meanwhile, Briffa had been surveying whatever it was that Wahl had sent him.

> Briffa: 21 July 2006: 1155402164
> Your comments have been really useful and reassuring that I am not doing MM [presumably Michael Mann, rather than McIntyre and McKitrick] a disservice. I will use some sections of your text in my comments that will be eventually archived so hope this is ok with you. I will keep the section in the chapter very brief – but will cite all the papers to avoid claims of bias.

The next day, Wahl wrote back, somewhat concerned:

> Wahl: 22 July 2006: 1155402164
> If I could get a chance to look over the sections of my text you would post to the comments before you do, I would appreciate it. If this is a burden/problem let me know and we'll work it out.
> If it is anything from the [CC] paper, of course that is fine to use at once since it is publicly available. There will only be exceedingly minor/few changes in the galleys, including a footnote pointing to the extended RE benchmarking analysis contained in the [revised comment].
> What I am concerned about for the time being is that nothing in the [revised comment] shows up anywhere. It is just going in, and confidentiality is important. The only exception to this are the points I make . . . concerning

the [McIntyre and McKitrick's] way of benchmarking the RE statistic. Those comments are fine to repeat at this point.

Fortunately for Wahl, Briffa appears to have been happy to play along:

> Briffa: 24 July 2006: 1155402164
>
> Here is where I am up to now with my responses (still a load to do) – you can see that I have 'borrowed (stolen)' from 2 of your responses in a significant degree – please assure me that this OK (and will not later be obvious) hopefully.
>
> You will get the whole text (confidentially again) soon. You could also see that I hope to be fair to Mike [Mann] – but he can be a little unbalanced in his remarks sometime – and I have had to disagree with his interpretations of some issues also.
>
> Please do not pass these on to anyone at all.

It seems fair to say then that Eugene Wahl provided Briffa with a copy of his unpublished *Climatic Change* comment in order to assist in rebutting McIntyre, and that contrary to IPCC rules, Briffa used this information to inform his drafting work. It also appears clear that Jones was aware of what had gone on.

On IPCC deadlines

There is an email from someone called 'Mel', working at the IPCC TSU to Overpeck and Jansen, the coordinating lead authors for the paleoclimate chapter which demonstrates that Hegerl's paper was accepted after the official cutoff date, that cutoff date being retrospectively changed to accommodate it and the other late breaking papers.

Mel: 10 August 2006: 1155497558

Although the deadline for additional accepted papers has now passed, this submission comes from a [chapter lead author] (Gabi Hegerl) so am forwarding on.

On data withholding

One interesting email exchange that is of direct relevance to the Hockey Stick story concerns McIntyre's attempts to get hold of Mann's code for the Hockey Stick papers.* The CRU email archive contains a message from one of the *Climatic Change* editorial board, Professor Christian Azar of the University of Goteborg, in which he indicates that he agrees with the rest of the board that unless what Mann has already posted is sufficient to allow reproduction of his results, then the code should be released. He adds that releasing it would anyway be beneficial to the debate.

The day after Azar's statement, Jones emailed the whole editorial board, with a long plea that Mann should be allowed to withhold his code.

Jones: 16 January 2004: 1074277559

The papers that [McIntyre and McKitrick] refer [to] came out in *Nature* in 1998 and to a lesser extent in GRL in 1999. These reviewers did not request the data . . . and the code. So, acceding to the request for this to do the review is setting a VERY dangerous precedent. Mike [Mann] has made all the data series [available] and this is all anyone should need. Making model code available is something else.

He goes on by explaining that the code is irrelevant to the debate and that sceptics are picking on Mann:

---

*   See page 120.

> Jones: 16 January 2004: 1074277559
>
> I'm not sure how many of you realise how vicious the attack on him has been. I will give you an example.

He then complains that the first Mann heard of McIntyre's 2003 paper in *Energy and Environment* was when the press told him about it. He claims that the peer review of McIntyre's paper was not independent and complains that McIntyre and McKitrick's paper had a figure labelled 'corrected version' which he felt contradicted their position that they weren't publishing their own reconstruction. Nothing of what Jones says, however, could conceivably be labelled a 'vicious attack'.

Closing off he addresses Schneider directly:

> Jones: 16 January 2004: 1074277559
>
> In trying to be scrupulously fair, Steve, you've opened up a whole can of worms.

An hour later he contacted Mann (capitals in original):

> Jones: 16 January 2004: 1074277559
>
> This is for YOURS EYES ONLY. Delete after reading – please !
> I'm trying to redress the balance. One reply . . . said you should make all available!!* . . . Told Steve separately . . . to get more advice from a few others as well as [the publisher] and legal.
>
> PLEASE DELETE – just for you, not even Ray and Malcolm.

The reasons why Jones asks so urgently for secrecy are not clear. It may be that he was asked not to discuss the decision with Mann.

---

*   Jones was not referring to the email from Azar, but to one of the other members of the board who had expressed similar sentiments.

It is clear from the email archive that requests for data are seen as burdensome and irritating by the Hockey Team. They believe that there is an attempt to prevent them from doing their work by tying them up in endless requests for information.

With the introduction of the UK Freedom of Information Act in 2005, the scientists were clearly worried. Tom Wigley wrote to Jones wondering if it meant he would have to release his computer code, a question that Jones said would only be answered in the fullness of time once legal precedents began to be set on the subject. He reassured Wigley that as an ex-employee of CRU he would probably not be covered by the Act, a theme on which he expanded in a later message:

> Jones: 21 January 2005: 1106338806
> I wouldn't worry about the code. If [the Freedom of Information Act] does ever get used by anyone, there is also [intellectual property rights] to consider as well. Data is covered by all the agreements we sign with [third parties], so I will be hiding behind them.

However, by 2007, in an email to Tom Karl, Jones was reporting that he had been able to persuade his FOI officers to help him out.

> Jones: 19 June 2007: 1182255717
> Think I've managed to persuade [University of East Anglia] to ignore all further [FOI] requests if the people have anything to do with Climate Audit.

And by 2008, Phil Jones was reporting to his colleagues that CRU was coping well.

> Jones: 20 August 2008: 1219239172
> [Keith Briffa and Tim Osborn are] still getting FOI requests as well as [the Hadley Centre at the Met Office and the University of Reading]. All our FOI officers have been in

discussions and are now using the same exceptions not to respond – advice they got from the Information Commissioner. . .

The . . . line we're all using is this. IPCC is exempt from any country's FOI – the skeptics have been told this.

Extraordinarily then, there is a strong hint that the Information Commissioner's office had been providing a variety of public bodies with advice on how to avoid public requests for information.

Meanwhile, other members of the Hockey Team were less happy with the way things were going. Ben Santer was one of these. He had been on the receiving end of one of McIntyre's requests for data, but his refusal to comply had cause some problems:

> Santer: 2 December 2008: 1228258714
>
> There has been some additional fallout from the publication of our paper in the *International Journal of Climatology*. After reading Steven McIntyre's discussion of our paper on climateaudit.com (and reading about my failure to provide McIntyre with the data he requested), an official at [The US Department of the Environment] headquarters has written to . . . [Lawrence Livermore National Laboratory, Santer's employer], claiming that my behavior is bringing [the lab's] good name into disrepute.

The next day he expanded on his concerns:

> Santer: 3 December 2008: 1228330629
>
> One of the problems is that I'm caught in a real Catch-22 situation. At present, I'm damned and publicly vilified because I refused to provide McIntyre with the data he requested. But had I acceded to McIntyre's initial request for climate model data, I'm convinced (based on the past experiences of [other Hockey Team members]) that I would have spent years of my scientific career dealing with

demands for further explanations, additional data, Fortran code, etc. (Phil has been complying with FOI requests from McIntyre and his cronies for over two years). And if I ever denied a single request for further information, McIntyre would have rubbed his hands gleefully and written: 'You see – he's guilty as charged!' on his website.

Jones tried to reassure the American:

> Jones: 3 December 2008: 1228330629
> When the FOI requests began here, the FOI person said we had to abide by the requests. It took a couple of half hour sessions – one at a screen, to convince them otherwise, showing them what [Climate Audit] was all about. Once they became aware of the types of people we were dealing with, everyone at UEA (in the registry and in the Environmental Sciences school – the head of school and a few others) became very supportive. I've got to know the FOI person quite well and the Chief Librarian – who deals with appeals. The VC* is also aware of what is going on – at least for one of the requests, but probably doesn't know the number we're dealing with. We are in double figures.

Meanwhile, it was clear that, just as they had done on the possibility of journals publishing sceptic papers, the Team were also going to use their collective influence to keep the journals in line on the subject of data availability. Having been refused data by Santer, McIntyre had taken the issue up with the *International Journal of Climatology* (IJoC). Unfortunately, the journal editor had said that data archiving was not required by the journal.

During the course of 2009, I corresponded with Professor Paul Hardaker, the chief executive of the Royal Meteorological Society (RMS), which publishes IJoC, on the question of why the journal

---

*   Apparently the Vice Chancellor, at that time Bill MacMillan.

had no policy on making data available. Professor Hardaker was very accommodating, and undertook to put the question of formulating a policy to the society's publications committee. However, shortly after I had made my first approaches to Professor Hardaker, the Hockey Team seem to have got in touch with him too. In March 2009. we see Phil Jones emailing Hardaker:

> Jones: 19 March 2009: 1237496573
> I had been meaning to email you about the RMS and IJoC issue of data availability for numbers and data used in papers that appear in RMS journals. This results from the issue that arose with the paper by Ben Santer et al in IJoC last year. Ben has made the data available that this complainant wanted. The issue is that this is intermediate data. The raw data that Ben had used to derive the intermediate data was all fully available.

Santer, meanwhile, was deeply unimpressed with the idea of having to make intermediate data available.

> Santer: 19 March 2009: 1237496573
> If the RMS is going to require authors to make ALL data available – raw data PLUS results from all intermediate calculations – I will not submit any further papers to RMS journals.

Jones also seems to have had another issue with the RMS.

> Jones: 19 March 2009: 1237496573
> I'm having a dispute with the new editor of *Weather*.* I've complained about him to [Hardaker]. If I don't get him to back down, I won't be sending any more papers to any RMS journals and I'll be resigning from the RMS.

---

*   Another RMS journal.

Since that time, the Royal Meteorological Society's publications committee has met twice, but to date there appears to be no new policy on data availability.

Other scientists couldn't quite believe the Hockey Team's approach to the subject. In 2006, Hans von Storch wrote to Keith Briffa having read an article that recounted some of McIntyre's problems with getting hold of Briffa's data. These requests, said von Storch, were entirely appropriate and he quoted what had been written in the article.

> Von Storch: 5 August 2006: 1155333435
> 'The issue of data access was discussed in the [dendroclimatology] conference in Beijing – some people suggesting that withholding data was giving the trade a black eye. Industry leaders, such as presumably Briffa, said that they were going to continue stonewalling'.

Von Storch was incredulous:

> Von Storch: 5 August 2006: 1155333435
> I can not believe this claim, and I would greatly appreciate if you would help me to diffuse any such suspicions . . . I am concerned if we do not apply a truly open data and algorithm-policy, our credibility will be severely damaged, not only in the US but also in Europe. 'Open' means also to provide data to groups which are hostile to our work – we have done so with our [own] data, which resulted in two hostile comments in *Science*, which were, however, useful as they helped to clarify some issues.

Briffa, however, merely brushed him aside:

> Briffa: 11 August 2006: 1155333435
> Just too bogged down with stuff to even read their crap –

but I have no intention of withholding anything. Will
supply the stuff when I get five minutes!!

As we have seen Briffa's data was only finally released three years
later.

Mann may have refused to send his residual series to McIntyre
but he was quite happy to send them to trusted colleagues. In July
2003 he sent the MBH99 figures to CRU's Tim Osborn (emphasis
added).

Mann: 31 July 2003: 1059664704

Tim,

Attached are the calibration residual series for experiments
based on available networks. . .

Basically, you'll see that the residuals are pretty red for
the first 2 cases, and then not significantly red for the 3rd
case – its even a bit better for the AD 1700 and 1820 cases,
but I can't seem to dig them up.* In any case, the
incremental changes are modest after 1600 – it's pretty
clear that key predictors drop out before AD 1600, hence
the redness of the residuals, and the notably larger
uncertainties farther back . . .

You only want to look at the first column (year) and
second column (residual) of the files. I can't even remember
what the other columns are! Let me know if that helps.

Thanks,

Mike

P.S. I know I probably don't need to mention this, but just
to insure absolutely clarity on this, I'm providing these for
your own personal use, since you're a trusted colleague. So
please don't pass this along to others without checking
[with] me first. *This is the sort of 'dirty laundry' one doesn't*

---

* Residuals should be 'white', which is to say entirely random, rather than
red. Mann is implying that there is a problem with the calibration.

want to fall into the hands of those who might potentially try to distort things. . .

On bristlecones

There is an interesting exchange of emails that was prompted by an series of exchanges on Climate Audit between paleoclimatologist Martin Juckes and McIntyre and his readers. Juckes had been trying to defend his use of bristlecones in the face of the NAS panel's conclusions that this was not advisable, and he indicated in an email that he was going to contact Gerry North to see if this was really what he meant. Unfortunately, North was not willing to come off the fence, deferring instead to the panel member who had written the paragraph on bristlecones. Juckes however felt that the evidence for carbon dioxide fertilisation in bristlecones was weak

Briffa on the other hand was quite convinced that there were problems with the species, if not necessarily to do with carbon dioxide fertilisation:

> Briffa: 6 November 2006: 1163715685
> In my opinion (as someone who has worked with the bristlecone data hardly at all!) there are undoubtedly problems in their use that go beyond the strip bark problem (that I will come back to later). . .
>
> The main one is an ambiguity in the nature and consistency of their sensitivity to temperature variations . . . The bottom line though is that these trees likely represent a mixed temperature and moisture-supply response that might vary on longer timescales.

He also said that stripbark problem meant that the bristlecones 'will have unpredictable trends as a consequence of aging and depending on the precise nature of each tree's structure', and referred to Mann's adjustment for carbon dioxide fertilisation as 'very arbitrary'. In fairness, he also noted that one author had suggested that there was in fact no carbon dioxide fertilisation at all.

Esper agreed, suggesting that the wider rings in recent years were due to physical damage rather than temperature:

> Esper: 6 November 2006: 1163715685
> I didn't visit the bristlecone sites yet, but the mechanism might be [physical damage]. I believe that over time the crown and root system are reduced, but not at the same rate than the reduction in circumference covered by the cambium. This would be the key for strip bark tree rings being wider than 'normal' rings.

Another paleoclimatologist, Rob Wilson, said that he had avoided using bristlecones after noting McIntyre's findings. He went on to note that there didn't seem to be a correlation between the bristlecones and temperature, although he thought there was at least some temperature response in the record.

On Yamal

Several of the emails touch on the subject of Yamal. It is revealed that Briffa was funding Stepan Shiyatov. In one email, Shiyatov asks Briffa to send multiple small payments to his personal bank account:

> Shiyatov: 7 March 1996: 0826209667
> It is important for us if you can transfer the ADVANCE* money on the personal accounts which we gave you earlier and the sum for one occasion transfer (for example, during one day) will not be more than 10,000 USD. Only in this case we can avoid big taxes and use money for our work as much as possible.

When Briffa's data was finally released, there clearly consternation among the scientists, a situation that was

---

*   This appears to be a funding programme of some kind.

exacerbated by the fact that Briffa was recovering from a serious illness. In his absence, Mann and Schmidt approached Briffa's colleague Tim Osborn for help in formulating a response.

But while Osborn had co-authored with Briffa in the past, he hadn't been involved with the Royal Society paper and wasn't able to help much.

> Osborn: 29 September 2009: 1254230232
> Regarding Yamal, I'm afraid I know very little about the whole thing – other than that I am 100% confident that 'The tree ring data was hand-picked to get the desired result' is complete crap. Having one's integrity questioned like this must make your blood boil (as I'm sure you know, with both of you having been the target of numerous such attacks) . . .
>
> Apart from Keith, I think Tom Melvin here is the only person who could shed light on the McIntyre criticisms of Yamal. But he can be a rather loose cannon and shouldn't be directly contacted about this . . .

Melvin, also from CRU, was a Briffa co-author on the Royal Society paper. It appears, however, that the Hockey Team did not consider him as one of their own. This must have been something of a problem, with Briffa out of action and Melvin not trusted. How would they respond? Fortunately, by the next day Briffa had agreed to pull himself from his sickbed, and Mann and Osborn agreed that a rebuttal would be issued, despite the fact that nobody had actually examined McIntyre's work at that point.

Mann set straight to work, responding to a request for information from the *New York Times*' Andy Revkin with this:

> Mann: 29 September 2009: 1254258663
> The preliminary information I have from others familiar with these data is that the attacks are bogus . . . even if there were a problem [with] these data, it wouldn't matter as far as the key conclusions regarding past warmth are

concerned. But I don't think there is any problem with these data, rather it appears that McIntyre has greatly distorted the actual information content of these data.

Again, it is not clear that any detailed examination of McIntyre's claims had yet taken place. It is also odd that Revkin would ask Mann for information about Briffa's paper. Mann had not been involved in Briffa's paper at all.

Meanwhile, other members of the Hockey Team were more concerned. Tom Wigley in particular was very worried:

> Wigley: 5 October 2009: 1254756944
>
> Keith [Briffa] does seem to have got himself into a mess. As I pointed out in emails, Yamal is insignificant. And you say that (contrary to what [McIntyre and McKitrick] say) Yamal is *not* used in MBH, etc.* So these facts alone are enough to shoot down [McIntyre and McKitrick] in a few sentences (which surely is the only way to go – complex and wordy responses will be counter productive).
>
> But, more generally, (even if it is irrelevant) how does Keith explain the McIntyre plot that compares Yamal-12 with Yamal-all? And how does he explain the apparent 'selection' of the less well-replicated chronology rather that the later (better replicated) chronology?
>
> Of course, I don't know how often Yamal-12 has really been used in recent, post-1995, work. I suspect from what you say it is much less often that [McIntyre and McKitrick] say – but where did they get their information? I presume they went thru papers to see if Yamal was cited, a pretty foolproof method if you ask me.

---

* Although some members of the press made this erroneous allegation, I have been unable to locate any instances of McIntyre doing so. It may be that Wigley is conflating McIntyre's comments with those of journalists.

Perhaps these things can be explained clearly and concisely – but I am not sure Keith is able to do this as he is too close to the issue and probably quite pissed off. And the issue of withholding data is still a hot potato, one that affects both you and Keith (and Mann). Yes, there are reasons – but many good scientists appear to be unsympathetic to these. The trouble here is that withholding data looks like hiding something, and hiding means (in some eyes) that it is bogus science that is being hidden.

I think Keith needs to be very, very careful in how he handles this. I'd be willing to check over anything he puts together.

On the Hockey Stick

There is one lovely email from John Mitchell, who we met earlier in his role of IPCC review editor. Here he is saying his piece to Eystein Jansen and Jonathan Overpeck on the subject of the second order draft comments.

> Mitchell: 21 June 2006: 1150923423
> I am in Geneva . . . so I have not had a lot of time to look at the [Second Order Draft] comments. I can not get to Bergen before Tuesday. I had a quick look at the comments on the Hockey Stick and include below the questions I think need to be addressed which I hope will help the discussions. I do believe we need a clear answer to the skeptics. I have also copied these comments to Jean [Jouzel, the other review editor] . . .
>
> 1. There needs to be a clear statement of why the instrumental and proxy data are shown on the same graph. The issue of why we don't show the proxy data for the last few decades (they don't show continued warming) but assume that they are valid for early warm periods needs to be explained.

This is an extraordinary statement. Clearly, senior IPCC scientists knew that the proxy records showed no warming in recent decades. Mitchell felt it needed to be explained. It is clear from the IPCC report however, that Jansen and Overpeck did not take him up on this suggestion. The information that proxy records do not now show any warming has been suppressed.

> Mitchell: 21 June 2006: 1150923423
>
> 2 . There are number of methodological issues which need a clear response. There are two aspects to this. First, in relation to the [Third Assessment Report and MBH98],* which seems to be the obsession of certain reviewers. Secondly (and this I believe this is the main priority for us) in relation to conclusions we make in the chapter we should make it clear where our comments apply to only MBH (if that is appropriate), and where they apply to the overall findings of the chapter. Our response should consider all the issues for both MBH and the overall chapter conclusions. a. The role of bristlecone pine data: Is it reliable? Is it necessary to include this data to arrive at the conclusion that recent warmth is unprecedented? b. Is the [PC analysis] approach robust? Are the results statistically significant? It seems to me that in the case of MBH the answer in each is no. It is not clear how robust and significant the more recent approaches are . . .

So amazingly, Professor Mitchell believed that the Hockey Stick used a biased methodology and gave results that were not statistically significant and yet signed off the paleoclimate chapter as 'a reasonable assessment' of the evidence. Yet here is how the final version of the IPCC report explained the Hockey Stick debate:

---

*   Mitchell wrote MBA, which I assume is an error. Presumably he meant MBH.

> McIntyre and McKitrick . . . raised . . . concerns about the details of the Mann et al. (1998) method, principally relating to the independent verification of the reconstruction against 19th-century instrumental temperature data and to the extraction of the dominant modes of variability present in a network of western North American tree ring chronologies, using Principal Components Analysis. The latter may have some theoretical foundation, but Wahl and Ammann (2006) also show that the impact on the amplitude of the final reconstruction is very small.[264]

Readers can judge for themselves whether Professor Mitchell should have accepted this as a fair reflection of the dispute.

Professor Mitchell's role in the IPCC review was to umpire disputes and ensure that both sides of any argument were fairly represented in the report. Yet here we see him engaged in an ongoing correspondence discussing not how to represent sceptic positions in the report, but how to give them 'a clear answer'. Disputes were supposed to be reported in an annex to the report, and yet there is no sign of this having been even considered by Mitchell and Briffa.

Where do we stand now?

The initial shock of the leaking of the CRU emails seems to be dying away, but it is clear that the reverberations will continue for months to come. The emails have been analysed by many sets of eyes, and the most obvious outrages are all now in the public realm. As those closer to the story survey the evidence more closely in coming weeks, there can be little doubt that further revelations will be made. Analysis has also begun on the files of data and code that were released along with the emails. Already we have seen the quality of computer programming come in for serious criticism.

The emails were released after completion of the text of this book. What is extraordinary to me as a writer is how much of the content of the emails entirely corroborates what I had written in the previous chapters. In light of all I had learned while researching this book, the emails read exactly as I might have imagined they would. The sceptic community, and particularly McIntyre and McKitrick, had been extraordinarily insightful in their analysis of what was happening behind the scenes of the global warming movement. This means that everything I wrote in Chapter 15 still stands. However, the CRU emails have shown us that the situation is even worse than was thought. For the purposes of this book there are two clear conclusions to be drawn from the emails. Firstly that senior climatologists have sought to undermine the peer review process and bully journals into suppressing dissenting views. This means that the scientific literature is no longer a representation of the state of human knowledge about the climate. It is a representation of what a small cabal of scientists feel is worthy of discussion. Secondly, the IPCC reports represent the outcome of a process in which a relatively small group of scientists produce a biased review of a literature they themselves have colluded to distort through gatekeeping and intimidation. The emails establish a pattern of behaviour that is completely at odds with what the public has been told regarding the integrity of climate science and the rigour of the IPCC report-writing process. It is clear that the public can no longer trust what they have been told. What is less clear is what we, as ordinary citizens, can do in the face of the powerful, relentless forces of corrupted science, to set things right. Awareness, however, is the essential first step.

Notes on the figures

All figures are original artwork prepared by the author, except as follows.

Chapter 1
1.1 The Medieval Warm Period. Adapted from the version published in the NAS report.
1.2 The Hockey Stick from *Nature*. This is the black and white version of the Hockey stick that caused the initial controversy.[14]

Chapter 3
3.1 The Gaspé series, Prepared by the author using data downloaded from http://www.uoguelph.ca/~rmckitri/research/trcsupp.html.
3.2 Twisted Tree, Heartrot Hill. Adapted from the version published in MM03.[37]
3.3 The Australia PC1. Adapted from the version published in MM03.[37]

Chapter 5
5.4 Sheep Mountain and Mayberry Slough. Adapted from the version in the *Nature* submission.[47]
5.5 The effect of short centring on red noise. Adapted from the version in the *Nature* submission.[47]

Chapter 6
6.1 Twelve Hockey Sticks. Adapted from the version shown by McIntyre at the AGU.
6.2 Outliers in the dataset. Adapted from the version published in MM05(EE).[79]
6.4 Gaspé – original and updated series. Adapted from the version published by McIntyre at Climate Audit.[265]

Chapter 8
8.1 Adapted from Huybers' comment in *Geophysical Research Letters*.[117]

Chapter 9

9.1 The average of Mann's proxies compared to the Hockey Stick. Adapted from McIntyre's NAS presentation.[155]

9.2 Mann at the centre of a paleoclimate web. Adapted from the Wegman Report.[15]

Chapter 10

10.1 A well-dated tree. Adapted from Climate Audit.[266]

10.2 Tree with uncertain dating. Adapted from Climate Audit.[266]

10.3 Yamal and the Polar Urals update. Adapted from Climate Audit.[267]

10.4 Some of the 64 flavours of temperature reconstruction. Downloaded from the supplementary information to Bürger and Cubasch at ftp://ftp.agu.org/apend/gl/2005GL024155/2005GL024155-fs02.jpg.

Chapter 11

11.1 Was a correction made to MBH99? Adapted from McIntyre's NAS presentation.[155]

Chapter 13

13.1 Analysis of rings of Almagre tree 84–55. Adapted from Climate Audit.[268]

13.2 Two cores from tree 84–56. Adapted from Climate Audit.[269]

13.3 The Ababich update to Sheep Mountain. Adapted from Climate Audit.[270]

13.4 Mann and Jones PC1 with different versions of Sheep Mountain. Adapted from Climate Audit.[270]

Chapter 15

15.1 Briffa's reconstruction from 2001. Prepared by the author using Steve McIntyre's R script (http://www.climateaudit.org/scripts/spaghetti/spaghetti.wilson07.txt). Data sources as per the script.

Chapter 16

16.1 The Yamal sensitivity test. Reproduced by the author using McIntyre's script at http://www.climateaudit.org/?p=7168. Data sources as per the script.

Sources

1. McKitrick R. What is the Hockey Stick debate about? 2005. Available at:http://www.friendsofscience.org/assets/files/documents/hockeystick.pdf

2. Weart SR. The discovery of global warming. Harvard University Press; 2004.

3. Hansen J, Johnson D, Lacis A, Lebedeff S, Lee P, Rind D, et al. Climate impact of increasing atmospheric carbon dioxide. *Science*. 1981; 213(4511): 957–966.

4. Hansen JE. The greenhouse effect: impacts on current global temperature and regional heat waves. Testimony to the United States Senate Committee on Energy and Natural Resources; June 1988. Available at http://image.guardian.co.uk/sys-files/Environment/documents/2008/06/23/ClimateChangeHearing1988.pdf.

5. Thatcher M. Speech to the Royal Society; 27 September 1988. Available at: http://www.margaretthatcher.org/speeches/displaydocument.asp?docid=107346.

6. Thatcher M. Speech to United Nations General Assembly (Global environment). 8 November 1989. Available at: http://www.margaretthatcher.org/speeches/displaydocument.asp?docid=107817.

7. Houghton J, Jenkins G, Ephraums J. Climate change: The IPCC scientific assessment (Contribution of Working Group I to the first assessment report of the Intergovernmental Panel on Climate Change). Cambridge: Cambridge University Press; 1990.

8. Lamb H. The early medieval warm epoch and its sequel. *Palaeogeography, Palaeoclimatology, Palaeoecology*. 1965; 1: 13–37.

9. Hughes M, Diaz H. Was there a Medieval Warm Period, and if so, where and when? *Climatic Change*. 1994; 26: 109–142.

10. Deming D. Climatic warming in North America: analysis of borehole temperatures. *Science*. 1995; 268(5217): 1576–1577.

11. Deming D. Global warming, the politicization of science, and Michael Crichton's State of Fear. *Journal of Scientific Exploration*. 2005; 19. Available at http://www.sepp.org/Archive/NewSEPP/StateFear-Deming.htm.

12. Lindzen R. Climate science: is it currently designed to answer questions? 29 November 2008. Available at http://arxiv.org/abs/0809.3762.

13. Houghton JT. Climate change 1995: the science of climate change. Cambridge: Cambridge University Press; 1996.

14. Mann ME, Bradley RS, Hughes MK. Global-scale temperature patterns and climate forcing over the past six centuries. *Nature*; 1998; 392: 779–787.

15. Wegman EJ, Scott DW, Said YH. Ad Hoc Committee report on the Hockey Stick global climate reconstruction. US Congress; 2006;

Available at: http: //www.uoguelph.ca/~rmckitri/research/Wegman Report.pdf.

16. University of Massachusetts Office of News & Information. UMass Amherst scientists lead team reconstructing global temperature over past six centuries. University of Massachusetts press release; 22 April 1998. Available at: http: //www.umass.edu/newsoffice/newsreleases/articles/12390.php.

17. Stevens WK. New evidence finds this is warmest century in 600 years. *New York Times*; 23 April 1998. Available at http://query.nytimes. com/gst/fullpage.html?res=950CEFDB1E3FF93BA15757C0A96E958260.

18. Manning A. 90s were warmest years in centuries. USA Today; 23 April 1998. Available at: http://www.usatoday.com/weather/climate/WC042398.HTM.

19. Llanos M. Millennium ending with record heat. MSNBC.com; 22 April 1998. Available at http://web.archive.org/web/19990222064034/msnbc.msn.com/news/160184.asp.

20. Notebook/planetwatch. It hasn't been this sizzling in centuries. Time.com; 4 May 1998. Available at http://web.archive.org/web/20021003055434/www.time.com/time/magazine/1998/dom/980504/notebook.planet_watch.it37.html.

21. Mann ME, Bradley RS, Hughes MK. Northern Hemisphere temperatures during the last millennium: inferences, uncertainties and limitations. *Geophysical Research Letters*; 1999; 26: 759–62.

22. University of Massachusetts Office of News & Information. 1998 was warmest year of millennium, UMass Amherst climate researchers report. 3 March 1999. Available at: http://www.umass.edu/newsoffice/newsreleases/articles/12577.php.

23. Daly J. The Hockey Stick. A new low in climate science. Available at http://www.john-daly.com/hockey/hockey.htm.

24. Fielden T. Report on the Hockey Stick controversy. BBC Radio 4; 24 February 2005. Audio available at http://www.bbc.co.uk/radio4/today/rams/ environment 20050224.ram.

25. Wilmking M, Juday GP, Barber VA, Zald HSJ. Recent climate warming forces contrasting growth responses of white spruce at treeline in Alaska through temperature thresholds. *Global Change Biology*. 2004; 10: 1724–1736.

26. Mann M. Climate over the past two millennia. Annual Review of Earth and Planetary Sciences. 2007; 35: 111–36. Available at: http://holocene.meteo.psu. edu/shared/articles/AREPS-preprint06.pdf.

27. Soon W, Baliunas S. Proxy climatic and environmental changes of the past 1000 years. *Climate Research*. 2003; 23: 89–110.

28. Mann M, Schmidt G. Peer review: A necessary but not sufficient condition II. RealClimate blog; 27 January 2005. Available at: http://www.realclimate. org/index.php?p=111.

29. Mann ME, Amman C, Bradley R, Briffa K, Jones P, Osborn T, et al. On past temperatures and anomalous late-20th century warmth. *Eos.* 2003; 84(27):256– 258.

30. Briffa K, Schweingruber F, Jones P, Osborn T, Shiyatov S, Vaganov E. Reduced sensitivity of recent tree-growth to temperature at high northern latitudes. *Nature.* 1998; 391(6668):678–682.

31. McIntyre S, Mann M. Correspondence related to the Hockey Stick. 2004. Available at http://web.archive.org/web/20031211211711/www.climate2003.com/file.issues.htm.

32. Essex C, McKitrick R. Taken by storm. Toronto; Key Porter Books; 2002.

33. McKitrick R, Wigle RM. The Kyoto Protocol: Canada's risky rush to judgment. 2002.

34. McKitrick R. Emission scenarios & recent global warming projections. Fraser Forum. 1 January 2003; p. 14–16.

35. McKitrick R. The Mann et al. Northern Hemisphere 'Hockey Stick' climate index. A tale of due diligence. In Michaels P, ed. Shattered consensus: the true state of global warming. Rowman and Littlefield; 2006. p. 20–49.

36. Appell D. Timing of M&M in E&E. Quark Soup blog; 1 December 2003. Available at http://web.archive.org/web/20040221195859/www.davidappell.com/archives/00000498.htm.

37. McIntyre S, McKitrick R. Corrections to the Mann et al. (1998) proxy data base and Northern Hemisphere average temperature series. *Energy & Environment.* 2003; 14:751–771.

38. Shultz N. Researchers question key global-warming study. USA Today; 28 October 2003. Available at: http://www.usatoday.com/news/opinion/editorials/2003-10-28-schulz_x.htm.

39. Appell D. E&E paper is 'wrong'. davidappell.com; 29 October 2003. http://web.archive.org/web/20040117141357/www.davidappell.com/archives/00000377.htm.

40. Appell D. M&M: the details. davidappell.com; 29 October 2003. http://web.archive.org/web/20040202010704/http://www.davidappell.com/archives/00000378.htm.

41. McIntyre S, McKitrick R. Letter to David Appell. 29 October 2003. http: //www.uoguelph.ca/~rmckitri/research/Response.Oct29.doc.

42. Mann M, Bradley R, Hughes M. Note on paper by McIntyre and McKitrick in *Energy and Environment.* Climatic Research Unit website. November 2003. Available from: http://web.archive.org/web/20031206180551/http://www.cru.uea.ac.uk/~timo/paleo/.

43. McIntyre S, McKitrick R. Reply to comments made by Professor Mann in response to our paper 'Corrections to the Mann et. al. (1998) Proxy Data Base and Northern Hemispheric Average Temperature Series'. *Energy and Environment* 14(6); 11 November 2003. Available at http://www.uoguelph.ca/~rmckitri/research/MM-nov12-part1.pdf.

44. McIntyre S, McKitrick R. Email to Michael Mann. 11 November 2003. Available at: http://web.archive.org/web/20071210020257/climate2003.com/ correspondence/mann.031111.htm.

45. Mann M. Email to McIntyre and McKitrick. 12 November 2003. Available at http://web.archive.org/web/20071214210857/climate2003.com/correspondence/mann.031112a.htm.

46. Mann M. Email to McIntyre and McKitrick (2). 12 November 2003. Available at http://web.archive.org/web/20071219231914/ climate2003.com/correspondence/mann.031112c.htm.

47. McIntyre S, McKitrick R. Global-scale temperature patterns and climate forcings over the past six centuries: A comment. (First unpublished submission to *Nature*). 2004. Available at http://www.uoguelph.ca/~rmckitri/research/fallupdate04/submission.1.final.pdf.

48. Cotter R. Email to McIntyre. 23 January 2004. Available at: http://www.climateaudit.org/correspondence/nature.040123.htm.

49. McIntyre S. Email to Rosalind Cotter. 7 February 2004. Available at: http: //www.climateaudit.org/correspondence/nature.040207.htm.

50. Mann M, Bradley R, Hughes M. Reply to: 'Global-scale temperature patterns and climate forcings over the past six centuries: a comment' by S McIntyre and R McKitrick. (First unpublished response to *Nature*); 2004.

51. McIntyre S. Letter to *Nature*. 8 April 2004. This is the covering letter that accompanied the third submission to *Nature* and includes the responses to Mann and to the peer reviewers. Available at http://www.climateaudit.org/ correspondence/nature.040408.htm.

52. McIntyre S. Letter to *Nature*. 21 March 2004. This is the covering letter that accompanied the second submission to *Nature* and includes the responses to Mann and to the peer reviewers. Available at http://www.climateaudit.org/ correspondence/nature.040321.htm.

53. Hughes MK, Funkhouser G. Frequency-dependent climate signal in upper and lower forest border tree rings in the mountains of the Great Basin. Climatic Change. 2003; 59(1):233–244.

54. Anonymous. *Nature* peer reviewers' comments on 'Global-scale temperature patterns and climate forcings over the past six centuries: A comment'. (This is the first round of review comments). 9 March 2004. Available from http://www.climateaudit.org/correspondence/nature.040309b.htm.

55. McIntyre S, McKitrick R. Global-scale temperature patterns and climate forcings over the past six centuries: A comment. (Second unpublished submission to Nature). 2004. Available at http://www.uoguelph.ca/~rmckitri/research/fallupdate04/MM.resub.pdf.

56. Cotter R. Email to McIntyre.26 March 2004. Available at: http://www.climateaudit.org/correspondence/nature.040326b.htm.

57. McIntyre S, McKitrick R. Global-scale temperature patterns and climate

forcings over the past six centuries: A comment. (Final unpublished submission to Nature); 2004. Available at http://www.uoguelph.ca/ ~rmckitri/research/fallupdate04/MM.short.pdf.

58. Jones P, Mann M. Climate over past millennia. *Reviews of Geophysics.* 2004; 42(2):1–42.

59. Cotter R. Email to McIntyre. 7 July 2004. Available at http://www.climateaudit.org/correspondence/nature.040707.htm.

60. Anonymous. *Nature* peer reviewers' comments on 'Global-scale temperature patterns and climate forcings over the past six centuries: A comment'. (This is the second round of review comments, which also contained the covering letter of rejection from Rosalind Cotter.); 3 August 2004. Available from http://www.climateaudit.org/ correspondence/nature.040803.htm.

61. Mann M. Responses to referees; 2004. These are the covering letter and responses to the peer reviewers that accompanied Mann's response to the final McIntyre submission to *Nature*. I am not aware of any online copy and I am therefore grateful to Ross McKitrick, who was able to locate a copy for me in his archives.

62. McIntyre S. Materials complaint to *Nature*. 17 November 2003. Available at: http://www.climateaudit.org/correspondence/nature .031117.htm.

63. Ziemelis K. Email to McIntyre. 9 December 2003. Available at: http://www. climateaudit.org/correspondence/nature.031209.htm.

64. McIntyre S. Second materials complaint; 17 December 2003. Available at: http://www.climateaudit.org/correspondence/nature.031217.htm.

65. Ziemelis K. Email to McIntyre; 18 December 2003. Available at: http://www. climateaudit.org/correspondence/nature.031218.htm.

66. Mann M. Reply to MBH98 materials complaint. February 2004. Available at http://www.climateaudit.org/correspondence/nature .complaint.reply. htm.

67. Langenburg H. Email to Steve McIntyre. 27 February 2004. Available at http://www.climateaudit.org/correspondence/nature.040227.htm.

68. McIntyre S. Email to Dinah Ashman. 17 March 2004. Available at http: //www.climateaudit.org/correspondence/nature.040317.htm.

69. Mann M, Bradley R, Hughes M. Corrigendum: Global-scale temperature patterns and climate forcing over the past six centuries. *Nature*. 2004; 430:105.

70. Langenburg H. Email to McIntyre. 26 March 2004. Available at: http:// www.climateaudit.org/correspondence/nature.040326.htm.

71. McIntyre S. Email to Heike Langenburg. 1 July 2004. Available at: http: //www.climateaudit.org/correspondence/nature.040701.htm.

72. Cotter R. Email to McIntyre. 4 July 2004. Available at: http://www. climateaudit.org/correspondence/nature.040707.htm.

73. 'Tamino'. PCA part 4: non-centered hockey sticks. Open Mind blog; 6 March 2008. Available at: http://tamino.wordpress.com/2008/03/06/ pca-

part-4-non-centered-hockey-sticks/.

74. Jolliffe I. Comments on principal components analysis. Open Mind blog; September 2008. Available at: http://tamino.wordpress.com/2008/08/10/open-thread-5-2/.

75. McIntyre S, McKitrick R. Hockey sticks, principal components, and spurious significance. *Geophysical Research Letters*; 2005; 32: L03710, doi:10.1029/2004GL021750.

76. McIntyre S. Errors matter #3: Preisendorfer's Rule N. Climate Audit blog; 13 February 2005. http://www.climateaudit.org/index.php?p=62.

77. Mann M. Email to Climatic Change 2004. Mann's email was forwarded by Schneider to McIntyre who quoted extracts from it at: http://www.climateaudit.org/?p=487.

78. Ziemelis K. Email to McIntyre. 7 September 2004. Available at: http://www. climateaudit.org/correspondence/nature.040907.htm.

79. McIntyre S, McKitrick R. The M&M critique of the MBH98 Northern Hemisphere climate index: update and implications. *Energy and Environment*. 2005; 16: 69–100.

80. Briffa K, Jones P, Schweingruber F. Tree-ring density reconstructions of summer temperature patterns across western North America since 1600. *Journal of Climate*. 1992; 5(7): 735–754.

81. Schoettle A. Ecological roles of five-needle pines in Colorado: potential consequences of their loss. *USDA Forest Service Proceedings*. 2004; RMRSP-32: 124–135. Available at: http://www.fs.fed.us/rm/pubs/rmrs p032/rmrsp032124135.pdf.

82. D'Arrigo RD, Kaufmann RK, Davi N, Jacoby GC, Laskowski C, Myneni RB, et al. Thresholds for warming-induced growth decline at elevational tree line in the Yukon Territory, Canada. *Global Biogeochemical Cycles*. 2004; 18: GB3021.

83. Jacoby G. Letter to *Climatic Change*. Quoted at http://www.climateaudit.org/index.php?p=79

84. McIntyre S. Jacoby's "lost" Gaspé cedars. Climate Audit blog; 25 April 2005. Available at: http://www.climateaudit.org/?p=182.

85. Muller R. Global warming bombshell; 15 October 2004. Available at: http: //www.technologyreview.com/Energy/13830/. MIT *Technology Review*.

86. Mann M. Myth vs. fact regarding the 'Hockey Stick'. RealClimate blog; 4 December 2004. Available at: http://www.realclimate.org/index.php/archives/2004/12/myths-vs-fact-regarding-the-hockey-stick/.

87. Connolley W. Posting to sci.environment; 15 October 2004. Available at: http://groups.google.com/group/sci.environment/msg/ 9a3673c6a0d64 c1f?hl=en&lr=&c2coff=1&rnum=5.

88. Annan J. Posting to sci.environment; 20 October 2004. Available at: http://groups.google.com/group/sci.environment/browse thread/thread/b3bbb3724c9c6eeb/9a3673c6a0d64c1f?hl=ena3673c6a0d64c1f.

89. Connolley W. William's page about MBH and M&M. Webpage; 2004.

Available at: http://www.wmconnolley.org.uk/sci/mbh/.

90. DeLong B. Medieval Warm Period. Brad DeLong's semi-daily journal: a weblog; 18 October 2004. Available at:http://www.j-bradford-delong.net/movable type/2004–2 archives/000406.html.

91. Lambert T. Deja Hockey Stick. Deltoid blog; 18 October 2004. Available at: http://scienceblogs.com/deltoid/2004/10/muller.php.

92. Anonymous. Welcome climate bloggers (Editorial). *Nature.* 2004; 432: 933. Available at: http://www.nature.com/nature/journal/v432/n7020/full/432933a.html.

93. Rutherford S, Mann M, Osborn T, Bradley R, Briffa K, Hughes M, et al. Proxy-based Northern Hemisphere surface temperature reconstructions: sensitivity to method, predictor network, target season, and target domain. *Journal of Climate.* 2005; 18(13): 2308–2329.

94. Mann M. Rutherford et al 2005 highlights. RealClimate blog; 22 November 2004. Available at: http://www.realclimate.org/index.php?p=10.

95. Mann M. False claims by McIntyre and McKitrick regarding the Mann et al. (1998) reconstruction. RealClimate blog; 4 December 2004. Available at: http://www.realclimate.org/index.php?p=8.

96. McIntyre S. Climate2003; 2003–2005. Available at: http://web.archive.org/web/*/http://www.climate2003.com.

97. Mann M. Peer review: a necessary but not sufficient condition. RealClimate blog; 20 January 2005. Available at: http://www.realclimate.org/index.php/archives/2005/01/peer-review-a-necessary-but-not-sufficient-condition/ langswitch lang/en.

98. Mann M, Crok M. Correspondence relating to the Hockey Stick. *Natuurwetenschaap & Techniek* website; April 2005. Available at: http://www.natutech. nl/00/nt/nl/49/nieuws/2299/index.html.

99. McKitrick R, Michaels PJ. A test of corrections for extraneous signals in gridded surface temperature data. *Climate Research.* 2004; 26(2): 159–173.

100. Quiggin J. McKitrick mucks it up. Crooked Timber blog; 25 August 2004. Available at: http://crookedtimber.org/2004/08/25/ mckitrick-mucks-it-up/.

101. Lambert T. McKitrick screws up yet again. Deltoid blog; 26 August 2004. Available at: http://scienceblogs.com/deltoid/2004/08/mckitrick6.php.

102. Crok M. Kyoto Protocol based on flawed statistics. *Natuurwetenshap & Techniek.* 2005; Feb. English translation available at http://www.uoguelph.ca/~rmckitri/research/Climate L.pdf.

103. Regalado A. In climate debate, the 'Hockey Stick' leads to a face-off. *Wall Street Journal* 14 February 2005. Available at: http://online.wsj.com/public/article/SB110834031507653590-DUadAZBzxH0SiuYH3tOdgUmKXPo 20060207.html?mod=blogs.

104. Editorial. Hockey Stick on Ice. *Wall Street Journal* 18 February 2005. Available at: http://online.wsj.com/article/SB110869271828758608.html.

105. Appell D. Behind the Hockey Stick. *Scientific American*. 2005; March. Available at: http://www.sciam.com/article.cfm?id=behind-the-hockey-stick.

106. BBC Radio 4 Today programme. Interview with Michael Mann; 24 February 2005. Audio file available at: http://www.bbc.co.uk/radio4/today/rams/ environment 20050224.ram.

107. Tennekes H. Email to McIntyre; 22 February 2005. Quoted by McIntyre at: http://www.climateaudit.org/?p=94.

108. Vranes K. Open season on Hockey Stick and peer review. Prometheus blog; 18 February 2005. Available at: http://sciencepolicy.colorado.edu/prometheus/archives/climate change/000355open season on hocke.html.

109. Cubasch U. Quote at SEPP website; 5 March 2005. Available at: http://www. sepp.org/Archive/weekwas/2005/Mar.%205.htm. The quotation is sourced from an article in German at: http://www.heise-medien.de/presseinfo/ bilder/tr/05/tr0503038.pdf.

110. Von Storch H. Quoted at 'von Storch on MBH "shoddiness"'. Posting on Climate Audit; 4 March 2005. Available at: http://www.climateaudit.org/index.php?p=127. The quotation is sourced from an article in German at: http://www.heise-medien.de/presseinfo/bilder/tr/05/tr0503038.pdf.

111. Von Storch H, Zorita E, Jones JM, Dimitriev Y, Gonzalez-Rouco F, Tett SFB. Reconstructing past climate from noisy data. *Science*. 2004; 306(5696): 679– 682.

112. Von Storch H, Zorita E. Comment on 'Hockey sticks, principal components, and spurious significance', by S. McIntyre and R. McKitrick. *Geophysical Research Letters*. 2005; 32: 20.

113. McIntyre S, McKitrick R. Reply to von Storch and Zorita. *Geophysical Research Letters*. 2005; 32(20): 20714. Available at: http://www. climateaudit.org/pdf/vz.rr.pdf.

114. McIntyre S. Reply to Huybers #1. Climate Audit blog; 2005. Available at: http://www.climateaudit.org/?p=369.

115. McIntyre S. Reply to Huybers #2: Re-scaling. Climate Audit blog; 2005. Available at: http://www.climateaudit.org/?p=370.

116. McIntyre S. Reply to Huybers #3: Principal components. Climate Audit blog; 2005. Available at: http://www.climateaudit.org/?p=405.

117. Huybers P. Comment on 'Hockey sticks, principal components, and spurious significance' by S. McIntyre and R. McKitrick. *Geophysical Research Letters*. 2005; 32: 735–738.

118. McIntyre S, McKitrick R. Reply to Huybers. *Geophysical Research Letters*. 2005; 32: L20713.

119. McIntyre S. UCAR, Ammann and Wahl and GRL. Climate Audit blog; 29 September 2005. Available at: http://www.climateaudit.org/?p=388.

120. UCAR press office. Media Advisory: The Hockey Stick controversy. New analysis reproduces graph of late 20th century temperature rise. UCAR website; 2005. Available at: http://www.ucar.edu/news/releases/2005/ ammann.shtml.

121. Amman C, Wahl E. Comment on 'Hockey sticks, principal components, and spurious significance' by S. McIntyre and R. McKitrick. Unpublished submission to *Geophysical Research Letters*; 2005. Ammann and Wahl's submissions to GRL are available as follows: Version 1 – http://www.climateaudit.org/pdf/others/AmmannWahl.2005.v1.pdf. Version 2 – http: //www.climateaudit.org/pdf/others/AmmannWahl2005.v2.pdf.

122. Wahl ER, Ammann CM. Robustness of the Mann, Bradley, Hughes reconstruction of northern hemisphere surface temperatures: Examination of criticisms based on the nature and processing of proxy climate evidence. *Climatic Change*. 2007; 85: 33–69. The various versions of Wahl and Amman are available as follows: Draft 1 – http://www.climateaudit.org/pdf/others/WahlAmmann2005.pdf. Draft 2 (including the R^2 statistics) – http://web.archive.org/web/20061206224952/www.cgd.ucar.edu/ccr/ammann/millennium/refs/WahlAmmann ClimaticChange inPress.pdf. Final accepted version – http://www.climateaudit.org/pdf/others/WahlAmmann2007.pdf.

123. McIntyre S. Peer reviewing Ammann and Wahl #1 -the correspondence. Climate Audit blog; 8 March 2006. Available at: http://www.climateaudit.org/?p=574.

124. McIntyre S. Letter to Steven Schneider; 5 July 2005. Available at: http: //www.climateaudit.org/pdf/wa.review.pdf.

125. Houghton J. Testimony to the US Senate; 21 July 2005. Available at: http://frwebgate.access.gpo.gov/cgi-bin/getdoc.cgi?dbname=109 senate hearings&docid=f:24631.pdf. Houghton's testimony was given on 21 July 2005. Ammann and Wahl's GRL comment was rejected on 6 June the same year.

126. Thacker P. How the Wall Street Journal and Rep. Barton celebrated a global warming skeptic. Environmental Science and Technology; 31 August 2005.

127. McIntyre S. Ammann at AGU: the answer. Climate Audit blog; 14 January 2006. Available at http://www.climateaudit.org/?p=492.

128. Schneider S. Interview in *Discovery* magazine; Oct 1989. This quotation is widely reproduced on the internet. See for example http://johnquiggin.com/index.php/archives/2003/09/19/honest-or-effective/.

129. McIntyre S, McKitrick R. Reply to Ammann and Wahl. (Unpublished); 2005. Available at: http://www.climateaudit.org/pdf/reply.ammann.pdf.

130. McIntyre S. IPCC 4AR and Ammann. Climate Audit blog; 10 March 2006. http://www.climateaudit.org/index.php?p=578.

131. McIntyre S. Ammann and Wahl little whopper rejected. Climate Audit blog; 16 March 2006. Available at: http://www.climateaudit.org/?p=589.

132. Anonymous. Bruiser Barton. *The Hill News*; 20 July 2005. Available at: http: //thehill.com/editorials/bruiser-barton-2005-07-20.html.

133. Barton J. Letters to climate scientists; 2005. See http://energycommerce.house.gov/reparchives/108/Letters/06232005 1570.htm.

134. Pease R. Politics plays climate 'hockey'. *BBC News Online*; 18 July 2005. Available at: http://news.bbc.co.uk/1/hi/sci/tech/4693855.stm.

135. Monastersky R. Congressman demands complete records on climate research by 3 scientists who support theory of global warming. *Chronicle of Higher Education*; 1 July 2005. Available at: http://chronicle.com/daily/2005/07/ 2005070101n.htm. (Subscription required.).

136. Brown P. Republicans accused of witch-hunt against climate change scientists. *Guardian*; 30 August 2005. Available at: http://www.guardian.co.uk/environment/2005/aug/30/usnews.research.

137. Eilperin J. GOP chairmen face off on global warming. *Washington Post*. 18 July 2005; p. A04. Available at: http://www.washingtonpost.com/wp-dyn/ content/article/2005/07/17/AR2005071701056.html.

138. Mann M. Letter to Barton; 15 July 2005. Available at: http://www.realclimate.org/Mann response to Barton.pdf.

139. Bradley R. Letter to Barton; 13 July 2005. Available at: http://www.realclimate.org/Bradley response to Barton.pdf.

140. Hughes M. Letter to Barton; 15 July 2005. Available at: http://www.realclimate.org/Hughes response to Barton.pdf.

141. Bloch E. Responsibilities of institutions and investigators in the conduct of research; 17 April 1989. Available at: http://128.150.4.107/pubs/stis1996/ iin106/iin106.txt.

142. Leinen M. Email to McIntyre regarding data; 28 July 2004. Available at: http://www.climateaudit.org/correspondence/nsf.040728.htm.

143. Cuffey K. There's no disguising it – global warming's no put-on. *San Francisco Chronicle*; 9 October 2005. Available at: http://www.sfgate.com/cgi-bin/article.cgi?f=/c/a/2005/10/09/ING5FF2U031.DTL.

144. McIntyre S. NAS news and schedule. Climate Audit blog; 24 February 2006. Available at: http://www.climateaudit.org/?p=551.

145. McIntyre S. Alley at the NAS panel. Climate Audit blog; 6 March 2006. Available at:http://www.climateaudit.org/?p=562.

146. McIntyre S. D'Arrigo: Making cherry pie. Climate Audit blog; 7 March 2006. Available at: http://www.climateaudit.org/?p=570.

147. Briffa KR. Annual climate variability in the Holocene: interpreting the message of ancient trees. *Quaternary Science Reviews*. 2000; 19(1-5): 87–105.

148. Briffa KR, Osborn TJ, Schweingruber FH, Harris IC, Jones PD, Shiyatov SG, et al. Low-frequency temperature variations from a northern tree ring density network. *Journal of Geophysical Research*. 2001; 106(D3): 2929–2941.

149. Hughes M. Statement to NAS panel; 2 March 2006. McIntyre quotes Hughes as replying to a question of whether tree rings were failing to pick up climatic warming: 'That's the third explanation that I wasn't going to discuss', apparently indicating that he wished to avoid the

subject. See: http://www.climateaudit.org/?p=573.

150. McIntyre S. Luterbacher and Hegerl at NAS. Climate Audit blog; 8 March 2006. Available at: http://www.climateaudit.org/?p=571.

151. McIntyre S. Mann at the NAS panel. Climate Audit blog; 16 March 2006. Available at: http://www.climateaudit.org/?p=588.

152. McIntyre S, McKitrick R. Presentation to the National Academy of Sciences expert panel, surface temperature reconstructions for the last 1,000–2,000 years. : Supplementary Comments; 3 April 2006. Available at: http://www. climateaudit.org/pdf/NAS-followup-M&M.pdf.

153. National Research Council. Surface temperature reconstructions for the last 2,000 years. North G, editor. National Acadamies Press; 2006. Available at: http://www.nap.edu/catalog/11676.html.

154. Osborn TJ, Briffa KR. The spatial extent of 20th-century warmth in the context of the past 1200 years. *Science*. 2006; 311(5762): 841–844.

155. McIntyre S, McKitrick R. Presentation to the National Academy of Sciences expert panel: surface temperature reconstructions for the past 1,000–2,000 years; 2 March 2006. Available at http://www.climateaudit.org/pdf/NAS. M&M.ppt.

156. National Academy of Science. Ensuring the integrity, accessibility, and stewardship of research data in the digital age; 2009. Available at: http://books.nap.edu/openbook.php?record id=12615&page=R1.

157. Pielke Jnr R. Quick reaction to the NRC hockey stick report. Prometheus blog; 22 June 2006. Available at: http://sciencepolicy. colorado.edu/prometheus/archives/climatechange/000859quick reaction to th.html.

158. Brumfiel G. Academy affirms hockey-stick graph. *Nature*. 2006; 441: 1032– 1033.

159. Anonymous. Backing the hockey-stick graph. *BBC News Online*; 23 June 2006. Available at: http://news.bbc.co.uk/nolpda/ukfs news/hi/newsid 5109000/5109188.stm.

160. Anonymous. Breaking the 'hockey stick'. *Washington Times*; 26 June 2006. Available at: http://www.washingtontimes.com/news/2006/jun/26/ 20060626-094411-2965r/.

161. Regalado A. Panel study fails to settle debate on past climates. *Wall Street Journal* 23 June 2006. Available at: http://online.wsj.com/article/ SB115098487133887497.html.

162. Regalado A. Hockey stick hokum. *Wall Street Journal* 14 July 2006. Available at: http://online.wsj.com/article/SB115283824428306460. html.

163. Questions surrounding the 'Hockey Stick' temperature studies: Implications for climate change assessments. Hearings before the Subcommittee on Oversight and Investigations of the Committee on Energy and Commerce, House of Representatives. (Transcript); July 19 and July 27, 2006. Available at: http://frwebgate.access.gpo.gov/cgi-bin/getdoc.cgi? dbname=109 house hearings&docid=f:31362.wais.

164. North G. A scientific graph stands trial. Chronicle of Higher Education Colloquy; 6 September 2006. Available at: http://chronicle.com/ live/2006/ 09/hockey stick/.

165. McIntyre S. Sciencemag on House E&C hearing. Climate Audit blog; 1 August 2006. Available at: http://www.climateaudit.org/?p=766.

166. Gore A. An inconvenient truth: the planetary emergency of global warming and what we can do about it. Rodale Books; 2006.

167. Thompson L, Mosley-Thompson E. Public lecture at Ohio State University; 11 January 2008. Quoted by Huston McCulloch. See: http://www.climateaudit. org/?p=2598.

168. RealClimate. Followup to the Hockeystick hearings; 31 August 2006. Available at: http://www.realclimate.org/index.php/archives/2006/08/ followup-to-the-hockeystick-hearings/.

169. Mann M. Responses of Dr. Michael Mann to questions propounded by the Committee on Energy and Commerce Subcommittee on Oversight and Investigations; 31 August 2006. Available at: http://www.meteo. psu.edu/~mann/house06/HouseFollowupQuestionsMann31Aug06.pdf.

170. North G. Lecture at Texas A&M University; 2006. Audio available at: http: //geotest.tamu.edu/userfiles/216/NorthH264.mp4. The quotation cited is at around 55 minutes in.

171. Jones P, Briffa K, Barnett T, Tett S. High-resolution palaeoclimatic records for the last millennium: interpretation, integration and comparison with general circulation model control-run temperatures. *The Holocene*. 1998; 8(4): 455.

172. Shiyatov S. Reconstruction of climate and the upper treeline dynamics. *Publications of the Academy of Finland*. 1995; 6: 144–147.

173. Naurzbaev MM, Hughes MK, Vaganov EA. Tree-ring growth curves as sources of climatic information. *Quaternary Research*. 2004; 62(2): 126–133.

174. Larson D. Letter to Steve McIntyre. Quoted at Climate Audit blog posting, Polar Urals #2: Broken core; 4 April 2005. Available at: http://www. climateaudit.org/?p=162.

175. Briffa K, Bartholin T, Eckstein D, Jones P, Karlen W, Schweingruber F, et al. A 1,400-year tree-ring record of summer temperatures in Fennoscandia. *Nature*. 1990; 346: 434–439.

176. Briffa K, Jones P, Bartholin T, Eckstein D, Schweingruber F, Karlen W, et al. Fennoscandian summers from AD 500: temperature changes on short and long timescales. *Climate Dynamics*. 1992; 7(3): 111–119.

177. Briffa K, Jones P, Schweingruber F, Karlen W, Shiyatov S. Tree-ring variables as proxy-climate indicators: problems with low-frequency signals. *NATO ASI Series – Global Environmental Change*. 1996; 41: 9–42.

178. Grudd H. Tree rings as sensitive proxies of past climate change. University of Stockholm; 2006. Available at: http://www.diva-portal.org/diva/ getDocument?urn nbn se su diva-1034-2 fulltext.pdf.

179. Crowley TJ, Lowery TS. How warm was the Medieval Warm Period? *Ambio*. 2000; 29: 51–54.

180. Crowley TJ. Causes of climate change over the past 1000 Years. *Science*. 2000; 289: 270–277.

181. Crowley T. Raising the ante on the climate debate. *Eos*. 2005; 86: 262–263.

182. McIntyre S, Crowley T. Correspondence relating to Crowley and Lowery 2000 data was posted by Steve McIntyre. The Crowley–McIntyre letters. Climate Audit blog; 1 July 2005. Available at: http://www.climateaudit. org/?p=246.

183. Zhu K, Chu K. A preliminary study on the climatic fluctuations during the past 5000 years in China. Scientia Sinica. 1973; 16: 226–256.

184. Zhang DE. Evidence for the existence of the medieval warm period in China. Climatic Change. 1994; 26: 289–297.

185. Anonymous. Dendroclimatologists answer back. Climate Audit blog; 29 March 2007. Available at: http://www.climateaudit.org/?p=1304.

186. McIntyre S. Thompson et al [1993] on Dunde. Climate Audit blog; 18 September 2005. Available at: http://www.climateaudit.org/?p=374.

187. Thompson L, Mosley-Thompson E, Davis M, Lin P, Yao T, Dyurgerov M, et al. 'Recent warming': ice core evidence from tropical ice cores with emphasis on Central Asia. *Global and Planetary Change*. 1993; 7(1-3): 145–156.

188. Esper J, Cook ER, Schweingruber FH. Low-frequency signals in long tree-ring chronologies for reconstructing past temperature variability. *Science*. 2002; 295: 2250–2253.

189. McIntyre S. Science – Email #39. Climate Audit blog; 11 May 2006. Available at: http://www.climateaudit.org/?p=668.

190. Esper J, Cook E, Krusic P, Peters K, Schweingruber F. Tests of the RCS method for preserving low-frequency variability in long tree-ring chronologies. Tree Ring Research. 2003; 59: 81–98. Available at:http://www.wsl.ch/staff/jan. esper/publications/TRR 2003.pdf.

191. Fletcher R. Unbalanced opinions. *UBC Thunderbird*; 2003. Available at: http://web.archive.org/web/20070320020115/www.journalism.ubc. ca/thunderbird/archives/2003.11/printer ready/climate print.html.

192. Moberg A, Sonechkin DM, Holmgren K, Datsenko NM, Karlen W. Highly variable Northern Hemisphere temperatures reconstructed from low-and high-resolution proxy data. *Nature*. 2005; 433(7026): 613–617.

193. Rincon P. Climate 'warmest for millennium'. *BBC News Online*; 10 Febuary 2006. Available at: http://news.bbc.co.uk/1/hi/sci/tech/ 4698652.stm.

194. Bürger G. Comment on 'The Spatial Extent of 20th-Century Warmth in the Context of the Past 1200 Years'. Science. 2007; 316: 1844. Available at: http://www.sciencemag.org/cgi/content/full/316/5833/1844a.

195. Stockwell D. Reconstruction of past climate using series with red noise. *Australian Institute of Geoscientists News*. 2006; 83: 14. Available at: http://landshape.org/enm/wp-content/uploads/2006/06/AIGNews Mar06%2014.pdf.

196. Osborn TJ, Briffa KR. Response to comment on 'The spatial extent of 20thCentury warmth in the context of the past 1200 years'. *Science.* 2007; 316: 1844. Available at: http://www.sciencemag.org/cgi/content/full/sci; 316/5833/1844b.

197. McIntyre S. Bürger comment on Osborn and Briffa 2006. Climate Audit blog; 30 June 2007. Available at: http://www.climateaudit.org/?p=1797.

198. Loehle C. A 2000-year global temperature reconstruction based on nontreering proxies. Energy & Environment. 2007; 18(7-8): 1049–1058.

199. Loehle C, McCulloch JH. Correction to: A 2000-year global temperature reconstruction based on non-tree ring proxies. *Energy & Environment.* 2008; 19(1): 93–100.

200. Bürger G, Cubasch U. Are multiproxy climate reconstructions robust? Geophysical Research Letters. 2005; 32: 1–4.

201. McIntyre S. Some thoughts on disclosure and due diligence in climate science. Climate Audit blog; 14 February 2005. Available at: http://www.climateaudit.org/index.php?p=66.

202. Pielke Jnr R. On the Hockey Stick; 6 July 2005. Available at: http://sciencepolicy.colorado.edu/prometheus/archives/climatechange/000480 on the hockey stick.html.

203. Vranes K. The Barton letters. Prometheus blog; 28 June 2005. Available at: http://sciencepolicy.colorado.edu/prometheus/archives/author vranes k/index.html\#000475.

204. Von Storch H. Hans von Storch on Barton. Prometheus blog; 8 July 2005. Available at: http://sciencepolicy.colorado.edu/prometheus/archives/climate change/000486hans von storch on b.html.

205. Various. IPCC Working Group I Fourth Assessment Report: Expert review comments on first-order draft; 16 November 2005. Available at http://www.climateaudit.org/pdf/ipcc/fod/AR4WG1 Ch06 FOR CommentResponses EDist.pdf.

206. Jansen E, Overpeck J, Briffa K, Duplessy JC, Joos F, Masson-Delmotte V. Palaeoclimate. In: Solomon S, Qin D, Manning M, Chen Z, Marquis M, Averyt KB, et al., editors. Climate change 2007: the physical science basis. Contribution of Working Group I to the Fourth Assessment Report of the Intergovernmental Panel on Climate Change (First order draft). IPCC; 2006. Available at: http://hcl.harvard.edu/collections/ipcc/.

207. Jansen E, Overpeck J, Briffa K, Duplessy JC, Joos F, Masson-Delmotte V. Palaeoclimate. In: Solomon S, Qin D, Manning M, Chen Z, Marquis M, Averyt KB, et al., editors. Climate change 2007: the physical science basis. Contribution of Working Group I to the Fourth Assessment Report of the Intergovernmental Panel on Climate Change (Second order draft). IPCC; 2006. Available at: http://hcl.harvard.edu/collections/ipcc/ A more user-friendly unofficial version is at: www.junkscience.com/draft AR4/Ch06 SOD Text TSU FINAL.pdf.

208. Various. IPCC Working Group I Fourth Assessment Report: Expert and government review comments on the second-order draft; 15 June 2006. http://www.climateaudit.org/pdf/ipcc/sod/AR4WG1 Ch06 SOR CommentResponses EDist.pdf.

209. IPCC. Deadlines for literature cited in the Working Group I Fourth Assessment Report; 2006. Available at: http://web.archive.org/web/20060207153611/ipcc-wg1.ucar.edu/wg1/PublicationDeadlines.pdf.

210. Wegman E. Response of Dr. Edward Wegman to questions posed by the Honorable Mr. Bart Stupak in connection with testimony to the Subcommittee on Oversight and Investigations; 2006. Available at: http://www.uoguelph. ca/~rmckitri/research/StupakResponse.pdf.

211. IPCC. Guidelines for inclusion of recent scientific literature in the Working Group I Fourth Assessment Report.;. Available at: http://web.archive.org/web/20070206012931/ipcc-wg1.ucar.edu/wg1/docs/PublicationDeadlines 2006-07-01.pdf.

212. IPCC. Appendix A to the Principles Governing IPCC Work: Procedures for the preparation, review, acceptance, adoption, approval and publication of IPCC reports; 2003. Available at:http://web.archive.org/web/ 20070202164941/www.ipcc.ch/about/app-a.pdf.

213. McIntyre S. IPCC schedule: WG1 report available only to insiders until May 2007. Climate Audit blog; 24 January 2007. Available at: http://www. climateaudit.org/?p=1101.

214. Borenstein S. Report on climate to include 'smoking gun' on global warming. This agency report was widely reproduced; 23 January 2007. See for example http://seattlepi.nwsource.com/national/300724 climate23.html.

215. Bhalla N. U.N. climate report will shock the world -chairman. *Reuters Alertnet*; 25 January 2007. Available at http://www.alertnet.org/thenews/newsdesk/DEL33627.htm.

216. Middlestaedt M. The fallout of global warming: 1,000 years. In stark terms, scientists confirm that climate change is 'unequivocal'. *Globe and Mail*; 31 January 2007. Available at http://www.theglobeandmail.com/servlet/ story/RTGAM.20070131.wclimate31/BNStory/National/home.

217. Alley RB, Berntsen T, Bindoff NL, Chen Z, Chidthaisong A, Friedlingstein P, et al. Summary for policymakers. In: Solomon S, Qin D, Manning M, Chen Z, Marquis M, Averyt KB, et al., editors. Climate Change 2007: The physical science basis. Contribution of Working Group I to the Fourth Assessment Report of the Intergovernmental Panel on Climate Change. IPCC; 2007.

218. Grønli KS. Hva skjedde med hockeykølla? *Forskning*; 9 February 2007. Available at http://www.forskning.no/Artikler/2007/februar/1170420490. 5. An English translation was posted at Climate Audit and can be seen at http://www.climateaudit.org/?p=1131.

219. McIntyre S. Where's Caspar? Climate Audit blog; 28 August 2007.

Available at: http://www.climateaudit.org/?p=1987.

220. McIntyre S. The dog that didn't bark. Climate Audit blog; 24 May 2008. Available at :http://www.climateaudit.org/?p=3111.

221. McIntyre S. Cunning IPCC bureaucrats. Climate Audit blog; 21 May 2007. Available at: http://www.climateaudit.org/?p=1589.

222. IPCC. Procedures for the preparation, review, acceptance, adoption, approval and publication of IPCC reports; Adopted 21 February 2003. Available at: http://www.climatescience.gov/Library/ipcc/app-a.pdf.

223. McIntyre S. Did IPCC review editor Mitchell do his job? Climate Audit blog; 30 January 2008. McIntyre quotes Mitchell's correspondence with Holland at this blog posting. Available at: http://www.climateaudit.org/?p=2678.

224. McIntyre S. Fortress CRU. Climate Audit blog; 20 June 2008. Available at: http://www.climateaudit.org/?p=3193.

225. Mann M. Comment on updating proxy records; December 2004. See http://www.realclimate.org/index.php?p=11.

226. Lamarche VC, Graybill DA, Fritts HC, Rose MR. Increasing atmospheric carbon dioxide: tree ring evidence for growth enhancement in natural vegetation. Science. 1984; 225: 1019–1021.

227. McIntyre S. Rocky Mountain high #2. Climate Audit blog; 7 April 2006. Available at: http://www.climateaudit.org/?p=622.

228. Ababneh L. Analysis of radial growth patterns of strip-bark and whole-bark bristlecone pine trees in the White Mountains of California: implications in paleoclimatology and archaeology of the Great Basin. University of Arizona; 2006. Available at: http://www.geo.arizona.edu/Antevs/Theses/AbabnehDissertation.pdf.

229. Salzer MW, Hughes MK. Bristlecone pine tree rings and volcanic eruptions over the last 5000 yr. *Quaternary Research*. 2007; 67(1): 57–68.

230. McIntyre S. Hughes and the Ababneh thesis. Climate Audit blog; 1 November 2007. McIntyre describes Ababneh's explanation in the comments to the posting at http://www.climateaudit.org/?p=2310.

231. Kaufman M. NASA revisions create a stir in the blogosphere. *Washington Post*; 15 August 2007. Available at: http://www.washingtonpost.com/wp-dyn/content/article/2007/08/14/AR2007081401677.html.

232. McLean D. NASA fixes data; 1934 ousts 1998 as hottest U.S. year; 14 August 2007. Available at: http://www.bloomberg.com/apps/news?pid=20601087&sid=aBBQO5XgLQu4&refer=worldwide.

233. Anonymous. NASA admits to flaw in warming data; 16 August 2007. Available at: http://timesofindia.indiatimes.com/articleshow/msid-2286256, prtpage-1.cms.

234. Mann ME, Zhang Z, Hughes MK, Bradley RS, Miller SK, Rutherford S, et al. Proxy-based reconstructions of hemispheric and global surface temperature variations over the past two millennia. Proceedings of the National Academy of Sciences. 2008; 105(36): 13252.

235. Penn State University Department of Public Information. News release:

global warming greatest in past decade; 1 September 2008. Available at: http://www.ems.psu.edu/NewsReleases/PDFArticles/ GlobalWarming91 08.pdf.

236. Black R. Climate 'hockey stick' is revived. *BBC News Online*; 1 September 2008. Available at http://news.bbc.co.uk/1/hi/sci/tech/ 7592575.stm.

237. Citizen O. Past decade warmest in 1,300 years; 2 September 2008. Available at: http://www.canada.com/topics/technology/story.html?id= f8d46935-258d-4234-bc7d-945940d92503.

238. Spotts PN. A gnarlier 'hockey stick,' the same message. Christian Science Monitor; 1 September 2008. Available at: http://features. csmonitor.com/environment/2008/09/01/a-gnarlier-hockey-stick-the-same-message/.

239. Cressey D. Jolly hockey sticks. The Great Beyond blog; 2 September 2008. Available at: http://blogs.nature.com/news/thegreatbeyond/ 2008/09/ jolly hockey sticks.html.

240. Tiljander M, Saarnisto M, Ojala A, Saarinen T. A 3000-year palaeoenvironmental record from annually laminated sediment of Lake Korttajärvi, central Finland. Boreas. 2003; 32(4): 566–577.

241. Rodrigo F, Esteban-Parra M, Pozo-Vazquez D, Castro-Diez Y. A 500-year precipitation record in Southern Spain. International Journal of Climatology. 1999; 19(11).

242. McIntyre S, McKitrick R. Proxy inconsistency and other problems in millennial paleoclimate reconstructions. Proceedings of the National Academy of Sciences of the United States of America. 2009; 106(6): E10.

243. Mann ME, Bradley RS, Hughes MK. Reply to McIntyre and McKitrick: Proxy-based temperature reconstructions are robust. *Proceedings of the National Academy of Sciences*. 2009; 106(6): E11.

244. Kaufman DS, Schneider DP, McKay NP, Ammann CM, Bradley RS, Briffa KR, et al. Recent warming reverses long-term Arctic cooling. *Science*. 2009; 325: 1236–1239.

245. Korhola A. Recession in climate science. CO₂ Rapporti blog; 27 September 2009. Korhola is professor of Arctic Global Change at the University of Helsinki. Excerpts from his blog posting were translated from the original Finnish by a Climate Audit reader. The original article can be seen at: http://www.co2-raportti.fi/index.php?page=blogi&news id=1370, and the translations at http://www.climateaudit.org/?p=7272.

246. Smith R. The trouble with medical journals. London: Royal Society of Medicine Press Ltd; 2006.

247. Godlee F, Gale CR, Martyn CN. Effect on the quality of peer review of blinding reviewers and asking them to sign their reports: a randomized controlled trial. *Journal of the American Medical Association*. 1998; 280(3): 237–240.

248. McCullough B, McKitrick R. Check the numbers: the case for due

diligence in policy formation. The Fraser Institute. 2009.

249. Zidek J. Editorial:(Post-normal) statistical science. *Journal of the Royal Statistical Society Series A.* 2006; 169(1): 1–4.

250. Buckheit J, Donoho D. Wavelab and reproducible research. In: Antoniadis A, Oppenheim G, editors. Wavelets and Statistics. Springer Verlag; 1995. p. 55–81. Available at: http:// eprints.kfupm.edu.sa/75983/1/75983.pdf.

251. Goodstein D. Conduct and misconduct in science; 2001. Available at: http: //www.physics.ohio-state.edu/~wilkins/onepage/conduct.html.

252. Altman D. In: The COPE report 2003. Committee on Publication Ethics; 2003. Available at: http://publicationethics.org/static/2003/ 2003pdfcomplete. pdf.

253. Pielke Jnr R. Invitation to McIntyre and Mann – So what? Prometheus blog; 31 October 2005. Available at: http://sciencepolicy.colorado.edu/ prometheus/archives/author pielke jr r/index.html.

254. Von Storch H. Comment on the importance of the Hockey Stick. Prometheus blog; 21 November 2005. Available at: http://sciencepolicy. colorado.edu/prometheus/archives/climate change/000641reflections on the c.html.

255. Bony S, Colman R, Kattsov VM, Allan RP, Bretherton CS, Dufresne JL, et al. How well do we understand and evaluate climate change feedback processes? Journal of Climate. 2006; 19(15): 3445–3482.

256. Randall DA, Wood RA, Bony S, Colman R, Fichefet T, Fyfe J, et al. Climate Models and Their Evaluation. In: Climate Change 2007: The Physical Science Basis. Contribution of Working Group I to the Fourth Assessment Report of the Intergovernmental Panel on Climate Change. Cambridge: Cambridge University Press; 2007. p. 636.

257. Hantemirov RM, Shiyatov SG. A continuous multimillennial ring-width chronology in Yamal, northwestern Siberia. The Holocene. 2002; 12: 717.

258. Briffa KR, Shishov VV, Melvin TM, Vaganov EA, Grudd H, Hantemirov RM, et al. Trends in recent temperature and radial tree growth spanning 2000 years across northwest Eurasia. Philosophical Transactions of the Royal Society B: Biological Sciences. 2008; 363(1501): 2269.

259. Royal Society Publishing. Policy on data. Available at: http://publishing. royalsociety.org/index.cfm?page=1684\#question10.

260. McIntyre S. Letter to the editors of Phil Trans Roy Soc B. Reproduced at Climate Audit blog; 17 July 2008. Available at: http://www.climateaudit. org/?p=3266\#comment-277137.

261. Royal Society Publishing. Reply to McIntyre. Quoted at Climate Audit blog; July 2008. Available at: http://www.climateaudit.org/?p=3352.

262. Hulme M. Newspaper scare headlines can be counter-productive. Nature. 2007; 445(7130): 818.

263. Information Commissioner's Office. Awareness Guidance 2: Information

provided in confidence; 12 September 2008. Available at: http://www.ico.gov.uk/upload/documents/library/freedom_of_informatio n/detailed_specialist_guides/confidentialinformation_v4.pdf.

264. Jansen E, Overpeck J, Briffa H, Duplessy JC, Joos F, Masson-Delmotte V, et al. Palaeoclimate. In: Climate Change 2007: The Physical Science Basis. Contribution of Working Group I to the Fourth Assessment Report of the Intergovernmental Panel on Climate Change. Cambridge: Cambridge University Press; 2007. p. 433–498.

265. McIntyre S. The updated Gaspé series. Climate Audit blog; 9 February 2005. Available at: http://www.climateaudit.org/index.php?p=41.

266. McIntyre S. Polar Urals #3: Crossdating. Climate Audit blog; 5 April 2005. Available at: http://www.climateaudit.org/?p=164.

267. McIntyre S. The Yamal substitution. Climate Audit blog; 12 February 2006. Available at: http://www.climateaudit.org/?p=528.

268. McIntyre S. A little secret. Climate Audit blog; 12 October 2007. Available at: http://www.climateaudit.org/?p=2183.

269. McIntyre S. Almagre strip bark. Climate Audit blog; 17 October 2007. Available at: http://www.climateaudit.org/?p=2214.

270. McIntyre S. The Sheep Mountain update. Climate Audit blog; 14 November 2007. Available at: http://www.climateaudit.org/?p=2371.

Index